OPTICS

Optics

An Introduction

Sarhan M. Musa, PhD

(Prairie View A&M University)

Mercury Learning and Information

Dulles, Virginia
Boston, Massachusetts
New Delhi

Publisher: David Pallai
MERCURY LEARNING AND INFORMATION
22841 Quicksilver Drive
Dulles, VA 20166
info@merclearning.com
www.merclearning.com
1-800-232-0223

S. M. Musa. *OPTICS: An Introduction.*
ISBN: 978-1-93854-970-0

Library of Congress Control Number: 2019940569

202122321   Printed on acid-free paper in the United States of America.

CONTENTS

PREFACE

Optics is known as the branch of science and engineering surrounding the physical phenomena and technologies connected with the generation, transmission, manipulation, detection, and utilization of light. It has acquired newer dimensions with the advent of lasers, fibers, and sensors which occupy a major position among the outstanding achievements of science and engineering in the present era. Optics has many applications in areas of information technology and telecommunications, health care and biotechnology, sensing, lighting, energy, and manufacturing. This book is a definitive guide to the fundamental principles and techniques of optics as well as their effective usage. Primarily intended as a textbook for courses in electrical engineering and physics, the book also serves as a basic reference and refresher for professionals in these areas. Mainly this book is organized into the thirteen chapters and two appendices.

Sarhan M. Musa
April 2020

CHAPTER 1

THEORY OF LIGHT

Chapter Outline

1.0   INTRODUCTION

Due to the importance of light in optics, we illustrate a brief review of the historical theory of light, the speed of light, classification, units and measures, a basic review of scientific notation, light pressure, and optics.

1.1   HISTORICAL THEORIES ABOUT LIGHT

1.1.1 Hindu Theories

Around the fifth or sixth century BC, the Hindu schools of Samkhya and Vaisheshika developed theories on light. According to the Samkhya school, light is one of the five fundamental "subtle" elements out of which emerge the gross elements.

The Vaisheshika school gives an atomic theory. The basic atoms are those of earth, water, fire, and air. These atoms are taken to form binary molecules that combine further to form larger molecules. Light rays are taken to be a high-velocity stream of fire atoms.

In 499 CE, Aryabhata, who proposed a heliocentric solar system of gravitation in his Aryabhatiya, wrote that the planets and the moon do not have their own light but reflect the light of the sun.

The Indian Buddhists viewed light as being an atomic entity equivalent to energy, similar to the modern concept of photons, and viewed all matter as being composed of these light/energy particles.

1.1.2 Greek and Hellenistic Theories

In the fifth century BC, Empedocles postulated that everything was composed of four elements: fire, air, earth, and water. He believed that Aphrodite made the human eye out of the four elements, and that she lit the fire in the eye which shone out from the eye, making sight possible. If this were true, then one could see during the night just as well as during the day; so, Empedocles postulated an interaction between rays from the eyes and rays from a source such as the sun.

In about 300 BC, Euclid wrote *Optica*, in which he studied the properties of light. Euclid postulated that light traveled in straight lines, and he described the laws of reflection and studied them mathematically. He questioned whether sight was the result of a beam from the eye, for he asked how if one closed one's eyes, then opened them at night, one saw the stars immediately. Of course, if the beam from the eye traveled infinitely fast, this was not a problem.

In 55 BC, Lucretius, a Roman who carried on the ideas of earlier Greek atomists, wrote in *On the Nature of the Universe*: "The light & heat of the sun; these are composed of minute atoms which, when they are shoved off, lose no time in shooting right across the interspace of air in the direction imparted by the shove." Despite being similar to later particle theories, Lucretius's views were not generally accepted, and light was still theorized as emanating from the eye.

Ptolemy (c. second century) wrote about the refraction of light in his book *Optics* and developed a theory of vision whereby objects are seen by rays of light emanating from the eyes.

1.1.3 Optical Theory

The Muslim scientist Ibn Al-Haytham (965–1040) developed a broad theory of vision based on geometry and anatomy in *Book of Optics*. Al-Haytham postulated that every point on an illuminated surface radiates light rays in all directions, but the only one that can be seen is the ray that strikes the eye perpendicularly. The other rays strike at different angles and are not seen. He described the pinhole camera and invented the camera obscura, which produces an inverted image, using it as an example to support his argument. This contradicted Ptolemy's theory of vision that objects are seen by rays of light emanating from the eyes. Ibn Al-Haytham held light rays to be streams of minute energy particles that traveled at a finite speed. He improved Ptolemy's theory of the refraction of light and went on to discover the laws of refraction.

He carried out the first experiments on the dispersion of light into its constituent colors. His major work, *Kitab al-Manazir (Book of Optics)* was translated into Latin in the Middle Ages, although his book dealt with the colors of the sunset. He dealt at length with the theory of various physical phenomena like shadows, eclipses, and rainbows. He attempted to explain binocular vision and gave an explanation of the apparent increase in size of the sun and the moon when near the horizon, known as the moon illusion. Because of his extensive experimental research on optics, Ibn Al-Haytham is considered the "father of modern optics."

Ibn Al-Haytham also correctly argued that we see objects because the sun's rays of light, which he believed to be streams of tiny energy particles traveling in straight lines, are reflected from objects into our eyes. He understood that light must travel at a large but finite velocity and that refraction is caused by the velocity being different in different substances. He also studied spherical and parabolic mirrors and understood how refraction by a lens will allow images to be focused and magnification to take place.

Abu Rayhan Al-Biruni (973–1048) also agreed that light has a finite speed, and he was the first to discover that the speed of light is much faster than the speed of sound. In the late thirteenth and early fourteenth centuries, Qutb Al-Din Al-Shirazi (1236–1311) and his student Kamal Al-Din Al-Farisi (1260–1320) continued the work of Ibn Al-Haytham, and they were the first to give the correct explanations for the rainbow phenomenon.

1.1.4 The "Plenum"

René Descartes (1596–1650) held that light was a disturbance of the plenum, the continuous substance of which the universe was composed. In 1637 he published a theory of the refraction of light that assumed, incorrectly, that light traveled faster in a denser medium than in a less dense medium. Descartes arrived at this conclusion by analogy with the behavior of sound waves. Although Descartes was incorrect about the relative speeds, he was correct in assuming that light behaved like a wave and in concluding that refraction could be explained by the speed of light in different media. As a result, Descartes's theory is often regarded as the forerunner of the wave theory of light.

1.1.5 Particle Theory

Ibn Al-Haytham proposed a particle theory of light in his *Book of Optics*. He held light rays to be streams of minute energy particles that travel in straight lines at a finite speed. He states in his *Optics* that "the smallest parts of light," as he calls them, retain only properties that can be treated by geometry and verified by experiment; they lack all sensible qualities except energy. Avicenna (980–1037) also proposed that "the perception of light is due to the emission of some sort of particles by a luminous source."

Pierre Gassendi (1592–1655), an atomist, proposed a particle theory of light which was published posthumously in the 1660s. Isaac Newton studied Gassendi's work at an early age and preferred his view to Descartes's theory of the plenum. He stated in his *Hypothesis of Light* of 1675 that light was composed of corpuscles (particles of matter) which were emitted in all directions from a source. One of Newton's arguments against the wave nature of light was that waves were known to bend around obstacles, while light traveled only in straight lines. He did, however, explain the phenomenon of the diffraction of light (which had been observed by Francesco Grimaldi) by allowing that a light particle could create a localized wave in the ether.

Newton's theory could be used to predict the reflection of light, but it could only explain refraction by incorrectly assuming that light accelerated upon entering a denser medium because the gravitational pull was greater. Newton published the final version of his theory in *Opticks (Optics)* in 1704. His reputation helped the particle theory of light to hold sway during the eighteenth century. The particle theory of light led Laplace to argue that a body could be so massive that light could not escape from it. In other words, it would become what is now called a black hole.

1.1.6 Wave Theory

In the 1660s, Robert Hooke published a wave theory of light. Christiaan Huygens worked out his own wave theory of light in 1678 and published it in his *Treatise on Light* in 1690. He proposed that light was emitted in all directions as a series of waves in a medium called the luminiferous ether. As waves are not affected by gravity, it was assumed that they slowed down upon entering a denser medium.

Thomas Young's sketches of the two-slit experiment showed the diffraction of light. Young's experiments supported the theory that light consists of waves.

Wave theory predicted that light waves could interfere with each other like sound waves (as noted around 1800 by Thomas Young), and that light could be polarized, if it were a transverse wave. Young showed by means of a diffraction experiment that light behaved as waves. He also proposed that different colors were caused by different wavelengths of light, and he explained color vision in terms of three-colored receptors in the eye.

Another supporter of wave theory was Leonhard Euler. He argued in *Nova theoria lucis et colorum* (1746) that diffraction could more easily be explained by wave theory.

Later, Augustin-Jean Fresnel independently worked out his own wave theory of light and presented it to the Académie des Sciences in 1817. Simeon Denis Poisson added to Fresnel's mathematical work to produce a convincing argument in favor of wave theory, helping to overturn Newton's corpuscular theory. By the year 1821, Fresnel was able to show via mathematical methods that polarization could be explained only by the wave theory of light and only if light was entirely transverse, with no longitudinal vibration whatsoever.

The weakness of wave theory was that light waves, like sound waves, would need a medium for transmission. A hypothetical substance called the luminiferous ether was proposed, but its existence was cast into strong doubt in the late nineteenth century by the Michelson-Morley experiment.

Newton's corpuscular theory implied that light would travel faster in a denser medium, while the wave theory of Huygens and others implied the opposite. At that time, the speed of light could not be measured accurately enough to decide which theory was correct. The first to make a sufficiently accurate measurement was Léon Foucault, in 1850. His result supported wave theory, and the classical particle theory was finally abandoned.

1.1.7 Electromagnetic Theory

A linearly polarized light wave frozen in time and showing the two oscillating components of light, an electric field and a magnetic field perpendicular to each other and to the direction of motion (a transverse wave), are shown in Figure 1.1, where λ is the wavelength, E is the amplitude of the electric field, and M is the amplitude of the magnetic field.

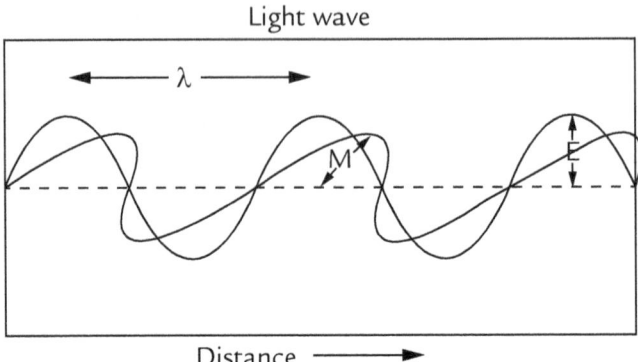

FIGURE 1.1. A linearly polarized light wave frozen in time.

In 1845, Michael Faraday discovered that the plane of polarization of linearly polarized light is rotated when the light rays travel along the magnetic field direction in the presence of a transparent dielectric, an effect now known as Faraday rotation. This was the first evidence that light was related to electromagnetism. In 1846, he speculated that light might be some form of disturbance propagating along magnetic field lines. Faraday proposed in 1847 that light was a high-frequency electromagnetic vibration that could propagate even in the absence of a medium such as the ether.

Faraday's work inspired James Clerk Maxwell to study electromagnetic radiation and light. Maxwell discovered that self-propagating electromagnetic waves would travel through space at a constant speed, equal to the speed of light. From this, Maxwell concluded that light was a form of electromagnetic radiation, first stated in 1862 in *On Physical Lines of Force*. In 1873, he published *A Treatise on Electricity and Magnetism*, which contained a full mathematical description of the behavior of electric and magnetic fields, still known as Maxwell's equations. Soon after, Heinrich Hertz confirmed Maxwell's theory experimentally by generating and detecting radio waves in the laboratory and demonstrating that these waves behaved exactly like visible light, exhibiting properties such as reflection, refraction, diffraction, and interference. Maxwell's theory and Hertz's experiments led directly to the development of modern radio, radar, television, electromagnetic imaging, and wireless communications.

1.1.8 The Special Theory of Relativity

Wave theory was widely successful in explaining nearly all optical and electromagnetic phenomena. However, a handful of experimental anomalies remained that could not be explained by or were in direct conflict with wave theory. One of these anomalies involved a controversy over the speed of light. The constant speed of light predicted by Maxwell's equations and confirmed by the Michelson-Morley experiment contradicted the mechanical laws of motion that had been unchallenged since the time of Galileo, which stated that all speeds were relative to the speed of the observer. In 1905, Albert Einstein resolved this paradox by revising the Galilean model of space and time to account for the constancy of the speed of light. Einstein formulated his ideas in his special theory of relativity, which advanced humankind's understanding of space and time. Einstein also demonstrated a previously unknown fundamental equivalence between energy and mass with his famous equation

$$E = mc^2 \tag{1.1}$$

where E is energy (in Joules, J), m is rest mass (in kilograms, kg), and c is the speed of light in a vacuum (2.998×10^8 meter/second (m/s)).

1.1.9 Particle Theory Revisited

Another experimental anomaly was the photoelectric effect, by which light striking a metal surface ejected electrons from the surface, causing an electric current to flow across an applied voltage. Experimental measurements demonstrated that the energy of individual ejected electrons was proportional to the frequency, rather than the intensity, of the light. Furthermore, below a certain minimum frequency, which depended on the particular metal, no current would flow regardless of the intensity. These observations appeared to contradict wave theory, and for years physicists tried in vain to find an explanation. In 1905, Einstein solved this puzzle as well, this time by resurrecting the particle theory of light to explain the observed effect. Because of the preponderance of evidence in favor of wave theory, however, Einstein's ideas were met initially by great skepticism among established physicists. But eventually Einstein's explanation of the photoelectric effect would triumph, and it ultimately formed the basis for wave-particle duality and much of quantum mechanics.

1.1.10 Quantum Theory

A third anomaly that arose in the late nineteenth century involved a contradiction between the wave theory of light and measurements of the electromagnetic spectrum emitted by thermal radiators, or so-called black bodies. Physicists struggled with this problem, which later became known as the ultraviolet catastrophe, unsuccessfully for many years. In 1900, Max Planck developed a new theory of black-body radiation that explained the observed spectrum correctly. Planck's theory was based on the idea that black bodies emit light (and other electromagnetic radiation) only as discrete bundles or packets of energy. These packets were called quanta, and the particle of light was given the name photon to correspond with other particles being described around this time, such as the electron and proton. A photon has an energy, E (in joules), proportional to its frequency, f (in hertz), by

$$E = hf = \frac{hc}{\lambda} \tag{1.2}$$

where h is Planck's constant (6.625×10^{-34} joule-second (J.s)), λ is the wavelength, and c is the speed of light. $f = \frac{1}{T}$, where T is the period (in seconds), and $\lambda = \frac{c}{f}$.

Often energy is given in terms of electron volts (eV). One eV is equal to 1.6×10^{-19} joules.

Likewise, the momentum p of a photon is also proportional to its frequency and inversely proportional to its wavelength:

$$p = \frac{E}{c} = \frac{hf}{c} = \frac{h}{\lambda} \tag{1.3}$$

As it originally stood, this theory did not explain the simultaneous wave- and particle-like natures of light, though Planck would later work on theories that did. In 1918, Planck received the Nobel Prize in Physics for his part in the founding of quantum theory.

Example 1.1

Determine the frequency and period of a visible light at free space wavelength 400×10^{-9}m (400 nm). Consider the speed of light in free space 3×10^8 m/s.

Solution:

$$\lambda = \frac{c}{f} \rightarrow f = \frac{c}{\lambda}$$

$$f = \frac{3 \times 10^8}{400 \times 10^{-9}} = 7.50 \times 10^{14} \text{ Hz}.$$

Example 1.2

Determine the energy of a photon from an Nd:YAG laser ($\lambda = 1.064 \times 10^{-6}$m) in joules and in terms of electron volts (eV).

Solution:

$$E = \frac{hc}{\lambda} = \frac{\left(6.625 \times 10^{-34}\,\text{J.s}\right)\left(2.988 \times 10^{8}\,\text{m/s}\right)}{1.064 \times 10^{-6}\,\text{m}} = 1.87 \times 10^{-19}\,\text{J}$$

$$E = 1.87 \times 10^{-19}\,\text{J}\left(\frac{1\text{eV}}{1.6 \times 10^{-19}\,\text{J}}\right) = 1.17\text{eV}.$$

1.1.11 Wave-Particle Duality

The modern theory that explains the nature of light includes the notion of wave-particle duality, described by Albert Einstein in the early 1900s, based on his study of the photoelectric effect and Planck's results. Einstein asserted that the energy of a photon is proportional to its frequency. The theory states that everything has both a particle nature and a wave nature, and various experiments can be done to bring out one or the other. It was not until 1924 that Louis de Broglie proposed to the scientific community that electrons also exhibited wave-particle duality. The wave nature of electrons was experimentally demonstrated by Davisson and Germer in 1927. Einstein received the Nobel Prize in 1921 for his work with wave-particle duality on photons (especially explaining the photoelectric effect thereby), and de Broglie followed in 1929 with his extension to other particles.

1.1.12 Quantum Electrodynamics

The quantum mechanical theory of light and electromagnetic radiation continued to evolve through the 1920s and 1930s, and it culminated with the development during the 1940s of the theory of quantum electrodynamics, or QED. This so-called quantum field theory is among the most comprehensive and experimentally successful theories ever formulated to explain a set of natural phenomena. QED was developed primarily by physicists Richard Feynman, Freeman Dyson, Julian Schwinger, and Shin-Ichiro Tomonaga. Feynman, Schwinger, and Tomonaga shared the 1965 Nobel Prize in Physics for their contributions.

1.2 SPEED OF LIGHT

The speed of light in a vacuum is presently defined to be exactly 299,792,458 m/s (about 186,282 miles per second). This definition for the speed of light means that the distance light in meter can travel in second of time through a given substance is now defined in terms of the speed of light. Light always travels at a constant speed, even between particles of a substance through which it is shining. Photons excite the adjoining particles that in turn transfer the energy to their neighbors. This may appear to slow the beam down through its trajectory in real time. The time lost between entry and exit is accounted for by the displacement of energy through the substance between each particle that is excited.

Different physicists have attempted to measure the speed of light throughout history. Galileo attempted to measure the speed of light in the seventeenth century. An early experiment to measure the speed of light was conducted by Ole Rømer, a Danish physicist, in 1676. Using a telescope, Ole observed the motions of Jupiter and one of its moons, Io. Noting discrepancies in the apparent period of Io's orbit, Rømer calculated that light takes about 22 minutes to traverse the diameter of Earth's orbit. Unfortunately, its size was not known at that time. If Ole had known the diameter of the Earth's orbit, he would have calculated a speed of 227,000,000 m/s.

Another, more accurate, measurement of the speed of light was performed in Europe by Hippolyte Fizeau in 1849. Fizeau directed a beam of light at a mirror several kilometers away. A rotating cog wheel was placed in the path of the light beam as it traveled from the source to the mirror and then returned to its origin. Fizeau found that at a certain rate of rotation, the beam would pass through one gap in the wheel on the way out and the next gap on the way back. Knowing the distance to the mirror, the number of teeth on the wheel, and the rate of rotation, Fizeau was able to calculate the speed of light as 313,000,000 m/s.

Léon Foucault used an experiment which used rotating mirrors to obtain a value of 298,000,000 m/s in 1862. Albert A. Michelson conducted experiments on the speed of light from 1877 until his death in 1931. He refined Foucault's methods in 1926 using improved rotating mirrors to measure the time it took light to make a round trip from Mt. Wilson to Mt. San Antonio in California. The precise measurements yielded a speed of 299,796,000 m/s.

1.3 CLASSIFICATION

All the known properties of light are described in terms of the experiments by which they were discovered and the many and varied demonstrations by which they are frequently illustrated. Numerous though these properties are, their demonstrations can be grouped together and classified under one of three heads: geometrical optics, wave optics, and quantum optics, each of which may be subdivided as follows:

Geometrical Optics

- Rectilinear propagation
- Finite speed
- Reflection
- Refraction
- Dispersion

Wave Optics

- Interference
- Diffraction
- Electromagnetic character
- Polarization
- Double refraction

Quantum Optics

- Atomic orbits
- Probability densities
- Energy levels
- Quanta
- Lasers

1.4   UNITS AND MEASURES

Light is measured with two main alternative sets of units: radiometry consists of measurements of light power at all wavelengths, while photometry measures light with wavelength weighted with respect to a standardized model of human brightness perception. Photometry is useful, for example, to quantify illumination intended for human use. The SI units for both systems are summarized in Tables 1.1 and 1.2.

Table 1.1. SI radiometry units.

Quantity	Symbol	SI unit	Abbreviation	Remark
Radiant energy	Q	Joule	J	energy

(continued)

(continued)

Quantity	Symbol	SI unit	Abbreviation	Remark
Radiant flux	ϕ	Watt	W	radiant energy per unit time, also called *radiant power*
Radiant intensity	I	watt per steradian	$W \cdot Sr^{-1}$	power per unit solid angle
Radiance	L	watt per steradian per square meter	$W \cdot Sr^{-1} \cdot m^{-2}$	power per unit solid angle per unit *projected* source area, called *intensity* in some other fields of study
Irradiance	E, I	watt per square meter	$W \cdot m^{-2}$	power incident on a surface, sometimes confusingly called "intensity"
Radiant exitance / Radiant emittance	M	watt per square meter	$W \cdot m^{-2}$	power emitted from a surface
Radiosity	J or J_λ	watt per square meter	$W \cdot m^{-2}$	emitted plus reflected power leaving a surface
Spectral radiance	L_λ or L_v	watt per steradian per meter^3 or watt per steradian per square meter per hertz	$W \cdot Sr^{-1} \cdot m^{-3}$ or $W \cdot Sr^{-1} \cdot m^{-2} \cdot Hz^{-1}$	commonly measured in $W \cdot Sr^{-1} \cdot m^{-2} \cdot nm^{-1}$

(continued)

Quantity	Symbol	SI unit	Abbreviation	Remark
Spectral irradiance	E_λ or E_v	Watt per meter^{-3} or watt per square meter per hertz	$W \cdot m^{-3}$ or $W \cdot m^{-2} \cdot Hz^{-1}$	commonly measured in $W \cdot m^{-2} \cdot nm^{-1}$

Table 1.2. The SI photometry units.

Quantity	Symbol	SI unit	Abbreviation	Notes
Luminous energy	Q_v	lumen second	lm . s	units are sometimes called talbots
Luminous flux	F	lumen (= cd.sr)	lm	also called *luminous power*
Luminous intensity	I_v	candela (= lm/sr)	cd	an SI base unit
Luminance	L_v	candela per square meter	cd/m^2	units are sometimes called "nits"
Illuminance	E_v	lux (= lm/m^2)	lx	used for light incident on a surface
Luminous emittance	M_v	lux (= lm/m^2)	lx	used for light emitted from a surface
Luminous efficacy	K	lumen per watt	Lm/W	ratio of luminous flux to radiant flux

The photometry units are different from most systems of physical units in that they take into account how the human eye responds to light. The cone cells in the human eye are of three types which respond differently across the visible spectrum, and the cumulative response peaks at a wavelength of around 555 nm. Therefore, two sources of light which produce the same intensity (W/m^2) of visible light do not necessarily appear equally bright. The photometry units are designed to take this into account and therefore are a better representation of how "bright" a light appears to be, rather than raw intensity. Photometry units relate to raw power by a quantity called luminous efficacy, which is used for purposes like determining how to best achieve sufficient illumination for various tasks in indoor and outdoor settings. The illumination measured by a photocell sensor does not necessarily correspond to what is perceived by the human eye without filters, which may be costly. Photocells and charge-coupled devices (CCDs) tend to respond to some light that is infrared, ultraviolet, or both.

1.5 BASIC REVIEW OF SCIENTIFIC NOTATION

Scientists, engineers, and technicians often use scientific notation in their work. Scientific notation makes a convenient way to write small or large numbers by using powers of ten. For example, the number 0.000234 becomes 2.34×10^{-4} and the number 234,000 becomes 2.34×10^5. Scientific notation involves a prefix, a decimal point, the number 10, and an exponent (or power). For example, the scientific notation 2.34×10^5 has the prefix = 2.34, the decimal point = the dot (.), base number = 10, and the exponent or power = 5. Each of the numbers in the prefix, 2, 3, or 4, is referred to as a digit. We can write any number as a prefix times a string of tens using scientific notation. For example:

$$300 = 3 \times 100 = 3 \times (10 \times 10) = 3 \times 10^2$$

$$4650 = 4.65 \times 1000 = 4.65 \times (10 \times 10 \times 10) = 4.65 \times 10^3$$

$$0.2 = 2 \times \frac{1}{10} = 2 \times \frac{1}{10^1} = 2 \times 10^{-1}$$

$$0.02 = 2 \times \frac{1}{100} = 2 \times \frac{1}{10^2} = 2 \times 10^{-2}$$

$$0.037 = 3.7 \times \frac{1}{100} = 3.7 \times \frac{1}{10^2} = 3.7 \times 10^{-2}$$

Scientific notation uses the power of ten. In scientific notation, a number is usually expressed as $x.yz \times 10^n$. Engineering notation expresses a number by using the power of 10 as shown in Table 1.3.

Table 1.3. The SI (metric) prefixes.

Power of 10	Prefix	Symbol
10^{24}	yotta	Y
10^{21}	zetta	Z
10^{18}	exa	E
10^{15}	peta	P
10^{12}	tera	T
10^{9}	giga	G
10^{6}	mega	M
10^{3}	kilo	k
10^{-3}	milli	m
10^{-6}	micro	μ
10^{-9}	nano	n
10^{-12}	pico	p
10^{-15}	femto	f
10^{-18}	atto	a
10^{-21}	zepto	z
10^{-24}	yocto	y

Example 1.3

Express each of the following numbers in scientific notation:

   a. 832,704

   b. 0.00000659

   c. 58,012,0000

   d. 0.0000365

Solution:

   a. The decimal point (not shown) is after 4, that is, 832,704.0. If we shift the decimal point to five places to the left, we obtain

     $832,704.0 = 8.32704 \times 10^{5}$, which is in scientific notation.

b. If we shift the decimal point six places to the right, we get
0.00000659 = 6.59×10^6, which is in scientific notation.

c. The decimal point (not shown) is after 0, that is, 58,012,0000.0. If we shift the decimal point to eight places to the left, we obtain
58,012,0000 = 5.8012×10^8, which is in scientific notation.

d. If we shift the decimal point five places to the right, we get
0.0000365 = 3.65×10^5, which is in scientific notation.

Example 1.4

Express each of the following numbers in scientific notation:

a. 68,300,000,000 m

b. 894,000,000 m

c. 0.0054 s

d. 0.0000932 s

Solution:

a. 68,300,000,000 m = 68.3×10^9 = 68.3Gm

b. 894,000,000 = 894×10^6m = 894 Mm

c. 0.0054s = 5.4×10^{-3}s = 5.4ms

d. 0.0000932s = 93.2×10^{-6}s = 93.2 μs

1.6 LIGHT PRESSURE

Light pushes on objects in its path, just as the wind would do. This pressure is most easily explainable in particle theory: photons hit and transfer their momentum. Light pressure can cause asteroids to spin faster, acting on their irregular shapes as wind would on the vanes of a windmill. The possibility to make solar sails that could accelerate spaceships in space is also under investigation.

Although the motion of the Crookes radiometer was originally attributed to light pressure, this interpretation is incorrect; the characteristic Crookes rotation is the result of a partial vacuum. This should not be confused with the Nichols radiometer, in which the motion is directly caused by light pressure.

1.7   OPTICS

Optics is the study of light and the phenomena associated with its generation, transmission, and detection. In a broader sense, optics includes all the phenomena associated with visible, infrared, and ultraviolet radiation. Geometrical optics assumes that light travels in straight lines and is concerned with the laws controlling the reflection and refraction of rays of light. Physical optics deals with phenomena that depend on the wave nature of light, for example, diffraction, interference, and polarization.

1.8   EXERCISES

1. According to the Samkhya School, what is one of the five fundamental "subtle" elements?

2. Who wrote that the planets and the moon do not have their own light, but reflect the light of the sun?

3. Empedocles postulated that everything was composed of what four elements?

4. Muslim scientist Ibn Al-Haytham developed a broad theory of vision based on _____ and _____ in *Book of* _____.

5. Who is considered the father of modern optics?

6. Qutb Al-Din Al-Shirazi and his student Kamal Al-Din Al-Farisi were the first to do what?

7. According to Descartes, what is plenum?

8. Descartes's theory is often regarded as the forerunner of the wave theory of _____.

9. What did Isaac Newton state in his *Hypothesis of Light*?

10. What are corpuscles?

11. Who published the wave theory of light in the 1600s?

12. Who was the first scientist to make a sufficiently accurate measurement about the speed of light?

13. What is Albert Einstein's famous equation?

14. According to Planck's theory, what name was the particle of light given?

15. The momentum p of a photon is also _____ to its _____ and _____ _____ to its _____.

16. In what year did Planck receive a Nobel Prize in Physics for his part in the founding of quantum theory?

17. How is the speed of light in a vacuum defined?

18. _____ includes all the phenomena associated with visible, infrared, and ultraviolet radiation.

In exercises 19–28, express the following numbers in scientific notation.

19. 73,400

20. 163,500

21. 800,000,000

22. 363

23. 0.0042

24. 0.831

25. 0.000036

26. 0.009

27. 3,452 m

28. 72.5 μs

In exercises 29–34, express the following scientific notation in decimal notation.

29. 3.16×10^{-1}

30. 84.35×10^{-4}

31. 295.8×10^{-4}

32. 7.321×10^{4}

33. 645×10^{-5}

34. 9×10^{8}

In exercises 35–38, express the following numbers in engineering notation.

35. 280×10^{-7}

36. 70×10^4

37. 9.2×10^{-3}

38. 0.7×10^{-9}

39. Determine the frequency and period of a visible light at free space wavelength 750×10^{-9} m (750 nm). Consider the speed of light in free space 3×10^8 m/s.

40. Determine the frequency and period of a visible light at free space wavelength 570×10^{-9} m (570 nm). Consider the speed of light in free space 3×10^8 m/s.

41. Determine the frequency, vacuum wavelength, and energy in joules of a photon having energy of 2.5 eV. Consider the speed of light in free space 3×10^8 m/s.

42. Determine the frequency, vacuum wavelength, and energy in joules of a photon having energy of 1.65 eV. Consider the speed of light in free space 3×10^8 m/s.

43. Find the momentum of a single photon of blue light ($f = 606$ THz) moving through free space.

44. Find the momentum of a single photon of yellow light ($f = 508$ THz) moving through free space.

45. Determine the photon energy and momentum values for a visible light at free space wavelength 590×10^{-9} m (590 nm). Consider the speed of light in free space 3×10^8 m/s.

46. Determine the photon energy and momentum values for a visible light at free space wavelength 620×10^{-9} m (620 nm). Consider the speed of light in free space 3×10^8 m/s.

CHAPTER 2

REFLECTION AND REFRACTION

2.0 INTRODUCTION

The term *light* in common phraseology is used to denote that aspect of radiant energy by which objects are made visible, due to stimulation of the retina of the eye. Nowadays it has become customary to include in this term certain other kinds of radiation called ultraviolet and infrared which, although incapable of exciting the sense of sight, nevertheless show other effects similar to those of visible light. For example, the radiation in the near ultraviolet region affects the photographic plate even more markedly than that in the visible region. The ultraviolet region lies beyond the violet end of the visible spectrum, while the infrared region lies beyond the red one.

The study of nature and properties of light forms a subject in physics called optics, which can be conveniently divided into three distinct branches: (a) geometrical optics, in which many facts concerning light can be investigated from the standpoint of the ray theory, that is, on the supposition that light travels in straight lines or rays, while no assumptions are made in it regarding the nature of light; (b) physical optics, in which certain phenomena like propagation, reflection, refraction, interference, diffraction, and polarization exhibited by light are studied from the standpoint of the wave theory; and (c) quantum optics, in which one studies the interaction of light with the atomic entities of matter.

Geometrical optics mainly deals with image formation by mirrors, lenses, and prisms. The whole subject can be developed on the basis of the following four fundamental properties of light and with the aid of geometrical or trigonometrical calculations:

1. In a homogeneous medium, light travels in straight lines.

2. Rays of light are independent, and any two of them can intersect each other without in any way affecting their future paths, which are the same as if each ray existed separately.

3. A beam of light is reflected from any optical surface in accordance with the following laws:

    a. The reflected ray lies in the plane of incidence, which is the plane containing the incident ray and the normal at the point of incidence.

    b. The angles of incidence and reflection are equal, the reflected ray and the incident ray being on the opposite sides of the normal.

4. The refraction of light from one medium to another is governed according to the following laws:

    a. The refracted ray lies in the plane of incidence.

    b. The sine of the angle of incidence bears a constant ratio to the sine of the angle of refraction.

The value of this ratio depends upon the media involved and on the wavelength of light. This law is called *Snell's law*, and in symbols it is expressed in the form

$$\frac{\sin i}{\sin r} = \frac{V_1}{V_2} = \frac{\mu_2}{\mu_1} = \mu_{21} \tag{2.1}$$

where V_1 is the velocity of light in the medium in which incident ray travels while V_2 is the velocity of light in the medium into which the light is refracted; μ_1 and μ_2, the indices of refraction of the first and the second medium with respect to a vacuum, respectively, are defined as

$$\mu_1 = \frac{\text{Velocity of light in vacuum}}{\text{Velocity of light in first medium}}$$

$$\mu_2 = \frac{\text{Velocity of light in vacuum}}{\text{Velocity of light in second medium}}$$

With the help of these fundamental laws it is possible to trace the path of light rays through any optical system and the formation of images studied. Accordingly, this branch of optics has important applications in the design, manufacture, selection, and use of excellent optical instruments.

Ray theory, however, fails to explain interference and diffraction phenomena. Moreover, a closer examination of the fact reveals that the rectilinear propagation of light is only approximate. These phenomena are explained on the basis of the wave nature of light. But owing to the small wavelength of visible light, the approximate limits being 4×10^{-5} cm and 7.5×10^{-5} cm, the rectilinear propagation of light is observed under normal circumstances, and hence assumptions made in geometrical optics are nearly true. Geometrical optics can therefore be regarded as an asymptotic form of wave optics when the wavelength of light is taken to be negligibly small, and under this limitation many results of wave optics pass over into those of ray optics.

The laws of geometrical optics, which have all been experimentally verified, can be derived from one basic principle called Fermat's principle of least time.

2.1   FERMAT'S PRINCIPLE

It states that "when light travels from one point to another it always follows the path along which the time taken is the least." In other words, the path of light between two given points is determined by the condition that the variation in the optical path is zero, that is

$$\delta \sum (\mu \cdot l) = 0$$

for any infinitesimal deviation from the actual path.

Stated in this form the law holds even if, in going from one point to another, light suffers one or more reflections or refractions. Moreover, it is only the optical path and not the direction of the propagation of light which is involved. The principle that the path of light is reversible as long as the media and the order in which light passes through these remains unchanged is, therefore, also included in this statement.

If we consider a ray of light traveling from one point to another in a homogeneous medium, Fermat's law implies that the length of its path between the two points should be the least. This path is clearly a straight line. It is therefore obvious that in an isotropic medium, light always travels in straight lines, and the law of rectilinear propagation results as a direct consequence of Fermat's principle.

Let us now apply the principle to deduce the laws of reflection and refraction.

2.1.1 Laws of Reflection

Let PQ in Figure 2.1 be a plane mirror. Suppose a ray of light starting from A reaches D after reflection from the mirror along the path ABD.

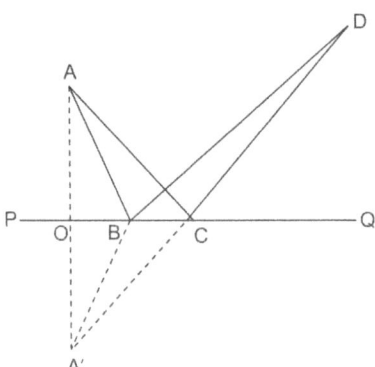

FIGURE 2.1. Law of reflection.

Draw AO normal to the mirror and produce it to A' such that AO = OA'. It follows, therefore, that whatever the position of B is we shall always have AB = A'B, and hence the length of the path ABD is always equal to A'BD, that is, ABD = A'BD.

Since the light is traveling in the same medium, the time taken by the light in going from A to D via the mirror will be the least if the length of the path itself is at a minimum.

If the light goes along ABD, the length of the path (as seen previously) is A'BD. It is clear from the triangle A'BD that

$$A'B + BD > A'D$$

or in other words, the minimum length from A' to D is A'CD. Hence, according to Fermat's principle, the light must follow the path ACD (since AC = A'C).

Now the triangles ACO and A'CO are congruent because
(*i*) AO = A'O, (*ii*) ∠AOC = ∠A'OC, and (*iii*) OC is common to both. It follows, therefore, that

$$\angle AOC = \angle OCA' \tag{2.2}$$

Since the straight lines PCQ and A'CD intersect at C,

$$\angle PCA' = \angle DCQ \tag{2.3}$$

So that the combination of Equations (2.2) and (2.3) give

$$\angle ACO = \angle DCQ$$

which is the law of reflection.

Example 2.1

Light is incident on a flat surface, making an angle of 20° with that surface, as shown in Figure 2.2. (a) What is the angle of incidence? (b) What is the angle of reflection? (c) Sketch the path of the reflected beam on the diagram.

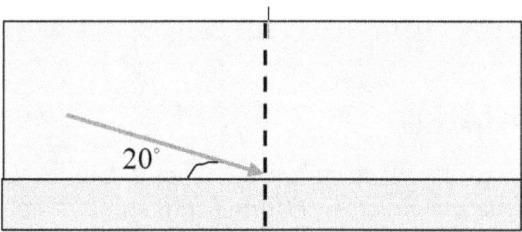

FIGURE 2.2. For Example 2.1.

Solution:

a. When the light makes an angle of 20° with the surface, it makes an angle of 70° with the normal to the surface. Thus, the angle of incidence (θ_i) is 70°.

b. Based on the *law of reflection*, the angle of reflection equals the angle of incidence. So the angle of reflection, θ_r (measured to the normal), is 70°.

c. The path of the light is shown in Figure 2.3 as follows:

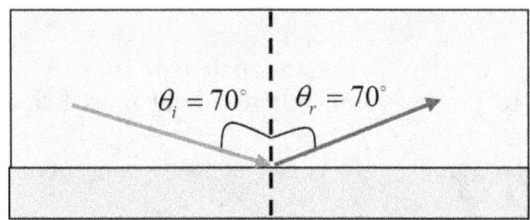

FIGURE 2.3. For solution to (c) of Example 2.1.

Example 2.2

A light ray strikes a reflective plane surface at an angle of 52° with that surface. (a) What is the angle of incidence? (b) What is the angle of reflection? (c) What is the angle made by the reflected ray and the surface? (d) What is the angle made by the incident and reflected rays?

Sketch the path of the reflected beam on the diagram.

Solution:

a. The angle of incidence is $\theta_i = 90° - 52° = 38°$.

b. The angle of reflection is $\theta_r = \theta_i = 38°$ (by the law of reflection).

c. The angle made by the reflected ray and the surface is $\theta_{rs} = 90° - 38° = 52°$.

d. The angle made by the incident and reflected rays is $\theta_i + \theta_r = 38° - 38° = 76°$.

2.1.2 Laws of Refraction

Suppose a ray of light starting from a point A in the first medium arrives at a point B in the second after suffering refraction at the plane surface of separation PQ as shown in Figure 2.4. If D is a point on PQ very near C, it follows from Fermat's principle that the time taken by the ray of light along the path ADB will, in the limit, be equal to that taken along ACB.

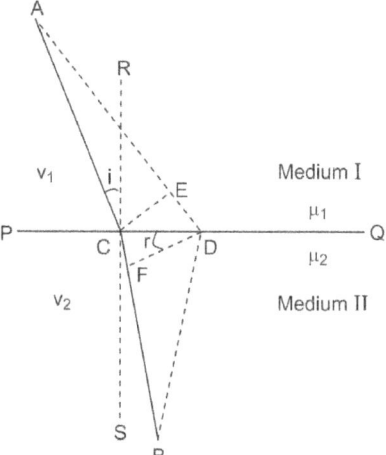

FIGURE 2.4. Law of refraction.

Draw CE perpendicular to AD and DF perpendicular to CB. Then as D approaches C, the length AE will approach AC, and similarly BD becomes more and more equal to BF. Hence, according to Fermat's principle, the time taken by light in covering the distance DE with a velocity V_1 will be the same as that taken by it in covering the distance CF with a velocity V_2. Therefore

$$\frac{ED}{V_1} = \frac{CF}{V_2} \qquad (2.4)$$

Or

$$\frac{ED}{CF} = \frac{V_1}{V_2} = \frac{\mu_2}{\mu_1}$$

where μ_1 and μ_2 are the refractive indices of media I and II respectively. This equation can be written as

$$\frac{ED/CD}{CF/CD} = \frac{\mu_2}{\mu_1}$$

Or

$$\frac{\sin ECD}{\sin CDF} = \frac{\mu_2}{\mu_1} \qquad (2.5)$$

Since

$$\angle ECD = \frac{\pi}{2} - \angle RCE = i \text{ the angle of incidence and}$$

$$\angle CDF = \frac{\pi}{2} - \angle DCF = r \text{ the angle of refraction.}$$

Equation (2.5) gives

$$\frac{\sin i}{\sin r} = \frac{\mu_2}{\mu_1}$$

If the first medium is air $\mu_1 = 1$, the previous equation reduces to

$$\frac{\sin i}{\sin r} = \mu$$

which is Snell's law. Here μ refers to the refractive index of the second medium with respect to air.

It has been experimentally observed that when light passes from a rarer medium to a denser medium, for example, from air to water, the refracted ray bends toward the normal, or the angle of refraction "r" is less than the angle of incidence i as shown in Figure 2.5, and hence μ is greater than 1.

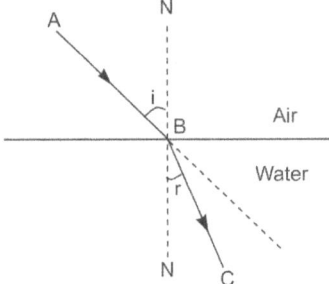

FIGURE 2.5. Refracted ray from air to water.

It has also been proven that the refractive index μ is the ratio of the velocities of light in the two media. Thus

$$\frac{\sin i}{\sin r} = \mu_{aw} = \frac{\text{Velocity of light in air}}{\text{Velocity of light in water}} = \frac{V_{air}}{V_{water}}$$

If instead of air the first medium is a vacuum, then

$$\mu_w = \frac{V}{V_{water}}$$

where "V" is the velocity of light in a vacuum and μ_w is the absolute refractive index of water. Also,

$$\mu_{aw} = \frac{V_{air}}{V_{water}} = \frac{V}{V_{water}} + \frac{V}{V_{air}} = \frac{\mu_w}{\mu_a}$$

In practice μ_a is taken as unity, and hence μ_{aw} does not appreciably differ from μ_w.

It can be readily seen that

$$\mu_{aw} = \frac{1}{\mu_{wa}}$$

or

$$\mu_{aw} \times \mu_{wa} = 1$$

In general, for more than two media

$$\mu_{12} \times \mu_{23} \times \mu_{34} \times \mu_{41} = 1$$

Example 2.3

Light travels from air into an optical fiber with an index of refraction of 1.40. If the angle of incidence on the end of the fiber is 25°, what is the angle of refraction inside the fiber?

Solution:

We know that $\mu_1 \sin i = \mu_2 \sin r$, so

$$(1)\sin 25° = (1.40)\sin r$$

$$\sin r = \frac{(1)\sin 25°}{(1.40)} \to r = \sin^{-1}\left(\frac{0.4226}{1.40}\right) = 17.6°.$$

The angle of refraction inside the fiber = 17.6°.

2.1.3 Breakdown of Snell's Law – Total Internal Reflection

It has been seen previously that $\sin i = \mu \sin r$, where μ is the refractive index of the denser medium with respect to air and is greater than unity. If light is now considered to be traveling from water to air as shown in Figure 2.6, the angle of incidence i will be less than the angle of refraction r and, consequently, μ_{wa} will be less than unity. As the angle of incidence i is increased, the refracted ray traveling in the rarer medium bends more and more away from the normal NN, till at a particular angle of incidence C, the refracted ray travels along the surface of separation of air and water or the angle of refraction $r = 90°$.

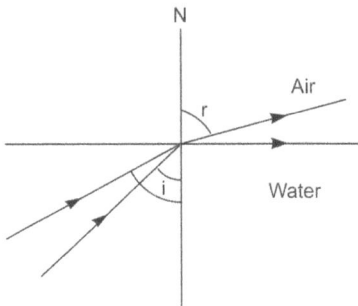

FIGURE 2.6. Critical angle.

In that case,

$$\mu_{wa} = \frac{\sin C}{\sin \pi / 2} = \sin C$$

The angle of incidence given by $\mu_{wa} = \sin C$ is known as the critical angle. If this angle is increased further, "$\sin r$" must be greater than unity, which is impossible. Consequently, there can be no refracted ray, and Snell's law breaks down. In this case light is reflected back in the (denser) medium, and the phenomenon is known as total internal reflection.

2.2 APLANATIC SURFACES

In the foregoing discussion the reflecting and refracting surfaces were considered plane. But if the surfaces are curved, it is no longer necessary that the length of the path followed by the light always be a minimum; it may sometimes be maximum as well. In such cases Fermat's principle of least time is not applicable. A consideration of aplanatic surfaces throws further light on such situations.

An aplanatic surface is such that from any point on it, the sum of the optical paths to any two points is a constant quantity. If P is any point such that

$$\mu_1 AP + \mu_2 BP = \text{constant} \qquad (2.6)$$

then the locus of the point P will be an aplanatic surface.

The curve represented by Equation (2.6) in one plane is called a Cartesian oval and is shown in Figure 2.7.

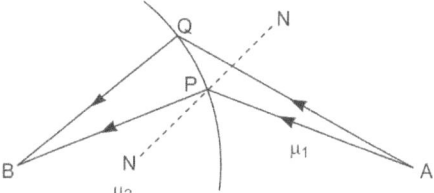

FIGURE 2.7. Definition of aplanatic surface.

If the figure is rotated about the axis AB, the corresponding aplanatic surface is obtained.

In the case of reflection, light is confined to a single medium, and Equation (2.6) reduces to

$$AP + BP = \text{constant} \qquad (2.7)$$

which is an ellipsoid of revolution with points A and B as foci as shown in Figure 2.8.

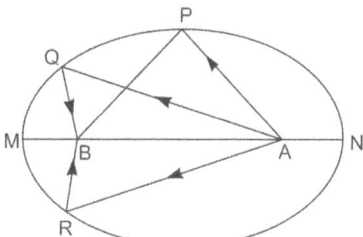

FIGURE 2.8. Aplanatic surface.

2.2.1 Ellipsoidal and Paraboloidal Mirrors

Let PQR as in Figure 2.8 be an arc of the ellipse whose major axis is MN and whose foci are A and B. If the arc is rotated about the major axis, a trace of the surface called the ellipsoid of revolution is obtained. The arc PQR may, therefore, be considered as the principal section of an ellipsoidal mirror.

Now, it is the property of an ellipse that the sum of the distances from any point on the ellipse to the two foci is a constant quantity:

i.e.,   *AP* + *BP* = constant

Hence, if a point source of light is placed at *A*, all the rays starting from *A* will, after reflection from *PQR*, pass through the point *B*. *PQR* is then an aplanatic surface, and the two conjugate points *A* and *B* are called aplanatic foci.

As the distance between the foci of the ellipse increases, it approximates more and more to a parabola. If, therefore, it is desired to converge all the rays coming from infinity to one single point, the mirror to be used must have a principal section which is a parabola. The mirror itself is called a paraboloidal mirror, as shown in Figure 2.9.

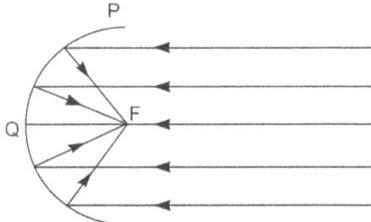

FIGURE 2.9. Paraboloidal mirror.

It may be mentioned here that a spherical mirror cannot be used for bringing a parallel beam of light to a point focus.

2.3 REFLECTION AT CURVED SURFACES – LAW OF EXTREME PATHS

Suppose the rays of light starting from a point object at *A* arrive at *B* after reflection from the mirror *MM*, as shown in Figure 2.10.

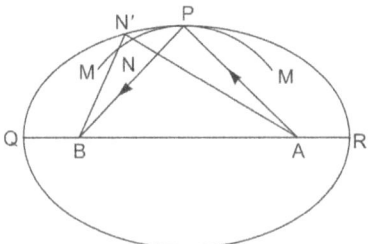

FIGURE 2.10. Law of extreme paths.

If an aplanatic surface is drawn with points A and B as its foci, its equation will be

$AP + BP$ = constant.

Such a surface will be tangential to the mirror at the point P. Take any point N on MM which is very near P. Join AN and produce it to meet the surface PQR at N'. Since N' lies on PQR, it must satisfy the condition

$$AN' + N'B = AP + BP \tag{2.8}$$

According to Fermat's principle

$AN + NB < AP + BP$

Now,

$$AN' + N'B = AN + NN' + N'B \tag{2.9}$$

Also, in the triangle $N'NB$

$$NN' + N'B > NB \tag{2.10}$$

Hence, it follows that

$AP + BP = AN + NN' + N'B > AN + NB$

It is thus seen that the time taken by light along the path APB, instead of being a minimum, is maximum.

It may be pointed out that if the curvature of MM is less than that of the aplanatic surface PQR, Fermat's principle of least time will hold.

Similarly, it can be shown that if a convex refracting surface is more convex than the corresponding aplanatic surface, the path followed by light in going from a less-refracting medium to a more-refracting medium is a maximum. On the other hand, if it is less convex or concave, the path is a minimum.

But whether the total optical path is a maximum or a minimum, its first derivative will always vanish, and hence the law can be written in a modified form, that is

$$\int_A^B \frac{ds}{V} = \text{max., min. or constant}$$

where "ds" is a small element of a path between any two points A and B, and "V" is the velocity of light in the medium. Since $\mu = c / V$, where "c" is the velocity of light in a vacuum, the previous equation may be written as

$$\int_A^B \mu \, ds = \text{max., min. or constant}$$

or

$$\delta \int_A^B \mu \, ds = 0$$

Fermat's law modified in this form is known as the *law of extreme paths* and may be stated as follows:

> Whenever light travels from one point to another (it may reach the second point directly or after one or more reflections and refractions) it always follows the path which is extreme, i.e., either maximum or minimum.

2.4 REFRACTION AT CURVED SURFACES

Let *RR*, as shown in Figure 2.11, be a curved refracting surface of radius *r* and center of curvature *C*. Suppose a ray of light *OP* starting from *O* placed at a distance "*u*" from *V* reaches *I* after suffering refraction at the surface.

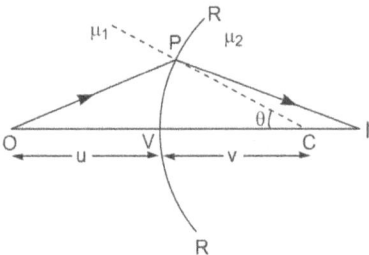

FIGURE 2.11. Refraction at curved surface.

Let *VI* = *v*. In the triangle *OPC*

$$OP^2 = r^2 + (u+r)^2 - 2r(u+r)\cos\theta \qquad (2.11)$$

Similarly, in the triangle *PCI*

$$PI^2 = r^2 + (v-r)^2 + 2r(v-r)\cos\theta \qquad (2.12)$$

According to the law of extreme paths, the first derivative of the total optical path should vanish, and hence

$$\delta(\mu_1 OP + \mu_2 PI) = 0 \qquad (2.13)$$

Equation (2.13) when combined with (2.11) and (2.12) yields

$$\delta[\mu_1\{r^2 +(u+r)^2 -2r(u+r)\cos\theta\}^{\frac{1}{2}}$$
$$+\mu_2\{r^2 +(v-r)^2 +2r(v-r)\cos\theta\}^{\frac{1}{2}}] = 0 \tag{2.14}$$

Differentiating partially with respect to θ (because θ varies if the path is changed), it is seen that

$$\frac{\mu_1 r(u+r)\sin\theta}{\left[r^2 +(u+r)^2 -2r(u+r)\cos\theta\right]^{1/2}}$$
$$+\frac{\mu_2[-r(v-r)\sin\theta]}{\left[r^2 +(v-r)^2 +2r(v-r)\cos\theta\right]^{1/2}} = 0 \tag{2.15}$$

If the angle θ is small, $\cos\theta$ will be nearly unity, and Equation (2.15) gives

$$\frac{\mu_1(u+r)}{u} = -\frac{\mu_2(v-r)}{v}$$

or

$$\frac{\mu_1}{u}+\frac{\mu_2}{v} = \frac{\mu_2 -\mu_1}{r} \tag{2.16}$$

In this case, however, "v" and "r" are positive and, therefore,

$$\frac{\mu_2}{v}-\frac{\mu_1}{u} = \frac{\mu_2 -\mu_1}{r} \tag{2.17}$$

If the medium to the left of the refracting surface is air, the previous equation simplifies to

$$\frac{\mu}{v}-\frac{1}{u} = \frac{\mu-1}{r}$$

2.5 REFRACTION AT A LENS

Suppose LAB is a convex lens whose refractive index is μ. Let P be a point object placed on the axis of the lens at a distance u from the optical center, that is, $OP = u$, as seen in Figure 2.12.

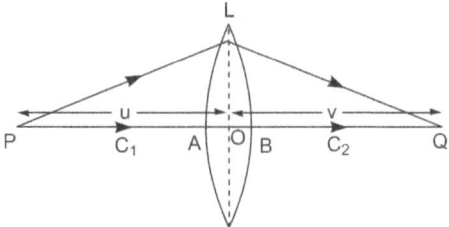

FIGURE 2.12. Refraction at a lens.

Any ray *PL* striking the lens at *L* will after refraction converge at the point *Q*. Similarly, another ray *PA* directed along the axis of the lens will go undeviated after refraction and cut the first refracted ray *LQ* at *Q* such that *OQ* = *v*. Then *Q* is the image of *P*.

Since the rays starting from *P* intersect at *Q*, the optical paths along the two directions according to the law of extreme paths must be the same, and hence

$$PL + LQ = PA + \mu AB + BQ = \text{max., min. or constant.} \qquad (2.18)$$

Since the ray *PA* travels a distance AB in glass, its equivalent optical path in air will be μAB. Also, in the triangle *PLO*

$$PL^2 = PO^2 + OL^2$$

and if *OL* is small as compared to *OP*

$$PL^2 = PO^2\left(1 + \frac{OL^2}{OP^2}\right)$$

or

$$PL \cong PO + \frac{OL^2}{2OP} \qquad (2.19)$$

Similarly, the triangle *OLQ* gives

$$OL \cong QO + \frac{OL^2}{2OQ} \qquad (2.20)$$

If these values of *PL* and *QL* are introduced in Equation (2.18), the latter reduces to

$$PO + QO + \frac{OL^2}{2}\left(\frac{1}{OP} + \frac{1}{OQ}\right) = PA + \mu AB + BQ$$

or

$$PO + QO + \frac{1}{2}OL^2 \left(\frac{1}{OP} + \frac{1}{OQ} \right) = PO - OB + QO - OA + \mu(AO + OB)$$

or

$$\frac{1}{2}OL^2 \left(\frac{1}{OP} + \frac{1}{OQ} \right) = (\mu - 1)(AO + BO)$$

Therefore,

$$\frac{1}{v} + \frac{1}{u} = 2(\mu - 1) \frac{AO + BO}{OL^2} \qquad (2.21)$$

For the surface whose radius of curvature is r_1 and center is C_1, it can be seen that

$$OL^2 = OB(2r_1 - OB)$$

If OB is small as compared to r_1, that is, if the lens is thin,

$$OL^2 = OB(2r_1) \qquad (222)$$

Similarly, it can be shown that for the surface of the radius of curvature r_2 and center C_2

$$OL^2 = OA(2r_2) \qquad (2.23)$$

On introducing the values of OL^2 from Equations (2.22) and (2.23) in Equation (2.21), it becomes

$$\frac{1}{v} + \frac{1}{u} = (\mu - 1) \left(\frac{1}{r_1} + \frac{1}{r_2} \right) \qquad (2.24)$$

where the proper signs are to be inserted.

2.6 EXERCISES

1. What are the three branches of optics?

2. In a homogenous medium, light travels in _____ lines.

3. Write the equation for *Snell's law*.

4. State Fermat's principle and use it to establish the laws of reflection and refraction of light.

5. State and explain Fermat's law of extreme paths and analyze a case where the actual path of light may be at a maximum. Use Fermat's law to deduce Snell's law of refraction.

6. Find angle $\theta°$ made by the system of the two mirrors (M_1 and M_2) shown in Figure 2.13 so that the incident ray at P_1 and the reflected ray at P_2 are parallel.

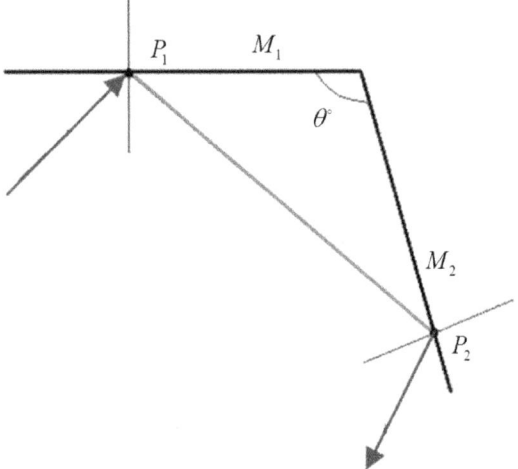

FIGURE 2.13. For Exercise 6.

7. Find angle θ of reflection at the point of incidence P_2 when a ray of light is reflected by the system of the two parallel mirrors (M_1 and M_2) at points P_1 and P_2, as shown in the following figure. The ray makes an angle of $30°$ with the axis of the two mirrors.

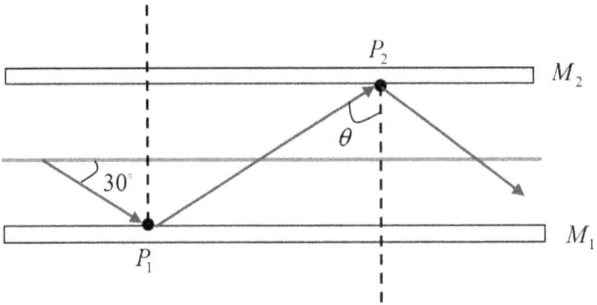

FIGURE 2.14. For Exercise 7.

8. Show that if the refracting angle of a glass prism is larger than twice the critical angle of refraction, no light can be passed through it by refraction.

9. Discuss the variation of deviation with the angle of incidence for a prism of lower refractive index than the surrounding medium.

10. Calculate the angle of minimum deviation for a 60° prism ($\mu = 1.5$) and the angle of incidence i at which it occurs. Hence, find the deviation for rays at incident angles $i + 15°$ and $i + 10°$ and then calculate the angular separation between the pair of rays on emergence. Also calculate the angular separation of rays on emergence which struck the prism at $i - 15°$ and $i - 10°$. Hence, explain Schuster's method of focusing the spectrometer.

11. Write notes on the following:

    a. Fermat's principle

    b. Dispersive power of a medium

    c. Irrationality of prismatic dispersion

CHAPTER 3

REFRACTION AT A SPHERICAL SURFACE

Chapter Outline

3.0 INTRODUCTION

The study of refraction of light through a single spherical surface separating two media of different indices of refraction is of great fundamental importance in geometrical optics. Therefore, we begin the study of this branch of optics by developing the theory of image formation by a single spherical refracting surface. In addition, the detailed discussion will be limited to *paraxial rays*, that is, rays inclined at small angles to the axes, for the reason that the results of paraxial theory also hold in the optics of "*corrected*" systems.

3.1 CONVENTION OF SIGNS

The derivations of various formulae in this branch of optics are based upon the measurement of angles and distances from a suitably chosen axis and origin in every optical system. It is therefore essential to adopt a convention of signs for distances and angles in order to ensure consistency in the derivations and use of various formulae. We shall adopt the following set of conventions which agree with the usual convention of the *Cartesian system of coordinates used in coordinate geometry*.

3.1.1 Convention of Distances

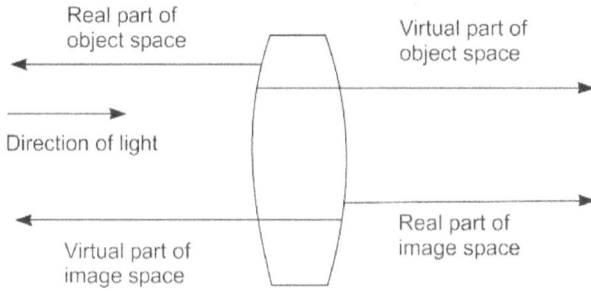

FIGURE 3.1. Object space and image space exist through all space.

1. Draw all figures with light advancing from left to right.

2. Consider all distances measured to the right of a suitably chosen origin *positive* and all distances to the left *negative*. Thus, all longitudinal distances measured from some origin will be considered positive when they are along the initial direction of the propagation of light, which is assumed to be always from left to right. In other words, according to this convention, the object distance *u* of a real object point is negative, whereas it is positive for a virtual object point. Similarly, the image distance *v* is positive for a real image point and negative for a virtual image point.

3. Consider a distance transverse to the axis *positive* when measured *upward* from it and *negative* when measured *downward*.

3.1.2 Convention for Angles

1. Consider the slope angle that a ray makes with the axis to be positive when an anticlockwise turn through this acute angle will bring a ruler from the principle axis direction into coincidence with that of the ray.

2. Consider angles of incidence and refraction positive when a ruler from a normal direction at the point of incidence on the surface must be rotated in the anticlockwise direction through this angle to bring it into coincidence with the ray.

It is advisable that this convention of signs should be memorized at this very stage, because this would make it easier to grasp and to derive consistent relations in this branch of optics.

3.2 REFRACTION AT A SPHERICAL SURFACE

Let us suppose that in Figure 3.2., the common boundary of two transparent homogeneous media of refractive indices μ_1 and μ_2 (where $\mu_1 < \mu_2$) is spherical in form, being convex toward the medium of index μ_1, in which is situated a luminous object point *P*.

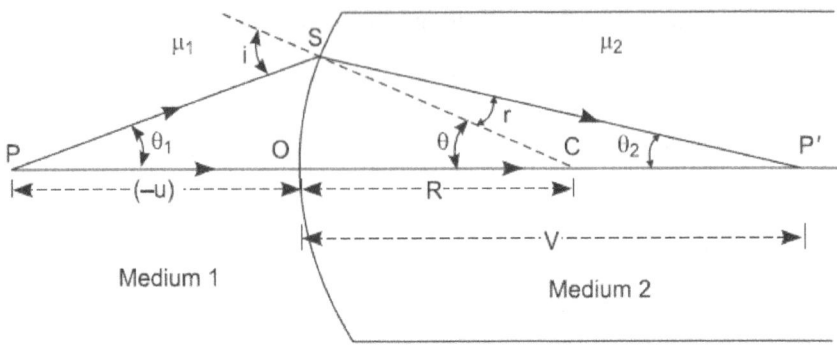

FIGURE 3.2. Refraction at convex surface forming a real image.

A line joining P and the center C of the spherical surface is called the axis of the system, and its intersect O with the refracting surface is called the pole of the surface. We shall consider this point O as the origin of the system. A ray PS from the axial object point P, inclined at an angle θ_1 with the axis, is incident on the surface at an angle i with the normal SC. Since $\mu_1 < \mu_2$, the incident ray PS on refraction at S bends toward the normal in accordance with the fundamental laws of refraction and pursues the path inclined at an angle r with the normal. The slope angle of the refracted ray SP' with the axis is θ_2. Another ray PO, which is coincident with the axis and therefore incident normally on the surface, enters undeviated in the second medium. The point of intersection P' of the ray SP' and the axial ray is, therefore, the image of the object point P. For some object positions with respect to the refracting surface, as illustrated in Figure 3.3, the refracted ray intersects the axial ray only when the former is produced backward. Accordingly, in such cases a virtual image is formed.

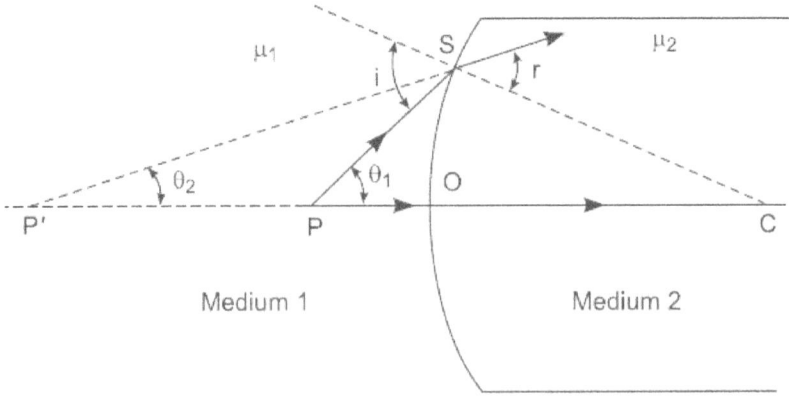

FIGURE 3.3. Refraction at convex surface forming a virtual image.

We can proceed to derive a relation between u and v, the distances of the object point P and its image P' from O, involving the constants of the system, namely μ_1, μ_2, and R, the radius of curvature of the refracting surface. It should be noticed that according to the convention of signs, in Figure 3.2 u is negative while v and R are positive, the pole O being the origin. The slope angle θ_1 of incident ray PS with the axis is positive, whereas the slope angle θ_2 of the refracted ray SP' is negative.

Now, in triangle PSC the law of sines gives

$$\frac{\sin\theta_1}{R} = \frac{\sin(\pi - 1)}{R + (-u)} = \frac{\sin i}{R - u}$$

Solving for i,

$$\sin i = \frac{R - u}{R}\sin\theta_1 \tag{3.1}$$

The angle of refraction r, conjugate to the angle of incidence i, may be found with Snell's law.

$$\frac{\sin i}{\sin r} = \frac{\mu_2}{\mu_1}$$

Solve for r,

$$\sin r = \left(\frac{\mu_1}{\mu_2}\right)\sin i \tag{3.2}$$

Slope angle θ_2 of the refracted ray should now be computed. In triangle PSP' the sum of the angles is π radians, that is

$$\theta_1 + (\pi - i) + r + (-\theta_2) = \pi$$

since θ_2 by convention is negative. On solving for θ_2

$$\theta_2 = r + \theta_1 - i \tag{3.3}$$

The image distance v from the pole O should now be computed. In triangle $P'SC$, the law of sines gives

$$\frac{\sin r}{v - r} = \frac{\sin(-\theta_2)}{R}$$

$$\sin\theta_2 = \left(\frac{R}{v - r}\right)\sin r \tag{3.4}$$

Solving for v,

$$v = R - \left(R \frac{\sin r}{\sin \theta_2} \right)$$

and with the help of Equations (3.1), (3.2), and (3.3), the previous equation can be easily given as

$$v = R - \left(\frac{\mu_1}{\mu_2} \right) \left(\frac{(R-u)\sin \theta_1}{\sin(r + \theta_1 - i)} \right)$$

Equations (3.1), (3.2), (3.3), and (3.4) are quite sufficient to determine the image distance v and the slope angle θ_2 conjugate to quantities u and θ_1 respectively. In general, the image distance v will be different for different slope angle θ_1 of the incident ray, because the ratio $\dfrac{\sin \theta_1}{\sin(r + \theta_1 - i)}$ is not constant for a spherical surface. Therefore, the image of the object point will not be a point image. This departure from the ideal point image of the axial object point is called *spherical aberration*.

3.2.1 First Order Theory

The sine of any angle can be expanded by *Maclaurin's theorem* into a power series in the angle, for example

$$\sin \theta = \theta - \frac{\theta^3}{3!} + \frac{\theta^5}{5!} - \frac{\theta^7}{7!} + \dots$$

For incident rays restricted to the region close to the axis, called the paraxial region, the slope angle θ_1 becomes very small, and as a consequence the slope angle θ_2 of the conjugate refracted ray, the angles i and r also become small to the same order of approximation. Rays which satisfy these restrictions are called paraxial rays, and for the refraction of these rays a very simple relationship between v, u, μ_1, μ_2, and R can be obtained by replacing, to a close approximation, the sine of the angle in Equations (3.1), (3.2), and (3.3) with the angle (in radians) itself. To assume $\sin \theta_1 = \theta_1$ is to neglect in the series all terms of higher order in θ_1 beyond the first order term and, therefore, the theory based on this approximation is called the first order theory.

Equations (3.1), (3.2), and (3.3) under paraxial conditions become

$$i = \left(\frac{R-u}{R} \right) \theta_1 \tag{3.5}$$

$$r = \left(\frac{\mu_1}{\mu_2}\right) i \tag{3.6}$$

$$\theta_2 = \left(\frac{R}{R-v}\right) r \tag{3.7}$$

Substituting the values of θ_2, r, and θ_1 in Equation (3.3), after canceling i and rearranging it becomes

$$\mu_1 v (R - u) = \mu_2 u (R - v)$$

Dividing by vuR and rearranging, we get

$$\frac{\mu_2}{v} - \frac{\mu_1}{u} = \frac{\mu_2 - \mu_1}{R} \tag{3.8}$$

or

$$\frac{f'}{v} + \frac{f}{u} = 1 \tag{3.9}$$

where

$$f = -\frac{\mu_1 R}{(\mu_2 - \mu_1)} \text{ and } f' = \frac{\mu_2 R}{(\mu_2 - \mu_1)}$$

Since the slope angles θ_1 and θ_2 do not appear in Equation (3.8), it follows that all rays diverging from an axial point P making small angles with the axis, after refraction at a spherical surface, cross the axis at a common point P' which is, therefore, the image corresponding to object point P. This equation, therefore, applies only to rays in the paraxial region, but it does apply to rays making large angles with the axis in optical systems which have been corrected for spherical aberration.

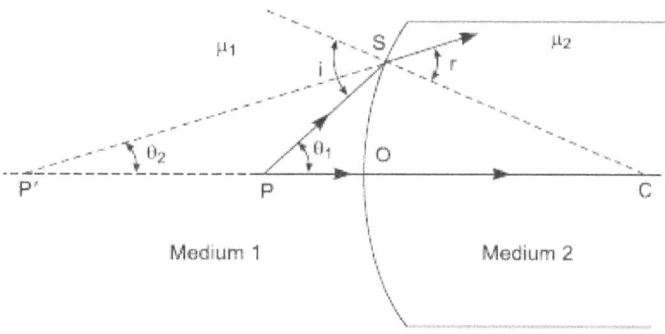

FIGURE 3.4. Refraction at convex surface forming a virtual image.

Although we have derived Equation (3.8) for the case of a convex refracting surface (R positive) when the object distance is such that a real image is formed, it holds equally well unchanged for object distance, as illustrated in Figure 3.3, which gives a virtual image, and also for the formation of a virtual image by a concave refracting surface (R negative), provided a consistent sign convention for u, v, and R is employed, that is, the correct sign is attached to the numerical values on the substitution for the symbols. Further, Equation (3.8) is of fundamental importance in *geometrical optics*, because it can be applied successively to each of the number of coaxial refracting surfaces, treating the image formed by the first surface as an object for the second and so on, in order to locate the final image of a given object.

In most of the practical cases of image formation, object point P is in air or a vacuum, that is, $\mu_1 = 1$ and the index of refraction of the second medium with respect to the first is $\mu = \dfrac{\mu_2}{\mu_1}$. Therefore, Equation (3.8) applied to this case becomes

$$\frac{\mu}{v} - \frac{1}{\mu} = \frac{\mu - 1}{R} \tag{3.10}$$

The conditions governing the object distance u for the formation of a real or virtual image due to refraction at a convex refracting surface may be very easily derived. Let u' be some object distance from the pole of the refracting surface. Putting $u = u'$ in Equation (3.8) and keeping the sign of R positive, we get

$$\frac{\mu_2}{v} = \frac{u'(\mu_2 - \mu_1) - \mu_1 R}{\mu' R} \tag{3.11}$$

Therefore, v is positive when

$$u' > \frac{\mu_1 R}{\mu_2 - \mu_1}$$

which is consequently the condition for the object distance for the formation of a real image. This condition corresponds to the case illustrated in Figure 3.2. The image distance v is negative and indicates a virtual image when

$$u' < \frac{\mu_1 R}{\mu_2 - \mu_1}$$

This condition corresponds to the case illustrated in Figure 3.3. However, when the object distance is

$$u = -u' = -\frac{\mu_1 R}{\mu_2 - \mu_1}$$

Equation (3.2 k) gives $v = \infty$, that is, the refracted rays are parallel to the axis and the image is formed at infinity.

3.3 PRINCIPAL FOCAL POINTS AND PRINCIPAL FOCAL LENGTHS

The axial object point which is imaged at infinity by a refracting surface is called the *principal focal point* of the *object space* or the *first principal focal point F* of the refracting surface. The distance of F from the pole of the surface is called the *first principal focal length* of the surface and is denoted by f. Therefore, f is the value of u conjugate to $v = \infty$, and hence its substitution in Equation (3.8) yields

$$f = -\frac{\mu_1 R}{\mu_2 - \mu_1} \qquad (3.12)$$

The axial image point conjugate to an infinitely distant axial object point is called the *principal focal point* of the *image space* or simply the *second principal focal point F'* of the system. Its distance from the pole of the surface is called the *second principal focal length* of the refracting surface and is denoted by f'. Accordingly, f' is the value of v conjugate to $u = -\infty$ and its substitution in Equation (3.8) gives

$$f' = \frac{\mu_2 R}{\mu_2 - \mu_1} \qquad (3.13)$$

Whence

$$-\frac{f'}{f} = \frac{\mu_2}{\mu_1} \quad \text{or} \quad -\frac{f'}{f} = \mu$$

Since μ, the index of refraction of the second medium with respect to the first, is always positive, it follows that the two focal lengths of the refracting surface are always of opposite signs.

This discussion obviously explains the physical significance of the symbols f and f' introduced in Equation (3.8).

3.4 LINEAR LATERAL MAGNIFICATION

Figure 3.5 illustrates linear lateral magnification. Let PQ be a small object of height y_1 placed transversely to the axis in front of the convex refracting surface. Ray PS from the foot of the object suffers refraction at S and follows the path SP' and intersects the axial ray PO, which suffers no deviation, at P'. Thus, P' is the image of P. Ray QS' is directed toward the center of curvature C and, therefore, it is incident at S' normally to the surface; as a consequence, it passes undeviated. Now a line drawn at P', perpendicular to the principal axis, intersects the ray QS' at the point Q', which is consequently the image of Q. Thus, $P'Q'$ is the image of PQ.

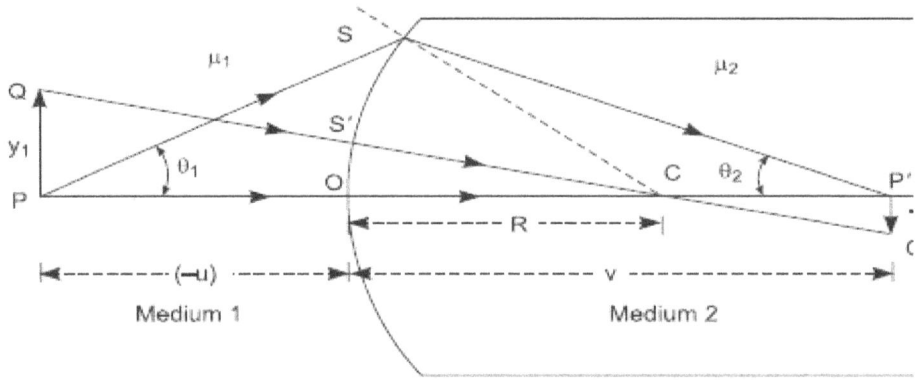

FIGURE 3.5. Linear lateral magnification.

Linear lateral magnification is defined by

$$m = \frac{P'Q'}{PQ} = \frac{y_2}{y_1} \tag{3.14}$$

where y_2 is the image height $P'Q'$. From similar triangles $CP'Q'$ and CPQ, we write

$$\frac{(-y_2)}{y_1} = \frac{CP'}{PC} = \frac{v-R}{R+(-u)} = \frac{v-R}{R-u}$$

$$\frac{y_2}{y_1} = \frac{v-R}{u-R} \tag{3.15}$$

Now, Equation (3.8) can be rewritten as

$$\mu_2\left(\frac{1}{v}-\frac{1}{R}\right)=\mu_1\left(\frac{1}{u}-\frac{1}{R}\right)$$

or

$$\left(\frac{v-R}{u-R}\right)=\frac{\mu_1 v}{\mu_2 u}$$

whence, Equation (3.15) becomes

$$m=\frac{y_2}{y_1}=\frac{\mu_1 v}{\mu_2 u} \qquad (3.16)$$

This relation determines the height y_2 of an image conjugate to height y_1 of the object formed by refraction at a single spherical surface, whether convex or concave, provided a consistent sign convention is used. This relation shows that if v and u are of opposite signs, y_1 and y_2 are also of opposite signs; that is, an inverted image of the object is formed, and magnification m is negative. On the other hand, if v and u have the same signs, the image is upright with respect to the object, and magnification m is positive.

Incidentally, we may derive the *Abbe sine condition* even at this stage very easily. From Equations (3.2 a), (3.2 b), and (3.2 d), we have the relation

$$\frac{v-R}{u-R}=\left(\frac{\sin\theta_1}{\sin\theta_2}\right)\times\left(\frac{\sin r}{\sin i}\right)=\left(\frac{\mu_1}{\mu_2}\right)\times\left(\frac{\sin\theta_1}{\sin\theta_2}\right)$$

Whence Equation (3.14) becomes

$$\mu_1 y_1 \sin\theta_1 = \mu_2 y_2 \sin\theta_2 \qquad (3.17)$$

which is known as the *Abbe sine condition*, and it holds good for all rays, whether paraxial or not.

3.5   LAGRANGE'S LAW AND HELMHOLTZ'S LAW

An important relation exists in the paraxial region between the object and image heights and the angles at which the conjugate rays are inclined to the principal axis. Let the incident ray PS in Figure 3.4 be inclined at an angle θ_1 with the axis and the conjugate refracted ray SP' inclined at θ_2 with the axis. According to sign convention, θ_2 is negative. Then, referring to Figure 3.4 and using Equation (3.16), we have

$$\frac{y_1}{(-y_2)} = \left(\frac{\mu_2}{\mu_1}\right) \times \left(\frac{(-\mu)}{v}\right)$$

In the paraxial region, the point S in Figure 3.4 is close to the pole O. We therefore write

$$\frac{\tan\theta_1}{\tan(-\theta_2)} = \frac{\left(\dfrac{-OS}{u}\right)}{\left(\dfrac{OS}{v}\right)}$$

On multiplying $\left(\dfrac{y_1}{y_2}\right) \times \left(\dfrac{\tan\theta_1}{\tan\theta_2}\right) = \dfrac{\mu_2}{\mu_1}$

or

$$\mu_1 y_1 \tan\theta_1 = \mu_2 y_2 \tan\theta_2 \tag{3.18}$$

This equation was first given by Lagrange and is frequently called *Lagrange's Law*. For the paraxial region, since θ_1 and θ_2 are small, Equation (3.18) reduces to a more familiar form

$$\mu_1 y_1 \theta_1 = \mu_2 y_2 \theta_2 \tag{3.19}$$

Now consider $(n-1)$ coaxial refracting surfaces separating n media of indices $\mu_1, \mu_2, \mu_3, \dots, \mu_n$ in succession. Then the image formed by the first surface becomes the object for the second surface and so on. Therefore, applying Equation (3.18) successively to each of $(n-1)$ refracting surfaces, we get

$$\mu_1 y_1 \tan\theta_1 = \mu_2 y_2 \tan\theta_2$$

$$\mu_2 y_2 \tan\theta_2 = \mu_3 y_3 \tan\theta_3$$

$$\overline{\mu_{n-1} y_{n-1} \tan\theta_{n-1} = \mu_n y_n \tan\theta_n}$$

Hence, on addition,

$$\mu_1 y_1 \tan\theta_1 = \mu_n y_n \tan\theta_n$$

This equation, therefore, holds good for a system of coaxial refracting surfaces. In this form it was first investigated by *Helmholtz*. Accordingly, it is now called *Helmholtz's law of magnification*.

3.6   REFLECTION AS A SPECIAL CASE OF REFRACTION

The only mathematical difference between the phenomena of refraction and reflection of light is that *Snell's Law*

$$\frac{\sin i}{\sin r} = \frac{\mu_2}{\mu_1}$$

replaces the law of reflection $i = r$ (here r is the angle of reflection). If we employ a consistent sign convention for angles, as stated in Section 3.1, in both the phenomena, the law of reflection should be written as

$$i = -r \qquad (3.20)$$

where r is the angle of reflection as shown in Figure 3.6.

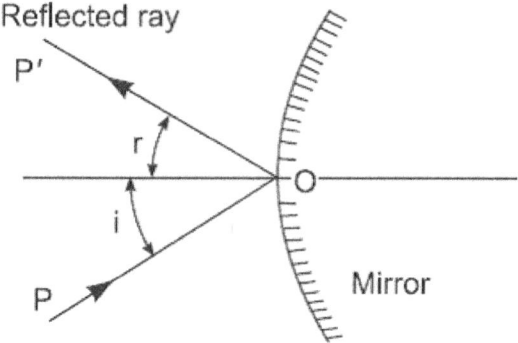

FIGURE 3.6. Reflection at a mirror.

It is therefore evident that the law of reflection represented by Equation (3.20) can be derived from Snell's law by using the artifice of replacing μ_2 by $-\mu_1$, that is, $\mu_2 = -\mu_1$, and interpreting the angle r as the angle of reflection. The steps of this procedure are

$$\mu_1 \sin i = \mu_2 \sin r = \left(-\mu_1\right) \sin r$$

$$\sin i = -\sin r$$

$$i = -r$$

As a consequence, all formulae for reflection of paraxial rays at a spherical mirror can be obtained from those for refraction of paraxial rays at a spherical refracting surface simply by putting $\mu_2 = -\mu_1$ in the formulae. Thus, Equation (3.8) becomes

$$\frac{\left(-\mu_1\right)}{v} - \frac{\mu_1}{u} = \frac{\left(-\mu_1\right) - \mu_1}{R}$$

or

$$\frac{1}{u} + \frac{1}{v} = \frac{2}{R} \tag{3.21}$$

which is therefore the relation between the conjugate object and image distances for a spherical mirror. Similarly, Equation (3.16) becomes

$$m = \frac{y_2}{y_1} = \frac{\mu_1 v}{\left(-\mu_1\right)u} = -\frac{v}{u} \tag{3.22}$$

where y_2 is the image height conjugate to height y_1 of the object. This expression for the magnification shows that if v and u are of the same sign, as in the case of a real image formed by reflection at a *concave mirror*, y_1 and y_2 are of opposite signs, that is, an inverted image of the object is formed. On the other hand, if v and u are opposite signs, as in the case of reflection at a *convex mirror*, y_1 and y_2 are of the same sign, that is, the image is upright with respect to the object.

Equations (3.21) and (3.22) are sufficient to locate the position and determine the size of the image of an object due to reflection at a spherical reflecting mirror.

3.7 LENSES

A lens is a portion of a refracting medium bounded by two curved surfaces which themselves may either be spherical or cylindrical. If both the bounding surfaces are spherical, the lens is said to be spherical. Similarly, if both the surfaces are cylindrical, the lens goes by the corresponding name. If one surface is spherical and the other cylindrical, the lens is said to be *spherocylindrical*. The spherical lenses are more commonly used, and hence they will be discussed in detail in this chapter.

In the case of spherical lenses, both the bounding surfaces may be curved, or one may be plane and the other curved (a plane surface may be taken as a curved surface of infinite radius of curvature). A straight line passing through the centers of curvature of the two surfaces is called the *principal axis of the lens*. The points of intersection of the two surfaces with this axis are called the *poles of the lens*. In thin lenses the distance between the poles is negligible as compared to the object and image distances, while in thick lenses this is not so.

If the periphery of the lens is circular in contour, its diameter is known as the aperture of the lens. Some of the more common types of lenses and their names are given in Figure 3.7.

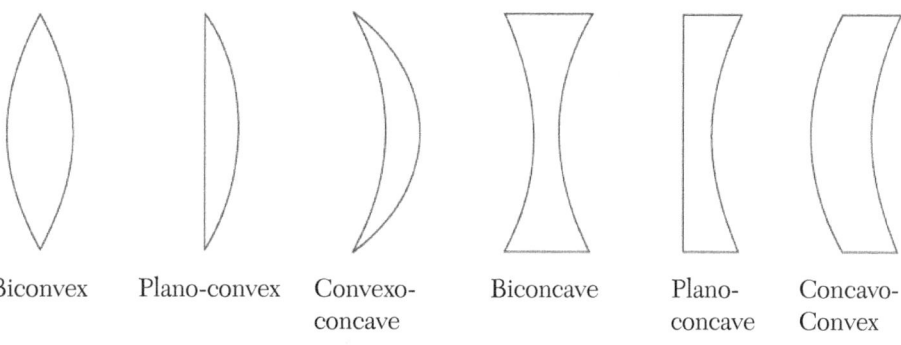

| Biconvex | Plano-convex | Convexo-concave | Biconcave | Plano-concave | Concavo-Convex |

FIGURE 3.7. Spherical lenses.

3.8   REFRACTION THROUGH A THICK LENS

Suppose an object O is placed on the axis of a thick lens at a distance "u" from the first surface as in shown in Figure 3.8.

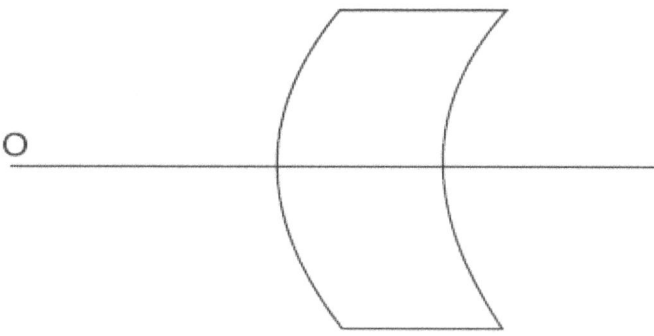

FIGURE 3.8. Thick lens.

Rays of light starting from O are refracted through the first surface and form an image, say at a distance v' from it. If μ is the refractive index of the lens and r_1 is the radius of curvature of the first surface, then

$$\frac{\mu}{v'} - \frac{1}{u} = \frac{\mu - 1}{r_1}$$

(3.23)

If in this expression $v' = \infty$, then $u = -\dfrac{r_1}{(\mu - 1)}$

If this value of u is denoted by f_1, then f_1 is called the first focal distance of the first surface of the lens and is given by

$$\frac{1}{f_1} = -\frac{(\mu - 1)}{r_1} \qquad (3.24)$$

The combination of equations (3.23) and (3.24) give

$$\frac{\mu}{v'} - \frac{1}{u} = -\frac{1}{f_1} \qquad (3.25)$$

This image at a distance v' from the first surface serves as an object for the second refracting surface of the lens. The distance of this object as measured from the second surface will be $(v' + t)$, where t is the thickness of the lens. If the final image is formed in air at a distance v from the second surface of the radius of curvature r_2, then

$$\frac{\left(\dfrac{1}{\mu}\right)}{v} - \frac{1}{v' + t} = \frac{\left(\dfrac{1}{\mu}\right) - 1}{r_2} \quad \text{(since light travels from glass to air in this case)}$$

This equation may be written as

$$\frac{1}{v} - \frac{\mu}{v' + t} = \frac{1 - \mu}{r_2} \qquad (3.26)$$

As before, if f_2 is the value of v for $v' = \infty$, f_2 is the second focal distance of the second surface and is given by

$$\frac{1}{f_2} = \frac{(1 - \mu)}{r_2} \qquad (3.27)$$

It can be seen from Equations (3.26) and (3.27) that

$$\frac{1}{v} - \frac{\mu}{v' + t} = \frac{1}{f_2} \qquad (3.28)$$

Eliminating v' from Equations (3.25) and (3.28) we get

$$\frac{\mu u f_1}{f_1 - u} = -t + \frac{\mu v f_2}{f_2 - v} \qquad (3.29)$$

which on simplification gives

$$\mu u f_1 (f_2 - v) + t(f_1 - u)(f_2 - v) - \mu v f_2 (f_1 - u) = 0$$

or

$$u(\mu f_1 f_2 - f_2 t) - v(\mu f_1 f_2 + f_1 t) + uv\left[\mu(f_2 - f_1) + t\right] + f_1 f_2 t = 0$$

Dividing throughout by $\left[\mu(f_2 - f_1) + t\right]$, it can be seen that

$$uv + \left(\frac{\mu f_1 f_2 - f_2 t}{\mu(f_2 - f_1) + t}\right)u - \left(\frac{\mu f_1 f_2 - f_1 t}{\mu(f_2 - f_1) + t}\right)v + \frac{f_1 f t}{\mu(f_2 - f_1) + t} = 0 \qquad (3.30)$$

Equation (3.30) is of the form

$$\left(\frac{1}{v - \beta}\right) - \left(\frac{1}{u - \alpha}\right) = \frac{1}{F} \qquad (3.31)$$

for on multiplying both the sides of Equation (3.31) by $F(u - \alpha)(v - \beta)$ and rearranging we get

$$uv - (F + \beta)u + (F - \alpha)v - F(\beta - \alpha) + \alpha\beta = 0 \qquad (3.32)$$

For Equations (3.30) and (3.32) to be identical, the following identities must hold

$$-(F + \beta) = \frac{\mu f_1 f_2 - f_2 t}{\mu(f_2 - f_1) + t} \qquad (3.33)$$

$$-(F - \alpha) = \frac{\mu f_1 f_2 + f_2 t}{\mu(f_2 - f_1) + t} \qquad (3.34)$$

and

$$\alpha\beta - F\beta + F\alpha = \frac{f_1 f_2 t}{\mu(f_2 - f_1) + t} \qquad (3.35)$$

The values of α, β, and F can be obtained by multiplying Equations (3.33) and (3.34) and adding the product to Equation (3.35). Thus

$$F^2 - \alpha\beta - F\alpha + F\beta = \frac{(\mu f_1 f_2 + f_1 t)(\mu f_1 f_2 - f_2 t)}{\left[\mu(f_2 - f_1) + t\right]^2} \qquad (3.36)$$

and

$$F^2 = \frac{(\mu f_1 f_2)^2 + \mu f_1 f_2 t(f_1 - f_2) - f_1 f_2 t^2}{\left[\mu(f_2 - f_1) + t\right]^2} + \frac{f_1 f_2 t}{\mu(f_2 - f_1) + t}$$

$$= \frac{\left(\mu f_1 f_2\right)^2 + \mu f_1 f_2 t \left(f_1 - f_2\right) - f_1 f_2 t^2 + f_1 f_2 t \left[\mu \left(f_2 - f_1\right) + t\right]}{\left[\mu \left(f_2 - f_1\right) + t\right]^2}$$

$$= \left(\frac{\mu f_1 f_2}{\mu \left(f_2 - f_1\right) + t}\right)^2$$

$$\therefore \ F = \pm \frac{\mu f_1 f_2}{\mu \left(f_2 - f_1\right) + t} \tag{3.37}$$

Now it has to be decided as to which of the two signs is admissible. It is clear that this equation will hold for all values of t, including $t = 0$, which is the case of a thin lens. In such a lens if a luminous object is placed at a distance f_1 from the first surface, the rays after refraction from the surface will be rendered parallel, and hence after refraction at the second surface of the lens they will come to a focus at a distance f_2 from the latter. Moreover, in the case of a thin lens, both α and β are zero. Thus, on putting $\alpha = \beta = 0$ and $v = f_2$ and $u = f_1$ in Equation (3.31), it reduces to

$$\frac{1}{f_2} - \frac{1}{f_1} = \frac{1}{F}$$

$$F = -f_1 f_2 \neq \left(f_2 - f_1\right) \tag{3.38}$$

It can at once be seen that the values of F obtained by putting $t = 0$ in Equation (3.37) will agree with that given by Equation (3.38) only when the negative sign is taken into account. As such the positive sign becomes inadmissible, and Equation (3.37) becomes

$$F = -\frac{\mu f_1 f_2}{\mu \left(f_2 - f_1\right) + t} \tag{3.39}$$

The values of α and β can now be easily obtained from Equations (3.33), (3.34), and (3.39).

Thus,

$$\alpha = \frac{f_1 t}{\mu \left(f_2 - f_1\right) + t} \tag{3.40}$$

$$\beta = \frac{f_2 t}{\mu \left(f_2 - f_1\right) + t} \tag{3.41}$$

such that,

$$\alpha f_2 = \beta f_1 \tag{3.42}$$

a relation which is of great importance.

If V is measured from a point at a distance β in the positive direction from the second surface and U from a point at a distance α in the positive direction from the first surface, then $u - \alpha = U$ and $v - \beta = V$.

Equation (3.31) may then be written as

$$\frac{1}{V} - \frac{1}{U} = \frac{1}{F} \tag{3.43}$$

a relation which is similar to that used in the case of a thin lens.

3.9 PRINCIPAL POINTS AND PRINCIPAL PLANES

In the case of thin lenses, the distance is always measured from the center of the lens. The distance from the center of the lens to either focus gives the focal length. But if a thick lens is used, the distance from its center to one focus is different from the distance to the other focus, and neither distance equals the focal length of the lens. Also, if an attempt is made to determine the focal length by, say, the U, V method, there is no fixed point from which these distances can be measured.

However, it has been shown previously that there are two points or planes from which the distances may be measured in all cases, and the simple formula for the lens may be applied as usual. These points are called *principal points*. Thus, the point at a distance α from the first surface is called the *first principal point* and, similarly, the point at a distance β from the second surface is called the *second principal point*. Planes drawn perpendicular to the axis and passing through these points are called *principal planes*.

In Equation (3.43) it is seen that if $V = \infty$, $U = -F$ so that $-F$ is the first focal distance of the lens. Its magnitude is given by Equation (3.39) and is measured from the first principal point. Similarly, if $U = \infty$, $V = F$ and then F is the second focal distance of the lens and is measured from the second principal point. A point at a distance F from the first principal point, as defined in Section 3.3, is called the first focal point of the lens. Similarly, the point at a distance F from the second principal point is termed as the second focal point of the lens.

If the lens is placed in a medium of a uniform refractive index, the first and second focal distances are found to be equal in magnitude but opposite in sign.

3.10 PROPERTIES OF PRINCIPAL PLANES AND PRINCIPAL POINTS

a. *Principal points are conjugate foci.* Equation (3.31) may be written as

$$v - \beta = \frac{F(u - \alpha)}{F + (u - \alpha)}$$

If $u = \alpha$, whatever the value of F is, $v = \beta$, so that if an object is placed at a distance α from the first surface or at the first principal point, its image will be formed at a distance β from the second surface, that is, at the second principal point, and vice versa. If the first principal point is outside the lens on the positive side, a real object can be placed, and if it is on the negative side, a virtual object must be placed.

In general, the two principal planes of a thick lens are conjugate planes.

b. *Principal planes are planes of unit magnification.* Let O as shown in Figure 3.9 be an object placed on the axis of a thick lens at a distance u from the first surface, and let AB be any ray passing through the first principal focus A of the first surface.

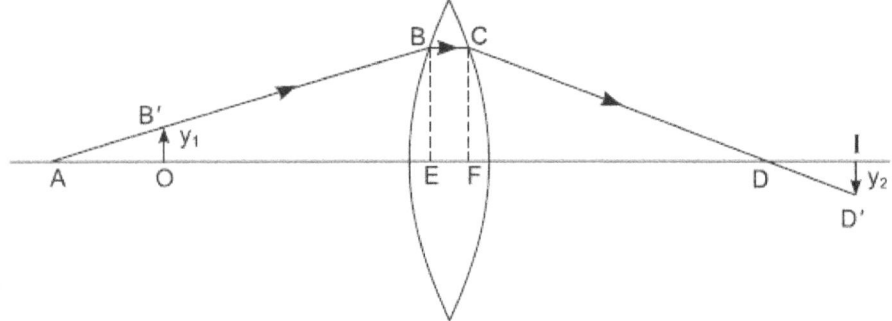

FIGURE 3.9. Planes of unit magnification.

This ray which touches the upper extremity of the object will after refraction from the first surface travel parallel to the axis inside the lens. After refraction from the second surface, this ray will pass through D, the second principal focus of the second surface. This ray, it is clear, will pass through the extremity of the image whose position is given by Equation (3.31). Let I be this image. Drop perpendiculars are BE and CF on the axis. When the rays AB and CD are nearly parallel to the axis, the points E and F coincide with the poles of the curved surfaces. Let the object distance $EO = -u$ and the image distance $v = +FI$. Also, let $EA = -f_1$ and $FD = +f_2$. If $BE = CF = h$, then $\left(\dfrac{h}{y_1}\right) = \left(\dfrac{f_1}{(f_1 - u)}\right)$ from the similar triangles ABE and $AB'O$ and $\left(\dfrac{h}{y_2}\right) = \left(\dfrac{-f_2}{+(v - f_2)}\right)$ from similar triangles CFD and DID' so that the magnification m is given by

$$m = \frac{y_2}{y_1} = \frac{f_1(f_2 - v)}{f_2(f_1 - u)} \tag{3.44}$$

This is the expression for magnification in any case. If the value of m is positive, the image I will be upright; if negative, the image I will be inverted. If now the object O is supposed to lie in the first principal plane $u = \alpha$, the image will lie in the second principal plane for which $v = \beta$. With these values the magnification becomes

$$m = \frac{f_1(f_2 - \beta)}{f_2(f_1 - \alpha)} = \frac{f_1 f_2 - f_1 \beta}{f_1 f_2 - f_2 \alpha} = +1 \tag{3.45}$$

since $f_1 \beta = f_2 \alpha$ from Equation (3.42).

Thus, the dimensions of the image seen in the second principal plane are equal to those of the object in the first principal plane, and also the lateral magnification is unity and positive.

3.11 LATERAL MAGNIFICATION REFERRED TO THE FOCAL POINTS

Figure 3.10 shows the lateral magnification. Let $Q_1 R_1$ and $Q_2 R_2$ be the principal planes of a thick lens, and let P_1, P_2 be the two principal points. Let an object $A_1 B_1$ of size y_1 be placed at a distance x_1 from the first focal point F_1. If $A_2 B_2$ is a plane conjugate to $A_1 B_1$, then the former will be the image of the latter.

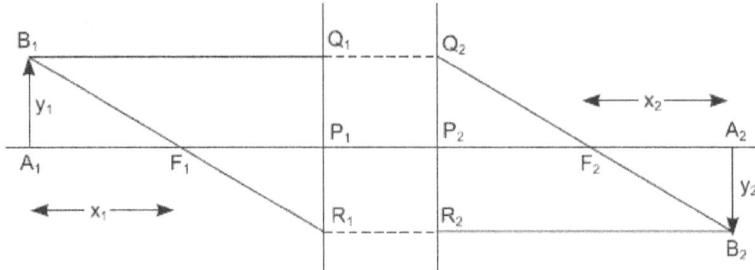

FIGURE 3.10. Lateral magnification.

Suppose the first and second focal distances P_1F_1 and P_2F_2 are respectively f_1 and f_2. Any ray B_1R_1 passing through F_1 and striking Q_1R_1 at R_1 will emerge in the same direction of the axis at the point R_2 such that $P_1R_1 = P_2R_2$ (lateral magnification is unity and positive for principal planes) and will be rendered parallel to the axis. Similarly, a ray B_1Q_1 parallel to the axis and striking Q_1R_1 at Q_1 will emerge on the same side of the axis at Q_2 such that $P_1Q_1 = P_2Q_2$. It is evident from the figure that the triangles $A_1B_1F_1$ and $P_1F_1R_1$ are similar and so are the triangles $P_2Q_2F_2$ and $F_2A_2B_2$. Therefore

$$\frac{y_1}{-P_1R_1} = \frac{x_1}{f_1} \text{ and } \frac{y_2}{P_2R_2} = \frac{x_2}{f_2} \tag{3.46}$$

Since $y_1 = P_1Q_1 = P_2Q_2$ and $y_2 = P_1R_1 = P_2R_2$, Equation (3.46) gives

$$\frac{P_1Q_1}{-P_1R_1} = \frac{x_1}{f_1} \text{ and } \frac{-P_2R_2}{P_2Q_2} = \frac{x_2}{f_2}$$

or

$$\frac{P_1Q_1}{P_1R_1} \times \frac{P_2R_2}{P_2Q_2} = \frac{x_1x_2}{f_1f_2}$$

$$\therefore \quad x_1x_2 = f_1f_2 \tag{3.47}$$

This is known as *Newton's formula*.

3.12 LATERAL MAGNIFICATION REFERRED TO PRINCIPAL POINTS

In the preceding section, the distances of the image and object were measured from the focal points. Sometimes it is convenient to measure the distances from the principal points P_1 and P_2 as in Figure 3.9.

Let $P_1A_1 = u$ and $P_2A_2 = v$. Then, according to our convention of signs

$$x_1 = -(u - f_1) \text{ and } x_2 = +(v - f_2)$$

Substituting these values of x_1 and x_2 in Equation (3.47) and remembering that f_1 is negative, it is seen that

$$(u - f_1)(v - f_2) = f_1 f_2$$

which on simplification yields

$$\frac{f_1}{u} + \frac{f_2}{v} = 1 \tag{3.48}$$

and the expression for lateral magnification becomes

$$m = -\frac{f_1}{u - f_1} = -\frac{v - f_2}{f_2} \tag{3.49}$$

A more useful expression for lateral magnification can be obtained by applying *Lagrange's law* to the principal planes. If θ_1 and θ_2 are the slope angles in the object and image spaces respectively, then

$$\frac{\mu_1}{\mu_2} = \frac{\theta_2}{\theta_1} \tag{3.50}$$

since lateral magnification is unity in principal planes. If θ_1 and θ_2 are small, as shown in Figure 3.11.

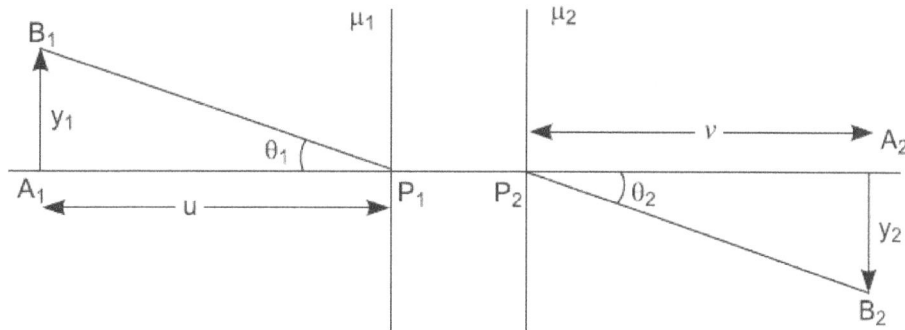

FIGURE 3.11. Lateral magnification referred to principal points.

$$-\theta_1 = \frac{y_1}{-u} \text{ and } -\theta_2 = \frac{y_2}{+v}$$

so that

$$m = \frac{y_2}{y_1} = +\left(\frac{v}{u}\right)\left(\frac{\theta_2}{\theta_1}\right) = +\left(\frac{\mu_1}{\mu_2}\right)\left(\frac{v}{u}\right) \tag{3.51}$$

because of Equation (3.50).

It will be found, on comparison, that Equations (3.16) and (3.51) are of the same form. It may be recalled that the distances u and v as in Equation (3.16) were measured from the vertex, while here these distances have been measured from the principal points. It is therefore inferred that the principal points of a single refracting surface coincide with its vertex.

3.13 DETERMINATION OF THE PRINCIPAL AND FOCAL POINTS

Equations (3.39) to (3.41) are used to determine the positions of the principal and focal points. According to Equation (3.39)

$$F = \frac{\mu f_1 f_2}{\mu(f_2 - f_1) + t}$$

Where $f_1 = \dfrac{r_1}{\mu - 1}$, $f_2 = \dfrac{r_2}{\mu - 1}$

If $\mu = 1.5$, then

$$F = \frac{1.5 f_1 f_2}{0.25\left[1.5 \times \left(\dfrac{1}{0.5}\right)(r_1 - r_2) + t\right]}$$

$$\therefore F = \frac{6 r_1 r_2}{t + 3(r_1 - r_2)} \tag{3.52}$$

Similarly,

$$\alpha = \frac{f_1 t}{\mu(f_2 - f_1) + t} = \frac{-\left(\dfrac{r_1}{0.5}\right)t}{1.5 \times \left(\dfrac{1}{0.5}\right)(r_1 - r_2) + t}$$

And if t is small,

$$\alpha = \frac{-2 r_1 t}{3(r_1 - r_2)} \tag{3.53}$$

and

$$\beta = \frac{-2r_2 t}{3(r_1 - r_2)} \qquad (3.54)$$

Let us calculate the values of α, β, and F for a biconvex lens.

Let $r_1 = r_2 = r$ for simplicity. If f and f' are the first and second focal distances

$$f = -f' = \frac{6r^2}{6r - t} = r$$

if t is small as compared to r,

$$\alpha = -\frac{t}{3} \text{ and } \beta = \frac{t}{3} \text{ if } t \text{ is small.}$$

Thus, if the thickness of the lens is known, the principal and focal points of a lens can be calculated. Further, if the lens is placed in air, the nodal points coincide with the principal points. Some typical cases are shown in Figure 3.12.

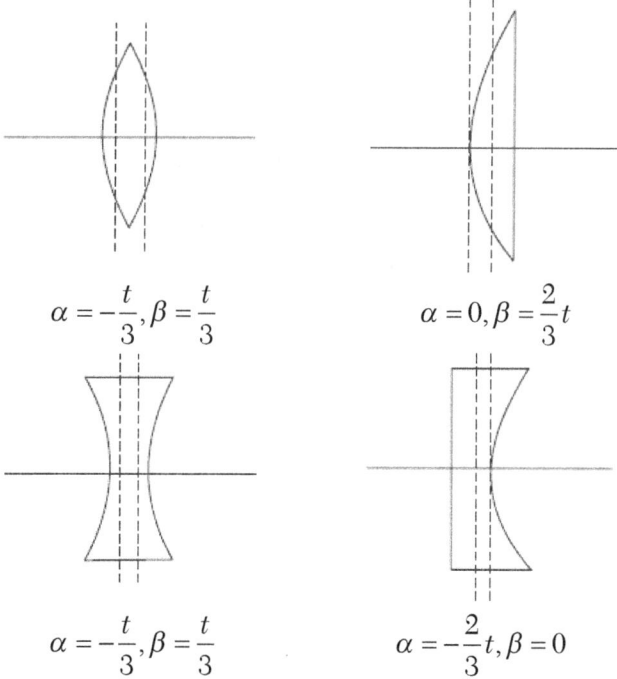

$$\alpha = -\frac{t}{3}, \beta = \frac{t}{3} \qquad\qquad \alpha = 0, \beta = \frac{2}{3}t$$

$$\alpha = -\frac{t}{3}, \beta = \frac{t}{3} \qquad\qquad \alpha = -\frac{2}{3}t, \beta = 0$$

FIGURE 3.12. Positions of principal points in thick lenses.

3.14 GRAPHICAL CONSTRUCTION OF IMAGES

Let LL' be a thick lens whose principal points are P_1 and P_2, as shown in Figure 3.13. These points can be calculated with the help of the previously given formula. Through these points, draw planes perpendicular to the axis. These will be the principal planes of the lens. Suppose an object O is placed at a certain distance from the lens.

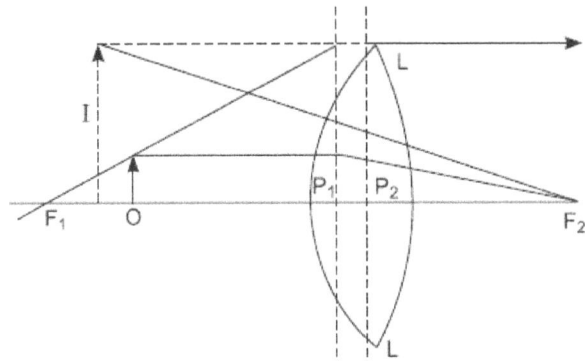

FIGURE 3.13. Construction of images.

To construct the image, take a ray passing through the upper extremity of the object and the first focal point F_1 of the lens. Then this ray will be rendered parallel to the axis after refraction by both surfaces of the lens. Also, this ray will emerge out of the second principal plane on the same side of the axis and as far from it as the point in which the incident ray cuts the first principal plane. Further, another ray from the object parallel to the axis will after refraction from the lens pass through F_2, the second principal focus. Draw a ray parallel to the axis through the upper extremity of O. The intersection of this ray with the first principal plane will be on the same side of the axis and at the same height as that of the refracted ray passing through F_2. Hence, the latter ray can be drawn. The point of intersection of both these rays gives the image I of the object O. The image can be constructed in a similar manner for a concave lens.

3.15 NODAL POINTS

Until now it has been assumed that the lens is placed in air or in one medium, but if the medium on one side of the lens is different from that on the other, it becomes necessary to consider two more points on the axis. These points are such that a ray of light directed from the positive side of the lens toward one of these points will after refraction from the lens appear to proceed in a parallel direction from the second. These points are respectively called the *first* and *second nodal points*.

Let *A* and *B* as in Figure 3.14 (a) be the centers of curvature of the two surfaces of a thick lens. From *A* draw a radius *AQ* of the front surface. From *B* draw a radius *BR* of the second surface parallel to *AQ*. Then it is clear that small elements of the surface at *Q* and *RA* will be parallel to each other (because the normal to these surfaces are parallel). Hence, a ray of light *PQ* directed toward N_1, which is incident at *Q*, will after refraction through the lens proceed along *RS* in a direction parallel to that of *PQ*. Produce *SR* back to cut the axis at N_2. Then N_1 and N_2 are the two nodal points.

In the previous construction, the media on both sides of the lens were the same. However, if the two media are different, two nodal points exist in this case as well.

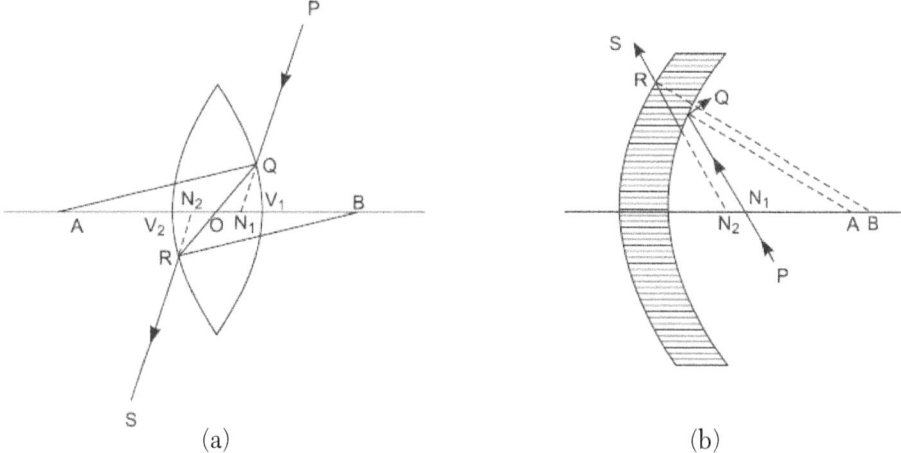

(a)   (b)

FIGURE 3.14. (a) Nodal points, (b) Nodal points.

Let P_1B and P_2G be the principal planes of a lens and let F_1 and F_2 be the first and the second principal focus respectively. Let a ray of light *AC* directed toward the first nodal point N_1 cut the first principal plane at *C*, as shown in Figure 3.15.

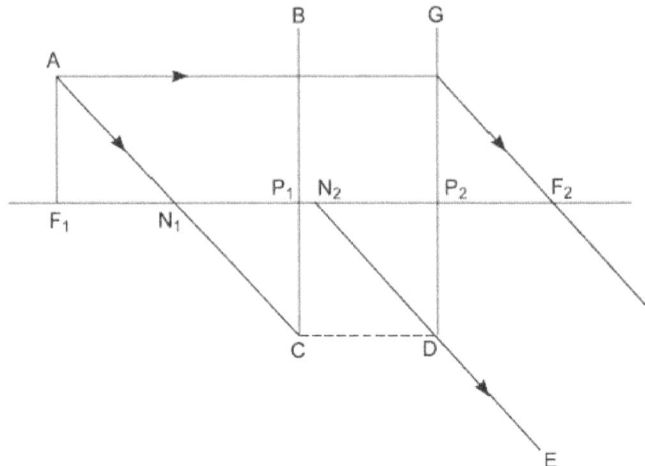

FIGURE 3.15. Nodal points.

It will then emerge at the second principal plane on the same side and at the same distance from the axis along DE such that $P_1C = P_2D$. Also, the ray DE will be parallel to the incident ray AC and will appear to proceed from the second nodal point N_2 (obtained by producing DE backward).

The two parallelograms N_1N_2DC and P_1P_2DC stand on the common base CD and between the two parallel lines P_1P_2 and CD, and hence they are equal in area.

Thus, Area P_1P_2DC = Area N_1N_2DC. Therefore, $N_1P_1 = N_2P_2$.

According to the previous equation, it is clear that the first nodal point is as much in advance of the first principal point as the second nodal point is in advance of the second principal point.

Again, draw a perpendicular F_1A cutting the incident ray AC in A. From A draw a ray AB parallel to the axis. This ray will emerge out at G such that $P_1B = P_2G$ and will after refraction through the lens pass through F_2. Since A is in the first focal plane, all the rays starting from A will after refraction through the lens be rendered parallel to each other. The rays DE and GF_2 will therefore be parallel, with the result that the triangles AF_1N_1 and GP_2F_2 will be equal in all respects.

$$P_2F_2 = N_1F_1$$

$$P_2F_2 = P_1F_1 - P_1N_1$$

or

$$P_1N_1 = n = P_1F_1 - P_2F_2 = -\left(F_1 + F_2\right)$$

which gives the distance between the nodal points and the corresponding principal points.

If the medium on both sides of the lens is air, $-F_1 = +F_2$, and in that case $n = 0$ or, in other words, the nodal points coincide with the principal points when the lens is placed in air. In the case of a thin lens, these points are evidently coincident with the center of the lens.

3.16 OPTICAL CENTER

In Figure 3.13 (a) the ray PQ after refraction by the surface follows the path QR and cuts the axis at O. This point O is called the optical center of the lens.

Since QN_1 is parallel to RN_2 and AQ is parallel to BR, the triangles AQN_1 and BRN_2 are similar and hence

$$\frac{OQ}{OR} = \frac{AQ}{RB}$$

If only the paraxial rays are considered

$OQ = OV_1$ and $OR = OV_2$, approximately

and if the radii of curvature AQ and RB are r_1 and r_2, then

$$\frac{V_1}{OV_2} = \frac{r_1}{r_2}$$

which gives the position of the optical center.

It will be noticed that the position of the optical center simply depends upon the radii of curvature of the surfaces and is independent of the refractive index of the lens. Hence, all the rays irrespective of their wavelengths or color pass through the optical center.

Since the ray PQ directed toward N_1 is refracted by the first surface along QR cutting the axis at O, it is clear that O is the image of N_1. Similarly, it can be seen that N_2 is the image of O formed by refraction at the second surface.

3.17 CARDINAL POINTS

Thus, it has been seen that there are certain points for a thick lens which simplify the consideration of its problem to a great extent. There are three pairs of these points: (a) principal points or Gauss points, (b) focal points, and (c) nodal points. These sets of points are called the *cardinal points* of a lens system.

A comparatively less important set of points is also occasionally encountered in literature. These points simplify the problems in rare cases. They are as follows:

1. *Anti-principal points*: They are conjugate points for which lateral magnification is unity and negative. They lie as far from the focal points as the corresponding principal points but on the negative side.

2. *Anti-nodal points*: They are points for which the angular magnification is unity and negative. They lie as far from the focal points as the corresponding nodal points but on the negative side.

3. *Bravais points*: They are self-conjugate points. These points do not exist at all in some cases. In some cases, one or even two pairs may exist.

3.18 THE THIN LENS

The focal length of a thick lens placed in air is given by Equation (3.39)

$$F = -\frac{\mu f_1 f_2}{\mu (f_2 - f_1) + t}$$

where

$$\frac{1}{f_1} = -\frac{\mu - 1}{r_1} \quad \text{and} \quad \frac{1}{f_2} = \frac{1 - \mu}{r_2}$$

With these values of f_1 and f_2, F becomes

$$f = F = -\frac{r_1 r_2}{(\mu - 1)\left[r_1 - r_2 + \frac{\mu - 1}{\mu} t \right]} \tag{3.55}$$

In a thin lens the thickness t is negligible. Hence, Equation (3.55) gives

$$\frac{1}{f} = (\mu - 1)\left(\frac{1}{r_1} - \frac{1}{r_2} \right) \tag{3.56}$$

Also, when t is small, it can be seen from Equations (3.40) and (3.41) that

$$\alpha = \beta = 0,$$

that is, in the case of a thin lens placed in air, the principal points as well as the nodal points coincide with the vertices of the surface, whose mutual separation can be neglected. As a consequence, a thin lens may be represented by a straight line perpendicular to the axis.

3.19 FOCAL LENGTHS OF THIN LENSES IN SPECIAL CASES

1. *Biconvex lens.* The radius of curvature r_1 of the surface facing the incident ray is positive, that of the other surface r_2 is negative, and hence

$$\frac{1}{f} = +(\mu - 1)\left(\frac{1}{r_1} + \frac{1}{r_2}\right)$$

The focal length of a biconvex lens is therefore positive.

2. *Plano-convex lens.* If the plane surface faces the incident light $r_1 = \infty$ and

$$\frac{1}{f} = +\frac{(\mu - 1)}{r_2}$$

If the convex surface faces the incident light, r_1 is positive. Even then the focal length is positive.

3. *Concavo-convex lens.* Here both the radii of curvature are on the same side so that if $r_1 > r_2$, the focal length will be negative, and if $r_2 > r_1$, it will be positive.

4. *Biconvex lens.* In this case the radius of curvature r_1 of the surface facing the incident ray is negative while r_2 is positive so that

$$\frac{1}{f} = -(\mu - 1)\left(\frac{1}{r_1} + \frac{1}{r_2}\right)$$

5. *Plano-concave lens.* If the plane surface faces the incident ray, $r_1 = \infty$, while r_2 is positive. Consequently

$$\frac{1}{f} = -\frac{(\mu - 1)}{r_2}$$

It can be shown that if the concave surface faces the incident light, the focal length will remain negative.

In general, if the thickness of a lens increases toward its periphery, the focal length is negative, and if it increases toward its central region, the focal length is positive, as shown in Figure 3.16.

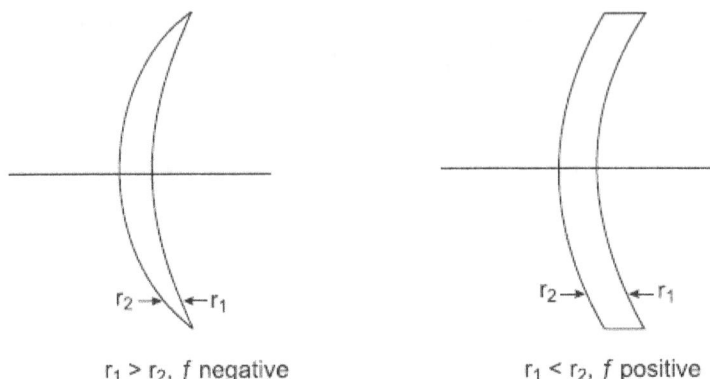

$r_1 > r_2$, f negative $r_1 < r_2$, f positive

FIGURE 3.16. Thickness of a lens.

3.19.1 Power of a Lens

The reciprocal of the focal length of a lens expressed in meters is called its power and is expressed in diopters. A diopter is defined as the power of a convex lens whose focal length is one meter, that is

$$+1D = +\frac{1}{100} \text{ cm}$$

Thus, the power of a convergent lens is positive, while that of a divergent lens is negative.

3.19.2 Principal Foci

Putting $t = 0$ in Equation (3.8 g), one gets

$$\frac{\mu u f_1}{f_1 - u} = \frac{\mu v f_2}{f_2 - v}$$

where

$$f_1 = -\frac{r_1}{\mu - 1} \text{ and } f_2 = -\frac{r_2}{\mu - 1}$$

With these values the previous equation becomes

$$\frac{1}{v} - \frac{1}{u} = (\mu - 1)\left(\frac{1}{r_1} - \frac{1}{r_2}\right) = \frac{1}{f} \tag{3.57}$$

This is the general equation for the conjugate distances in the case of a thin lens placed in air. The distances u and v are measured from the optical center of the lens, which coincides with the two principal points for all practical purposes.

If in Equation (3.57) $v = \infty$, then

$$\frac{1}{u} = -(\mu - 1)\left(\frac{1}{r_1} - \frac{1}{r_2}\right) = \frac{1}{f} \tag{3.58}$$

Expressed in words it means that if rays starting from an object are rendered parallel after refraction through the lens, the particular object distance given by Equation (3.58) is termed as the first principal focal distance of the lens and is given by

$$\frac{1}{f_1} = -(\mu - 1)\left(\frac{1}{r_1} - \frac{1}{r_2}\right)$$

The point on the axis at a distance f_1 or, in other words, the point at which the object is placed in this particular case, is called the *first principal focus* of the lens.

If $u = \infty$, that is, parallel rays are incident on the lens, the image distance v is given by

$$\frac{1}{v} = (\mu - 1)\left(\frac{1}{r_1} - \frac{1}{r_2}\right) \tag{3.59}$$

This particular value of v at which parallel rays converge or appear to converge after refraction through the lens is called the second principal focal length of the lens and is given by

$$\frac{1}{f_2} = (\mu - 1)\left(\frac{1}{r_1} - \frac{1}{r_2}\right)$$

The point at which the parallel rays converge or appear to converge after refraction is called the second principal focus of the lens. Thus, for a thin lens placed in air

$$f_1 = -f_2$$

3.20 THE CARDINAL POINTS OF THIN LENSES

A method for locating the cardinal points of a combination of lenses may also be given. Before considering the combination of two lenses, we shall show that the deviation of any ray due to refraction through a thin lens is independent of the angle of incidence and depends upon the height h of the point where total deviation occurs above the principal axis and the value of the second principal focal length. In Figure 3.17, I is the real image of the axial point object O. A ray OA on refraction through the lens undergoes a deviation, which is evidently given by

$$\delta = \alpha + (-\beta)$$

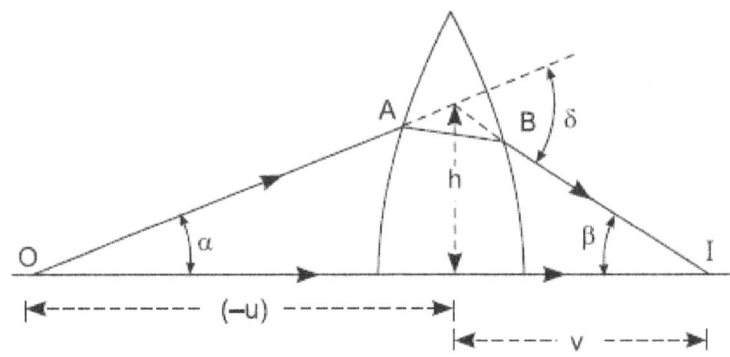

FIGURE 3.17. Cardinal points of a thin lens, where I is the real image of the axial point object O.

In the paraxial region, since inclinations of the incident and refracted rays to the principal axis are small, we can write

$$\alpha = \frac{h}{(-u)} \text{ and } (-\beta) = \frac{h}{v}$$

Hence, $\delta = h\left(\dfrac{1}{v} - \dfrac{1}{u}\right) = \dfrac{h}{f'}$ $\hspace{3cm}$ (3.60)

We therefore conclude that all rays incident at the same point of the lens, from whatever directions, undergo the same deviation on refraction through it. This statement is true for a divergent lens also.

Consider now two thin convergent lenses 1 and 2, as shown in Figure 3.18 separated by a distance d. Let f_1 and f_1' be the first and the second principal focal lengths of the first lens, and let f_1 and f_2' denote the same quantities for the second lens.

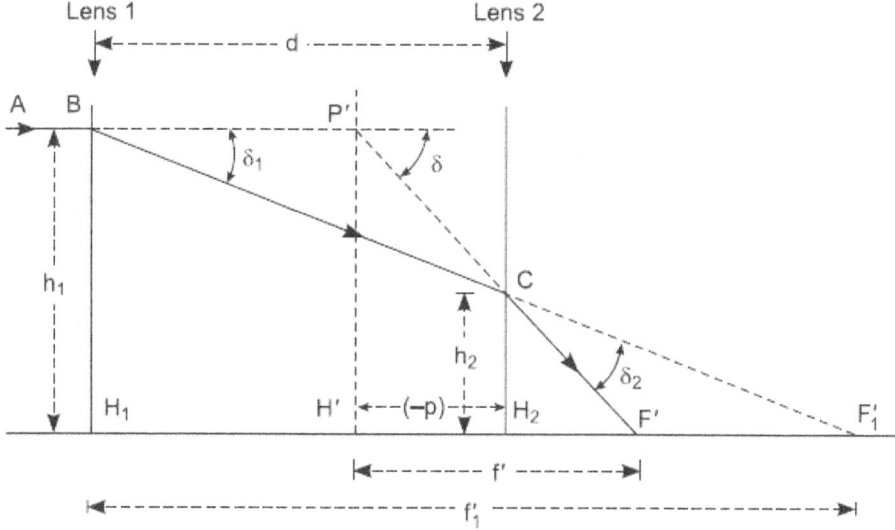

FIGURE 3.18. Location of second principal plane *P'H'* of the combination of two lenses.

A ray of monochromatic light, parallel to the principal axis, is shown incident at the height h_1 above the axis of the first lens. Due to refraction through the first lens, it is deviated toward the second principal focal point F_1' of this lens. The deviation δ_1 thus produced is given by

$$\delta_1 = \frac{h_1}{f_1'}$$

The emergent ray from the first lens, before reaching the focal point F_1', meets the second lens at height h_2 above the principal axis and undergoes deviation δ_2 in refraction through it, and finally crosses the principal axis at point F'. Evidently, the deviation δ_2, for reasons similar to those used in deriving Equation (3.60), is given by

$$\delta_2 = \frac{h_2}{f_2'}$$

The total deviation δ of the incident ray due to refraction through the combination of lenses is equal to the sum of δ_1 and δ_2, because the deviations are in the same direction.

The ray *AB* incident on the optical system is parallel to the principal axis; hence, F', the point where the emergent ray *CF'* intersects the principal axis, is termed as the second principal focus of the optical system. The conjugate rays *AB* and *CF'* when produced intersect at the point *P'*. As a consequence, it would appear that the combination of lenses functions as if the deviation of the incident ray occurs at P'. We may therefore replace the optical system by a single lens of second principal focal length equal to $H'F'$, placed in position $P'H'$, where $P'H'$ is a plane through P' transverse to the principal axis. Under this condition rays parallel to the principal axis, after refraction through this hypothetical lens, would also come to focus at F'. Accordingly, this lens is called the equivalent lens for parallel rays incident on lens 1. However, for parallel rays incident or emergent from the second lens, we shall have another equivalent lens in another position. $H'F'$ is called the *equivalent focal length of the combination* and $P'H'$ is called the *second principal plane of the optical system*.

Now

$$\delta = \delta_1 + \delta_2$$

Substituting the values of δ, δ_1, and δ_2, we get

$$\frac{h_1}{f'} = \frac{h_1}{f_1'} + \frac{h_2}{f_2'} \tag{3.61}$$

From the similar triangles CH_2F_1' and BH_1F_1', we conclude

$$\frac{h_2}{h_1} = \frac{f_1' - d}{f_1'} \tag{3.62}$$

or

$$h_2 = h_1 - \frac{h_1 d}{f_1'}$$

Substituting the value of h_2 from Equation (3.61), we get

$$\frac{1}{f'} = \frac{1}{f_1'} + \frac{1}{f_2'} - \frac{d}{f_1'f_2'} \tag{3.63}$$

or

$$f' = -\frac{f_1'f_2'}{d - f_1' - f_2'} = -\frac{f_1'f_2'}{\Delta} \tag{3.64}$$

where, we have written here

$$\Delta = d - f_1' - f_2' \tag{3.65}$$

or

$$\Delta = d + f_1 + f_2 \tag{3.66}$$

To derive the first principal focal length f of the combination of lenses, we proceed as shown in Figure 3.19.

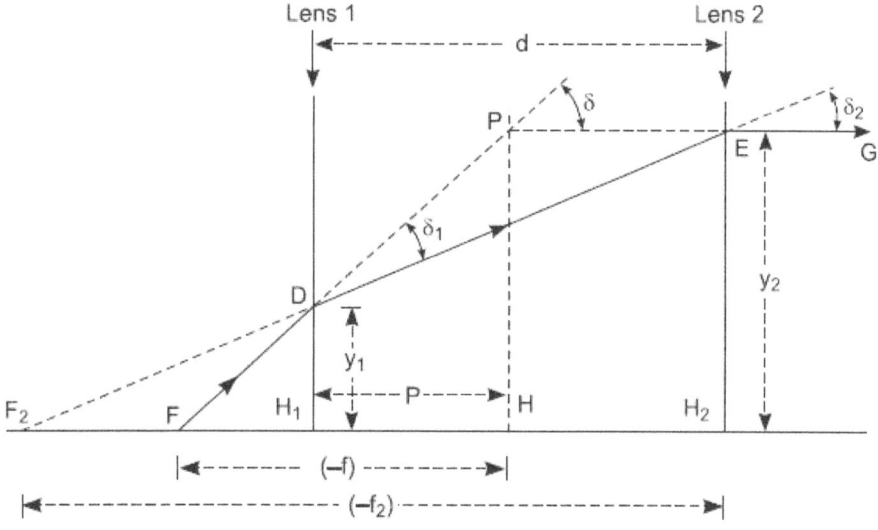

FIGURE 3.19. Location of first principal plane *PH*.

A ray of light from an axial point F after suffering deviations δ_1 and δ_2 due to refraction through the first and the second lens respectively is rendered parallel to the principal axis. As a consequence, F is the first principal focal point of the optical system. The total deviation δ is the sum of δ_1 and δ_2, and the point where the deviation appears to have occurred is P, the intersection of the emergent ray parallel to the principal axis and its conjugate incident ray on the optical system. *PH* is called the first principal plane, and *HF* is called the first principal focal length of the optical system. By convention of signs it is negative, since F is to the left of H, the first principal point of the optical system.

Now

$$\delta = \delta_1 + \delta_2 \tag{3.67}$$

But

$$\delta = \frac{y_2}{(-f)}, \ \delta_1 = \frac{y_1}{(-f_1)}, \ \delta_2 = \frac{y_2}{(-f_2)}$$

On substituting the values of δ, δ_1, and δ_2 in Equation (3.67), we get

$$\frac{y_2}{f} = \frac{y_1}{f_1} + \frac{y_2}{f_2} \tag{3.68}$$

From the similar triangles DH_1F_2 and EH_2F_2, we have

$$\frac{y_1}{y_2} = \frac{(-f_2)-d}{(-f_2)} = \frac{f_2+d}{f_2} \tag{3.69}$$

or

$$y_1 = y_2 + y_2\left(\frac{d}{f_2}\right)$$

Substituting the value of y_1 in Equation (3.68), we get

$$\frac{1}{f} = \frac{1}{f_1} + \frac{1}{f_2} + \frac{d}{f_1f_2} \tag{3.70}$$

or

$$f = \frac{f_1f_2}{f_1+f_2+d} = \frac{f_1f_2}{\Delta} \tag{3.71}$$

where we have written here

$$\Delta = d + f_1 + f_2$$

3.20.1 Situation of Principal Points

Let p denote the distance of the first principal plane PH from the first lens and let p' denote that of the second principal plane $P'H'$ from the second lens. Owing to the reason that the principal planes are located from the given position of the lens, we should assign signs to p and p' according to our sign convention, with reference to lens positions as origins. Accordingly, the numerical value of H_2H' is assigned a negative sign, since H' is to the left of H_2. But H is to the right of H_1, and hence we have shown H_1H as $(+p)$.

To evaluate the distance of the second principal plane from the second lens, we refer to Figure 3.18. Now, from similar triangles $P'H'F'$ and CH_2F', we have

$$\frac{h_2}{h_1} = \frac{f'-(-p')}{f'} = \frac{f'+p'}{f'} \tag{3.72}$$

Equating the value of $\left(\frac{h_2}{h_1}\right)$ as given by Equation (3.62) and Equation (3.72), we get

$$\frac{f'+p'}{f'} = \frac{f_1'-d}{f_1'} \text{ or } p' = -\frac{f_1'}{f_1'}(d) \tag{3.73}$$

Substituting the value of f' from Equation (3.64) in Equation (3.73),

$$p' = \frac{f_2'd}{\Delta} \tag{3.74}$$

To evaluate p we refer to Figure 3.19, in which from similar triangles PHF and DH_1F we have

$$\frac{y_1}{y_2} = \frac{(-f)-p}{(-f)} = \frac{f+p}{f} \tag{3.75}$$

Equating the value of $\left(\dfrac{y_1}{y_2}\right)$ as given by Equations (3.69) and (3.75), we get

$$\frac{f+p}{f} = \frac{f_2+d}{f_2}(d)$$

or

$$p = +\frac{f}{f_2}(d) = +\frac{f'}{f_2'}(d) \tag{3.76}$$

Substituting the value of f from Equation (3.71) in Equation (3.76), we get

$$p = \frac{f_1d}{\Delta} \tag{3.77}$$

Example 3.1

Two convex lenses each of 20 cm focal length are placed 5 cm apart. A tower of height 100 meters and 200 meters distant is viewed through them. Find the position and the size of the image.

Solution:

Suppose light from the tower is incident on the given optical system from the left. The lenses are situated in air.

Here, $f_1' = +20$ cm, $f_2' = +20$ cm, and $d = 5$ cm

Hence

$$\frac{1}{f'} = \frac{1}{f_1'} + \frac{1}{f_2'} - \frac{d}{f_1'f_2'} = \frac{2}{20} - \frac{5}{20 \times 20} = \frac{7}{80}$$

or

$$f' = \frac{80}{7} \text{ cm}$$

Similarly, $f = \dfrac{80}{7}$ cm

and

$$p' = -\frac{df'}{f_1'} = -\frac{5 \times 80}{20 \times 7} = -\frac{20}{7} \text{ cm}$$

and

$$p = +\frac{df'}{f_2'} = +\frac{20}{7} \text{ cm.}$$

Thus, the first principal point H is $\left(\dfrac{20}{7}\right)$ cm to the right of the first lens, and the second principal point H' is $\left(\dfrac{20}{7}\right)$ cm to the left of the second lens. The tower is 20000 cm in front of the first lens or $20000 + \left(\dfrac{20}{7}\right) = \dfrac{140020}{7}$ cm in front of the first principal plane. Hence, for the tower, $u = -\left(\dfrac{140020}{7}\right)$ when referred to principal points.

Now,

$$\frac{f'}{v} + \frac{f}{u} = 1 \rightarrow \left(\frac{80}{7 \times v}\right) + \left(\frac{80 \times 7}{7 \times 140020}\right) = 1 \rightarrow v = +11.4 \text{ cm}$$

The positive sign indicates that the image of the tower is formed at 11.4 cm to the right of the second principal point H' or at $11.4 - \left(\dfrac{20}{7}\right) = 8.54$ cm to the right of the second lens. Also, since the lenses are situated in air, we have

$$\frac{y'}{y} = \frac{v}{u} = -\frac{11.4 \times 7}{140020} \rightarrow y' = -\frac{11.4 \times 7 \times 10000}{140020} = -5.699 \text{ cm.}$$

The image of the tower is of height 5.699 cm and the negative sign indicates that the image is inverted with respect to the object.

Example 3.2

Obtain an expression for the focal length of the combination of two thin coaxial lenses of focal lengths f_1 and f_2 separated by a distance "a," the refractive index of the medium enclosed between the lenses being "n."

Solution:

It will be recalled, as proved in Section 3.3, that the ratio of the second principal focal length to the first principal focal length of a thin lens is numerically equal to the ratio of the refractive index of the real image space to the index of the real object space, that is,

$$f_1' = n f_1$$

where n is the refractive index of the real image space with respect to the index of the real object space. Since we assume $n > 1$, hence $f_1' > f_1$. A reverse of this holds good if the refractive index of the real object space with respect to that of the real image space is n, that is, in this case the second principal focal length will be smaller than the first principal focal length.

In our problem of two lenses separated by a medium of index "n," we assume light to be incident on the combination from the left in air. Then, for the first lens, the real object space is an air medium, while for the second lens, the real image space is an air medium. We further assume that the given focal lengths $\mathbf{f}_1$ and $\mathbf{f}_2$ are the numerical values of the smaller focal lengths of the two lenses. Accordingly, the second principal focal length, f_1', of the first lens is given by

$$f_1' = n\mathbf{f}_1$$

and the second principal focal length of the second lens is given by

$$f_2' = n\mathbf{f}_2$$

and its first principal focal length is given by

$$f_2' = nf_2' = n\mathbf{f}_2$$

Now, the optical separation, Δ, between the two lenses is given by

$$\Delta = a - f_1' + f_2 = a - n\mathbf{f}_1 - n\mathbf{f}_2$$

Therefore, the second principal focal length f' of the combination is

$$f' = -\frac{f_1' f_2'}{\Delta} = -\frac{n\mathbf{f}_1\mathbf{f}_2}{a - n\mathbf{f}_1 - n\mathbf{f}_2} = \frac{\mathbf{f}_1\mathbf{f}_2}{\mathbf{f}_1 + \mathbf{f}_2 - \left(\dfrac{a}{n}\right)}$$

and the first principal focal length of the combination is given by

$$f = \frac{f_1 f_2}{\Delta} = \frac{(-\mathbf{f}_1)(-n\mathbf{f}_2)}{a - n\mathbf{f}_1 - n\mathbf{f}_2} = -\frac{\mathbf{f}_1\mathbf{f}_2}{\mathbf{f}_1 + \mathbf{f}_2 - \left(\dfrac{a}{n}\right)}$$

3.21 EXERCISES

1. The study of the refraction of light through a single spherical surface is of great fundamental importance in what?

2. What are the derivations of various formulae in this branch of optics based upon?

3. What is the equation of the angle of refraction r conjugate to the angle of incidence i according to Snell's Law?

4. The theory based upon the approximation that $\sin\theta_1 = \theta_1$ beyond the first order term is referred to as what?

5. The axial object point which is imaged at infinity by a refracting surface is called the _____ _____ _____ point F.

6. What is the distance of F from the pole of the surface called?

7. What variable represents the first principle focal length of the surface?

8. What is the principal focal point of the image space?

9. The principal focal point of the image space is also referred to as what?

10. The two focal lengths of the refracting surface are always of opposite signs.

    True or False

11. What does the Abbe sine condition do for all rays?

12. What is the equation given by Lagrange called?

13. If one surface is spherical and the other cylindrical, the lens is said to be spherical.

    True or False

Fill in the Blanks:

14. A _____ passing through the centers of curvature of the two surfaces is called the principal axis of the lens.

15. The distance from the center of the lens to either focus gives the _____.

16. Planes drawn perpendicular to the axis and passing through these points are called what?

17. The first and second focal distances are found to be _____ in magnitude but _____ in sign.

18. What are the three cardinal points of a lens system referred to as?

Match the following points to their meanings:

19. Anti-Principle Points   self-conjugate points

20. Bravais Points   angular magnification

21. Anti-Nodal Points   lateral magnification

Fill in the Blanks:

22. The focal length of a biconvex lens is _____.

23. If the convex surface faces the incident light, r_1 is _____; the focal length is _____.

24. In general if the _____ _____ __ _____ increases toward its periphery, the _____ _____ is negative, and if it increases towards its central region, the focal length is _____.

25. PH is called the _____ _____ _____ and HF is called the _____ _____ _____ _____ of the optical system.

26. Define the cardinal points of a system of coaxial lenses. Describe how you would determine the principal planes of a combination of two thin lenses separated by a distance.

27. What are the properties of the cardinal points of a coaxial lens system? Plot the cardinal points of a Huygens eyepiece. How can they be determined experimentally?

28. Show that for a coaxial lens system, $xx' = ff'$ where x and x' are the respective distances of the object and image from the first and second focal points, and f and f' are two focal lengths. What form does the expression take when the media on the two sides of the system are the same?

29. Find the expression for the equivalent focal length of two thin lenses separated coaxially by a finite distance.

30. How would you determine experimentally the positions of the focal and principal points of a coaxial lens system? Explain and illustrate how to trace the image of an object formed by a lens system, given the positions of the cardinal points.

31. Give the graphical construction of the formation of an image by a system of lenses. How are the nodal points of a system of lenses determined experimentally?

32. Two thin lenses of focal lengths 12 cm and 4 cm are kept separated by a distance of 8 cm. Plot the positions of the cardinal points for the combination. Derive the formula used.

33. A telephoto lens consists of a convergent lens of focal length 12 cm facing the object and a divergent lens of focal length 5 cm placed 8 cm behind the former. Find the position where the plate should be placed to photograph a distant object.

34. What are different kinds of magnifications associated with a pair of conjugate positions of a coaxial optical system? Establish a relation between them.

35. Two thin convergent lenses each of 20 cm focal length are set coaxially 5 cm apart. An image of an upright pole 200 meters distant and 10 meters high is formed by the combination. Find the position of the unit and focal planes and the image. Also find the size of the image.

36. Prove for a combination of two thin lenses of focal lengths f_1 and f_2 separated by a distance d that the focal length of the combination is given by

$$\frac{1}{f} = \frac{1}{f_1} + \frac{1}{f_2} - \frac{d}{f_1 f_2}$$

37. Two convex lenses of focal lengths 20 cm and 5 cm are 10 cm apart. Calculate the power of such a combination.

38. Derive an expression for the focal length of a system of two thin lenses separated by a distance and calculate the position of the principal points. A thin converging lens and a thin diverging lens are placed coaxially in air at a distance of 5 cm. If the focal length of each is 10 cm, find for the combination (a) the focal length, (b) the power, and (c) the position of the principal points.

39. Derive an expression for the equivalent focal length of two thin lenses of focal lengths f_1 and f_2 arranged coaxially at a distance apart.

40. Two thin convex lenses of focal lengths 12 cm and 6 cm are placed coaxially 8 cm apart. Determine the position of the cardinal points of the system.

41. An object is placed at a distance 60 cm from a thin convex lens of focal length 20 cm. There is a second thin convex lens of focal length 30 cm that is 10 cm away from the first. Calculate the distance of the final image from the second lens.

42. Two plano-concave lenses of refractive index 1.50 have radii of curvature of 20 cm and 30 cm respectively. They are placed in contact with the curved surfaces facing each other, and the space between them is filled with a liquid of refractive index 1.33. What is the focal length of the combination?

43. A concavo-convex lens has an index of refraction 1.5, and the radii of curvature of its surface are 10 cm and 20 cm. The concave surface is upward and filled with an oil of refractive index 1.60. Calculate the focal length of the oil-glass combination.

44. What do you mean by conjugate planes? How many conjugate planes are there for an optical system?

45. Define principal planes. Where do these principal points lie in the case of a glass sphere?

46. Define nodal planes. Where do the nodal planes lie in the case of a glass hemisphere?

47. A refracting surface separates two media, one called an incident medium having $(r)(i)(n_1)$, the other having $(r)(i)(n_2)$. If f_1 and f_2 are image side focal length and object side focal length respectively, then prove

$$\left(\frac{f_1}{n_1}\right) + \left(\frac{f_2}{n_2}\right) = 0.$$

CHAPTER 4

ABERRATIONS OF OPTICAL IMAGES

Chapter Outline

4.0 INTRODUCTION

The theory of image formation by mirrors and lenses, developed in the previous chapters, was based on two fundamental assumptions: (a) the object and the image points were considered as if they were situated on or very close to the principal axis, and (b) the rays of light diverging from the object point were assumed to be confined to a narrow cone of a small angular opening so as to allow us to replace the sine of the slope angle θ of the ray by the angle θ. These paraxial conditions lead to a point image of a point object in the *first order theory*, so called because of the inclusion of the first order of the slope angle only.

In reality objects have finite dimensions and, moreover, considerations of the brightness of the image demand the use of reflecting and refracting systems having large apertures. In actual practice, therefore, the rays from object points are confined not simply to a region close to the principal axis, termed the paraxial region, but they form a cone of such a wide angular opening that in the expansion of the sine of the slope angle θ into a series,

$$\sin\theta = \theta - \frac{\theta^3}{3!} + \frac{\theta^5}{5!} - \frac{\theta^7}{7!} + ...$$

and it becomes essential to take into account, in our discussions of the image formation, at least the first two terms of this series. This theory, called the *third order theory* due to the inclusion of the third order term in θ, was developed by *Ludwig Von Seidal* in 1855 and gives various departures of the actual image from the one predicted by the first order theory. These departures, the so-called Seidal aberrations, may be classified as (a) spherical aberration, (b) coma, (c) astigmatism, (d) curvature, and (e) distortion. These aberrations are present in the images formed by ordinary lenses, even when the light of only one wavelength is employed. They are, therefore, also known as *monochromatic aberrations* (monos is a Greek word for single, and chroma is a Greek word for color). Each of the five monochromatic aberrations depends upon the wavelength of light but, leaving aside spherical aberration, the variation is always negligibly small in the others.

Furthermore, the focal length of the lens depends upon its index of refraction, which varies with wavelength of light. Therefore, even in the first order theory, if the light diverging from the object is not monochromatic, a lens will form a number of colored images of different sizes and at different positions. Thus, we shall encounter two new aberrations known as the *lateral chromatic aberration* and the *longitudinal chromatic aberration*. These aberrations are, however, absent in the images formed by mirrors.

A rigorous mathematical study of these aberrations is beyond the scope of this book but, wherever possible, simple mathematical deductions will be given while describing them in detail.

4.1 SPHERICAL ABERRATION OF A LENS

When the ratio of the aperture to the focal length of the lens is relatively large, a cone of rays from any axial object point *P*, after refraction, is not focused at a single point in a manner demanded by the first order theory, but produces an image as illustrated in Figure 4.1.

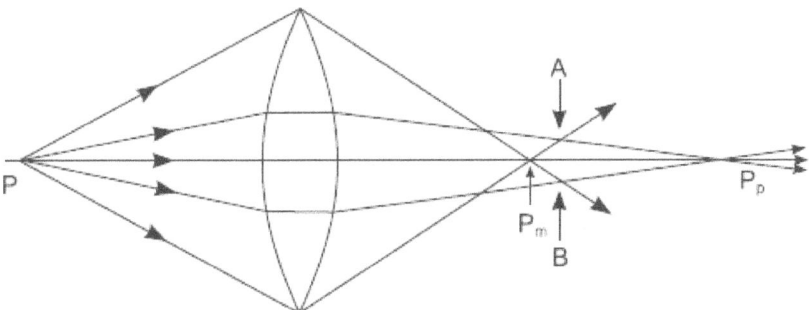

FIGURE 4.1. Spherical aberration of a convergent lens for an axial point object.

This effect is known as spherical aberration and arises due to the fact that the different annular zones of the lens have different focal lengths—the greater the radius of the zone, the smaller the focal length along the axis. Consequently, in Figure 4.1 the marginal rays, that is, those which are refracted at the boundary of the lens, converge to an image point P_m which is considerably closer to the lens than the point P_p, the focus of the paraxial rays. In general, the rays which are refracted at any other circular zone cross the axis at one image point between P_m and P_p. However, a different axial image point corresponds to each annular zone. There is evidently no plane in which a sharp image of *P* is formed. The cross-section of the refracted beam is circular everywhere, and it will be seen from the figure that there is one plane, *AB*, at which the cross-section has the least diameter. This smallest cross-section is known as the *circle of least confusion*, and the best image of *P* is obtained if a screen is placed perpendicular to the lens axis at this point.

The spherical aberration of a biconcave lens for a parallel incident beam is illustrated in Figure 4.2, while that for a virtual object is shown in Figure 4.3.

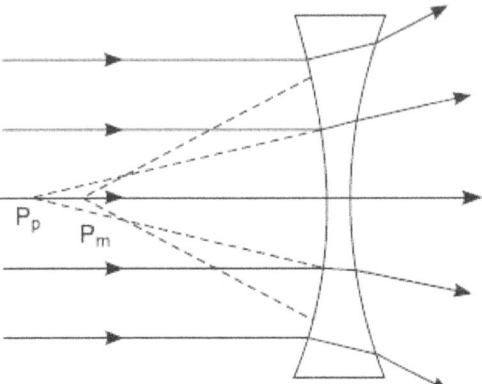

FIGURE 4.2. The spherical aberration of a biconcave lens for a parallel incident beam.

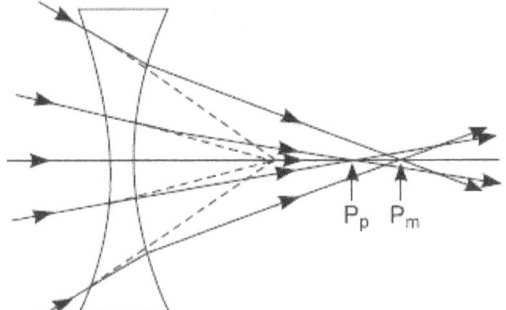

FIGURE 4.3. The spherical aberration of a biconcave lens for a virtual object.

The separation between the marginal focus P_m and the paraxial focus P_p is taken as a measure of the *longitudinal spherical aberration* of the lens, while the radius of the circle of least confusion is taken as a measure of the *lateral* or *transverse spherical aberration*. The spherical aberration varies approximately as the square of the height of the incident ray above the axis, and it also depends upon the distance of the object point. The longitudinal spherical aberration is considered positive or negative according to whether the marginal ray's focus P_m lies on the left or on the right of the paraxial ray's focus P_p. A convergent lens produces a positive aberration, while a divergent lens produces a negative aberration.

4.1.1 Reduction or Elimination of Spherical Aberration

The following devices are usually employed for reducing spherical aberration in images:

1. The spherical aberration of a given lens depends on the radius of the lens aperture. Therefore, it may be reduced by *stopping down* the lens aperture with the help of a coaxial aperture stop, thus using only the central portion of the lens. Of course, this would reduce the amount of light transmitted by the lens.

2. A single lens of a given focal length can also be designed so that the spherical aberration is at a minimum. The longitudinal spherical aberration (x) of a thin lens for parallel incident rays is given by

$$x = +k\frac{h^2}{f'}\left[\frac{\beta^2\mu^3 + \beta\left(\mu + 2\mu^2 - 2\mu^3\right) + \mu^3 - 2\mu^2 + 2}{2\mu(\mu-1)^2(1-\beta)^2}\right] \tag{4.1}$$

where h is the radius of the lens aperture, f' is its second principal focal length to be substituted with proper sign, and $\beta = \left(\dfrac{R_1}{R_2}\right)$ is called the shape factor of the lens. For a given value of h, f', and μ, the condition for minimum spherical aberration is $\dfrac{dk}{d\beta} = 0$, which leads to

$$\beta = \frac{2\mu^2 - \mu - 4}{\mu(1+2\mu)} \tag{4.2}$$

Since β has only one turning value, it obviously gives a minimum value of spherical aberration k, because when $\beta = 1$, k becomes infinite. If $\mu = 1.5$, Equation (4.2) gives $\beta = \left(\dfrac{R_1}{R_2}\right) = -\left(\dfrac{1}{6}\right)$. Thus, the form of a lens giving minimum spherical aberration is biconvex or biconcave, and the radius of curvature R_1 of the surface facing the object is about one sixth of that of the other face. In general, the more curved face of the lens should face toward the incident or emergent rays, whichever are more nearly parallel to the lens axis. A lens having $\left(\dfrac{R_1}{R_2}\right) = -\left(\dfrac{1}{6}\right)$ is termed as the *crossed lens* as shown in Figure 4.4, and this process of changing the shape of the lens without a change of focal length is known as *bending the lens* for minimum spherical aberration. It should be emphasized that the spherical aberration cannot be completely removed by any bending of the lens with spherical surfaces.

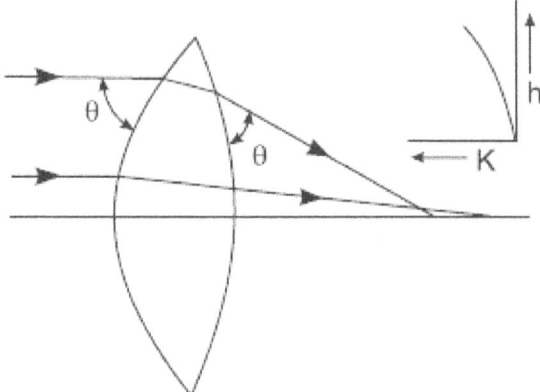

FIGURE 4.4. Crossed lens.

Spherical aberration can also be reduced in a single lens of a given focal length by constructing it from material of a high refractive index. For example, suppose $f' = 100$ cm, $h = 10$ cm, and $\mu = 1.5$. For minimum spherical aberration $\beta = -\left(\dfrac{1}{6}\right)$ and from Equation (4.1), $k = 1.07$ cm. Similarly, with $\mu = 2$, $\beta = +\left(\dfrac{1}{5}\right)$, and k is diminished to 0.44 cm.

3. In some optical instruments, plano-convex lenses are employed because they are cheaper to make, and when their curved side faces the parallel incident (or emergent) light, the spherical aberration is very nearly the same as that of a crossed lens. However, if the plane side is turned toward the object, the spherical aberration is very large. The spherical aberration being the result of the greater deviation of the marginal rays as compared to that of the paraxial rays, it is obvious that if the deviation of the marginal rays are made minimum, the marginal focal point P_m will be at its farthest to the right, and thus the least value of the spherical aberration will result. Just as in the case of a prism, the deviation is at a minimum when the incident and emergent rays make equal angles with its faces; similarly, in the case of a lens, the deviation of the marginal rays will be at a minimum when they enter the first lens surface and leave the second surface

at more or less equal angles. Thus, as a general rule, spherical aberration will be minimized when a lens is so designed or used that the total deviation of a given ray is divided equally between the two refractions. In a plano-convex lens with plane side toward the distant object, as shown in Figure 4.5, the deviation of a given ray is simply produced at the curved surface, and therefore the condition of minimum spherical aberration is violated, thus accounting for the presence of a large spherical aberration. But when the convex side faces the distant object, as shown in Figure 4.5, the deviation is divided between the two surfaces, thus accounting for the presence of minimum spherical aberration.

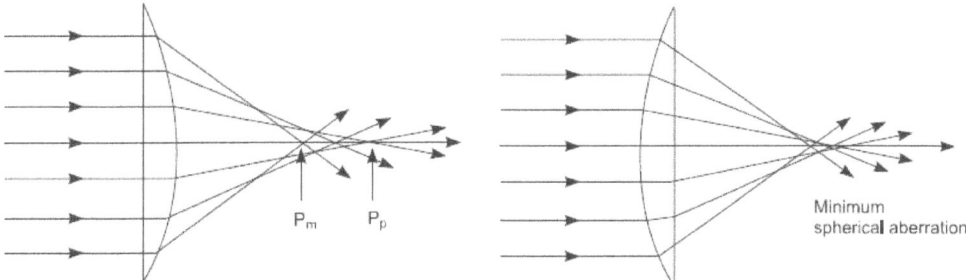

FIGURE 4.5. Spherical aberration is minimized if the deviation is equally shared between the surfaces.

The previous principle can be applied for finding the distance apart of two convergent lenses so that the resultant spherical aberration is a at a minimum. Consider a ray PQ, parallel to the principal axis, incident on the lens L_1 and traversing the optical system as shown in Figure 4.6.

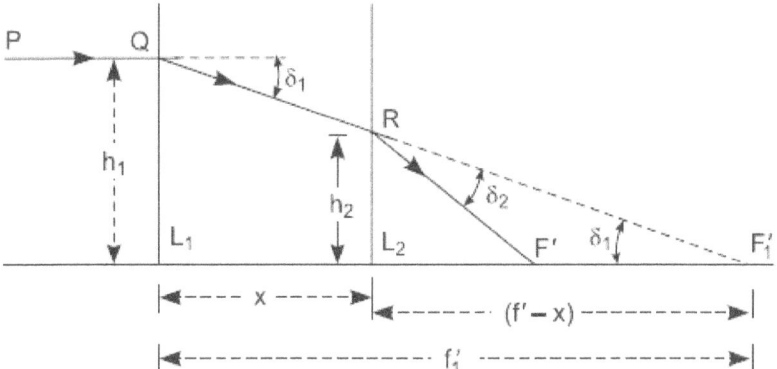

FIGURE 4.6. Distance apart of two convergent lenses so that the resultant spherical aberration is a at a minimum.

It will be seen from Figure 4.6 that the deviation produced in the ray by L_1 is δ_1 while that produced by L_2 is δ_2. Hence, the spherical aberration will be at a minimum when

$$\delta_1 = \delta_2$$

or

$$\frac{h_1}{f_1'} = \frac{h_2}{f_2'} \tag{4.3}$$

also, from similar triangles QL_1F_1' and $R_2L_2F_1'$, we have

$$\frac{h_2}{f_1' - x} = \frac{h_1}{f_1'} \tag{4.4}$$

Equating the values of $\left(\dfrac{h_1}{h_2}\right)$ as given by Equations (4.3) and (4.4), we get

$$x = f_1' - f_2' \tag{4.5}$$

Thus, the spherical aberration of a combination of two convergent lenses is at a minimum when their distance apart is equal to the difference of their principal focal lengths. It is essential in this case that the incident ray should suffer refraction first through the lens of the larger focal length and then through the one of the smaller focal length.

4. The spherical aberration of a single convergent lens for object positions beyond the first principal focus is positive, the marginal focus P_m being nearer to the lens than the paraxial focus P_p. A concave lens exhibits negative spherical aberration for a virtual object as shown in Figure 4.2, the paraxial focus P_p being nearer to the lens than the marginal focus P_m. Therefore, by combining a convergent and a divergent lens of proper shapes, it is possible to compensate the spherical aberration of a single zone of one lens with that of the other. If the two components are in contact and the combination is to act as a convergent lens, the principal focal length of the convergent component should be smaller than that of the divergent one. Since the spherical aberration increases with decrease in focal length, it is obvious that the spherical aberration of the convergent lens will be greater than that of the divergent one. Consequently, the convergent lens should be so shaped as to have minimum spherical

aberration, while the shape of the divergent one should be such as to have increased aberration by an amount just to make the resultant aberration zero. Thus, the system is said to be corrected if the aberration is zero for one zone and negligible for others. Figure 4.7 gives a graph and the spherical aberration of a cemented doublet corrected for marginal zone. For other zones the focal length is slightly less than the paraxial focal length. Combinations of lenses of this type are usually made in such a way as to eliminate both the spherical aberration and the chromatic aberration.

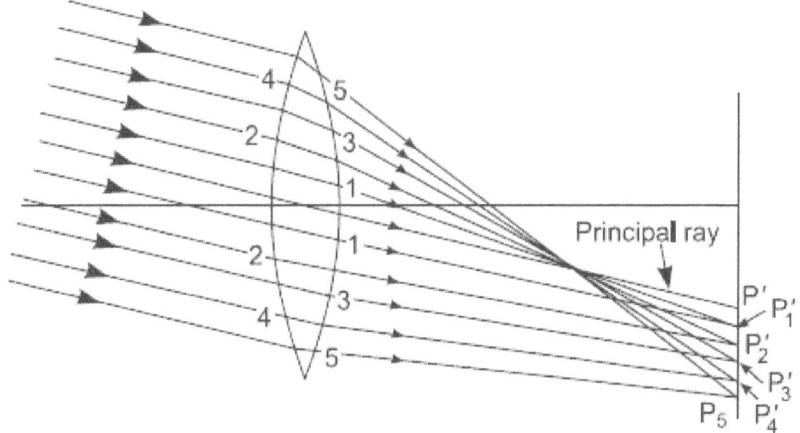

FIGURE 4.7. Spherical aberration of a cemented doublet corrected for the marginal zone.

5. Although *spherical aberration* cannot be eliminated, in general, from a single lens, it is absent for one pair of conjugate points called the aplanatic points of the lens. For this pair of points, the lens is also free from another aberration called *coma*. We shall refer to this point in the next section.

4.2   COMA

The aberration known as *coma* affects rays that come from object points even if located very close to the axis of the lens. It arises from the fact that for these non-axial point objects, there is either an increase as shown in Figure 4.8 (a) or a decrease as in Figure 4.8 (b) of lateral magnification with the height of the narrow circular zone of the lens through which the rays are refracted. In order to discuss the comatic defect in the image, it would be assumed that the lens system is free from other aberrations.

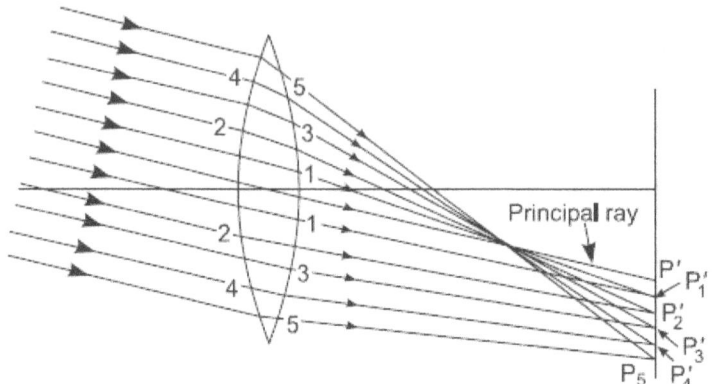

FIGURE 4.8 (a). Illustrating the formation of "*positive coma*" in the meridional plane.

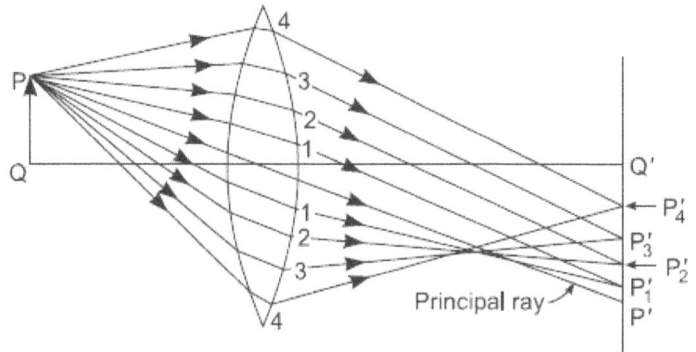

FIGURE 4.8 (b). Illustrating the formation of "*negative coma*" in the meridional plane.

Figure 4.8 (a) illustrates, in a lens, the effect of coma in the image of an object point situated close to the axis but at infinity. It would be observed that the ray below the optical center strikes the equivalent prism in any zone of the lens approximately in a symmetrical condition of refraction (equal angles of incidence and emergence) when compared to the ray above it; therefore, the former ray has nearly minimum deviation, while the latter has comparatively greater deviation. As a consequence, the rays from the upper and the lower portions of the same zone, after refraction, intersect below the principal ray, and thus the lateral magnification for the outer zone is greater than that for the central zone, and as explained in the sequence, the "*coma*" is said to be positive.

For the non-axial object point *P* at a finite distance from the lens, it would be observed, as shown in Figure 4.8 (b), that the ray above the optical center strikes the equivalent prism in any zone of the lens approximately in the symmetrical condition of refraction than the ray below it.

As a consequence, the rays from the upper and lower portions of the same zone, after refraction, intersect above the principal ray, and thus the lateral magnification for outer zones is less than that for the central zone. The coma under this condition, as explained in the sequence, is said to be negative. In reality, a cone of rays diverges from the point P. Therefore, we should also consider the refraction of rays not lying in the meridional plane. This is illustrated by Figure 4.9, which shows the refraction of rays through different points of a single zone.

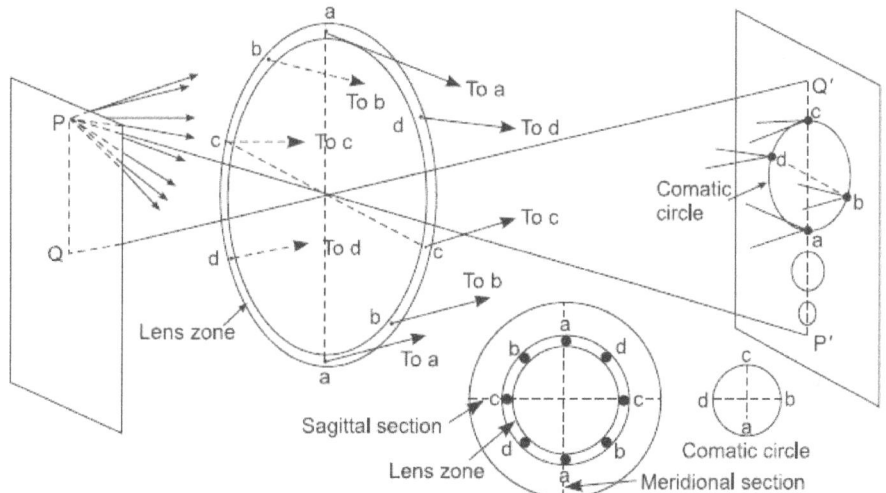

FIGURE 4.9. Each zone of the lens forms a ring-shaped image called a *comatic circle*.

By tracing the rays it is found that the meridional rays P_a, after refraction through the points a, i of the lens zone, come in focus at a point a of the image plane, while the rays P_c, in a plane perpendicular to the meridional plane, after refraction through the points c, c of the lens zone come in focus at the point c of the image plane, and so on. The annular cone of rays from P, which is refracted through the lens zone, thus comes to a focus in a circle a, b, c, d, called H. In the image plane, there will be such a *comatic circle* corresponding to each zone of the lens. The radius and the distance of the center of the *comatic circle* from the paraxial image point is proportional to the square of the radius of the lens zone. The image of the non-axial point P, therefore, consists of an expanding series of overlapping *comatic circles*. The totality of these circles produces an image having a balloon type of flare of light like a comet, with a bright nucleus at P' and with the tail becoming fainter and fainter as it grows, as shown in Figure 4.10—hence the name *coma*.

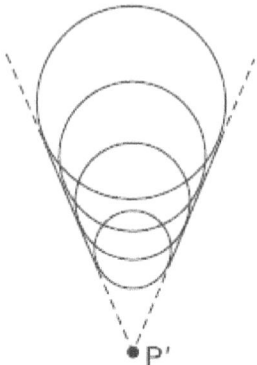

FIGURE 4.10. Comatic flare.

If the lateral magnification for outer zones is smaller than that for the central zone, as shown is Figure 4.8 (b), the tail of the comatic image extends inward toward the principal axis, and the coma is said to be negative. But if the lateral magnification for outer zones is greater than that for the central zone, as illustrated in Figure 4.8 (a), the tail of the comatic image extends outward from the principal axis, and the coma is said to be positive.

4.2.1 Reduction or Elimination of Coma

We see that a coma is the result of the variation of the lateral magnification for rays passing through different zones of a lens. Therefore, to eliminate the coma the lens should be so designed that the lateral magnification $\left(\dfrac{y'}{y}\right)$ is the same for all zones. The *Abbe sine condition* is shown in Figure 4.11.

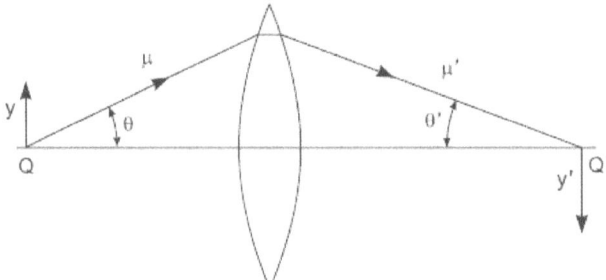

FIGURE 4.11. Abbe sine condition.

From the *Abbe sine condition*, which is presented in Equation (3.4), $\mu y \sin \theta = \mu' y' \sin \theta'$, it follows that an optical system is free from coma when the ratio

$$\frac{\sin \theta}{\sin \theta'} = \frac{\mu' y'}{\mu y} = \text{Constant} \qquad (4.6)$$

for all values of θ, where θ in the previous relation is the angle subtended by the zone radius at the axial point in the object plane.

For a very distant object near the axis, $\sin \theta$ is proportional to h, the height of the incident ray. In this case, the condition for freedom from coma is then

$$\frac{h}{\sin \theta'} = \text{Constant} \qquad (4.7)$$

From Figure 4.12 it is evident that the ratio $\left(\dfrac{h}{\sin \theta'} \right)$ is the effective focal length f_x' of the zone, which is the distance measured along the ray from the focal point F' to the point A where the ray deviation appears to occur.

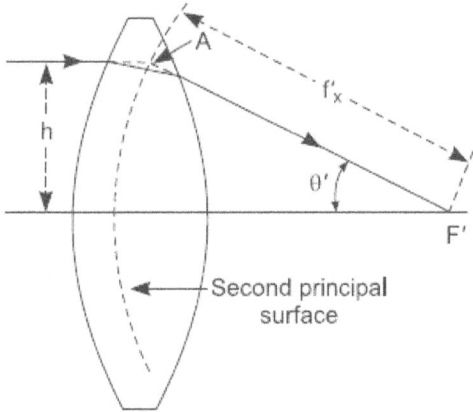

FIGURE 4.12. Illustrating the ratio $\left(\dfrac{h}{\sin \theta'} \right)$ is the effective focal length $f'x$ of the zone.

To eliminate the coma, the effective focal lengths of the zones must be identical. This condition demands that for a lens to be free of coma, the principal surface of the lens should be a spherical one with a center at F' and a radius f_x'. This ideal condition also ensures that the marginal focal length is the same as the paraxial focal length, and so spherical aberration is also absent.

The *coma*, like spherical aberration, may be minimized by the proper choice of radii of curvature of the lens surface, and this is investigated by the ray tracing method. But a lens designed for a minimum spherical aberration, however, cannot be free from coma, because the necessary shape of the lens for a minimum spherical aberration is not the same as required for zero coma. However, the so-called *aplanatic lens*, as illustrated in Figure 4.13, is simultaneously free from coma and spherical aberration for one particular pair of object and image points called the *aplanatic points* of the lens.

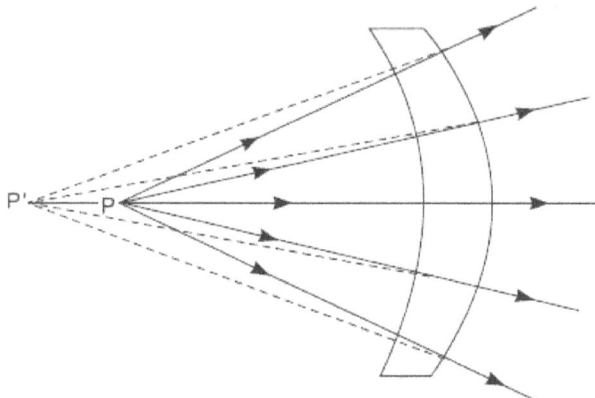

FIGURE 4.13. Aplanatic lens.

All rays diverging from an object placed at the center of curvature of the first surface, after refraction at the second surface, appear to diverge from the image point P' with coma and spherical aberration missing. This also holds good for a small element of object perpendicular to the lens axis at the *aplanatic point*.

A lens is free from spherical aberration for two neighboring axial points if the condition

$$\frac{\mu \sin\left(\dfrac{\theta}{2}\right)}{\mu' \sin\left(\dfrac{\theta'}{2}\right)} = \text{Constant} \tag{4.8}$$

is satisfied. Evidently, this condition and the *Abbe sine condition* cannot be satisfied simultaneously. Therefore, a lens system can be corrected to form a true image of either a small element along the axis (absence of spherical aberration) or a small element perpendicular to the principal axis (absence of coma), but not both simultaneously.

4.3   ASTIGMATISM

Even if the lens system is corrected for *spherical aberration* and *coma*, the image of an object point situated at an appreciable distance from the lens axis is not a point but a pair of mutually perpendicular lines some distance apart. This aberration is therefore known as *astigmatism*, as shown in Figure 4.14.

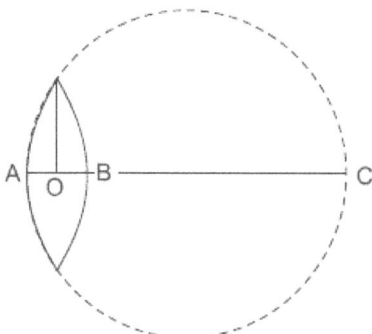

FIGURE 4.14. Astigmatism.

To explain the cause of astigmatism, we shall first express the focal length of a thin convergent lens

$$\frac{1}{f'} = +(\mu - 1)\left(\frac{1}{R_1} + \frac{1}{R_2} \right) \tag{4.9}$$

in terms of the diameter d of the lens aperture and the axial thickness t of the lens. By elementary geometrical considerations applied to Figure 4.14, we have

$$OE^2 = (OA) \times (OC)$$

$$\frac{1}{4} d^2 = t_1 \left(2R_1 - t_1 \right) \cong \frac{2R_1}{t_1}$$

Hence,

$$t_1 = \left(\frac{d^2}{8R_1} \right)$$

Similarly,

$$t_2 = \left(\frac{d^2}{8R_2} \right)$$

where

$$t = OA + OB = t_1 + t_2$$

whence

$$\frac{1}{R_1} + \frac{1}{R_2} = \frac{8t}{d^2}$$

Hence,

$$f' = +\frac{d^2}{8(\mu - 1)t} \tag{4.10}$$

Thus, the focal length of a thin convergent lens is proportional to $\left(\dfrac{d_2}{t}\right)$.

Now, in Figure 4.15, Q' is the image formed by a convergent lens of an axial object point Q.

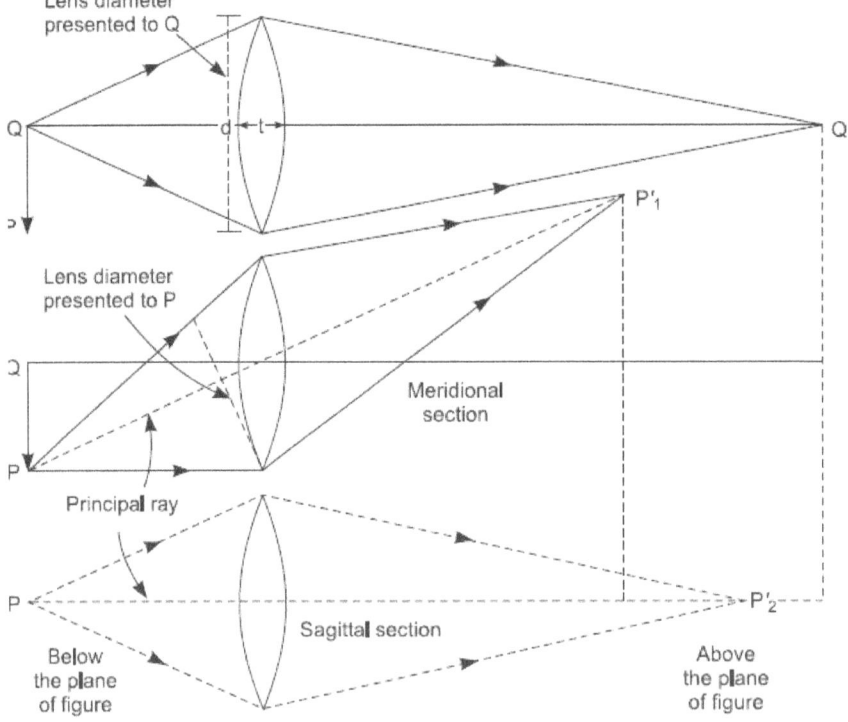

FIGURE 4.15. Illustration of the origin of astigmatism.

Owing to the symmetry of the lens with respect to Q, every section of the lens through its axis presents the same diameter to the axial point Q, and so the focal length is the same in all sections through the axis, and hence the refracted beam converges to one point. However, for the object point P, far away from the axis, the lens is not perfectly symmetrical. In the *meridional section*, the lens diameter presented to P is less than d. Moreover, the principal ray now passes inclined to the principal axis of the lens and therefore traverses a thickness greater than the axial thickness t of the lens. As a consequence of both the foregoing reasons, the focal length of lens in the *meridional section* is decreased for an off axial object point. Consequently, the image of P in this section is at P_1', nearer to the lens than the image point, which would be obtained by assuming the focal length in this section to be the same as along the axis. In the *sagittal section* of the lens, that is, the section by a plane through the principal ray and perpendicular to the *meridional section*, the lens diameter presented to non-axial object point P is the same as for the axial point, but the principal ray traverses greater thickness of the lens. The focal length in the *sagittal section* is therefore smaller than the axial focal length, but the decrease is not so much as in the *meridional section*, because in the former section the lens diameter presented to P remains d. The image of P, due to the refraction of rays in this section, is formed at P_2', which is slightly farther away from the lens than P_1' but closer to the lens than the image point, which would be given by the axial focal length. These considerations show that for an off-axial object point, the focal length of any narrow coaxial circular lens zone is not the same at every point of it, being least in the *meridional section* and maximum in the *sagittal*, but in each case less than the axial focal length.

The effect of astigmatism in the image of an off-axial object point P is illustrated in Figure 4.16, where it is assumed that other aberrations are absent.

The *meridional section* and the *sagittal section* of the cone of rays diverging from P are shown by vertical and horizontal shading respectively, before and after refraction through the lens. The rays in the meridional plane GHQ, owing to shorter focal length of the lens in this section, after refraction come to focus at P_1', while those in the sagitta plane PAB come to focus at another point P_2', farther away from the lens than P_1'. Similar considerations also show that all rays diverging from P, after refraction through the lens, do not pass through a single point but pass through a horizontal line ST, which is perpendicular to the principal ray $PP_1'P_2'$, and later on pass through a vertical line UW. Thus, a small refracted beam is, in general, not stigmatic but astigmatic.

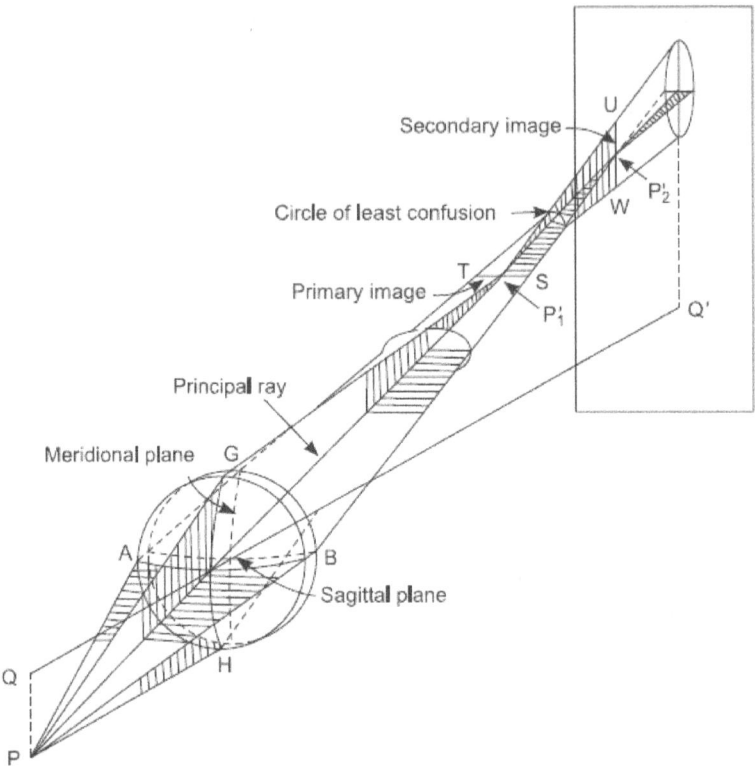

FIGURE 4.16. A diagram that shows two focal lines which constitute the image of an off-axial object point *P*.

The meridional line focus $SP_1'T$, which is perpendicular to the meridional plane, is called the *primary image*, and the sagittal line focus $UP_2'W$, which is vertical in the meridional plane, is called the *secondary image*. The cross-section of the refracted beam is, in general, elliptical, the ellipse degenerating into straight lines at positions of the primary and secondary foci and into a circle somewhere between these positions, where the major and minor axes of the ellipse are equal. This circle is known as the *circle of least confusion* and gives the best focus of the astigmatic pencil. For a simple convergent lens, the meridional focus P_1' is nearer to the lens than the sagittal focus P_2' as shown in Figure 4.16, and the astigmatism is said to be positive. But for a simple divergent lens as shown in Figure 4.17 (which illustrates only the refracted rays when produced backward) the sagittal focus P_2' is nearer to the lens than the meridional focus P_1', and astigmatism is said to be negative.

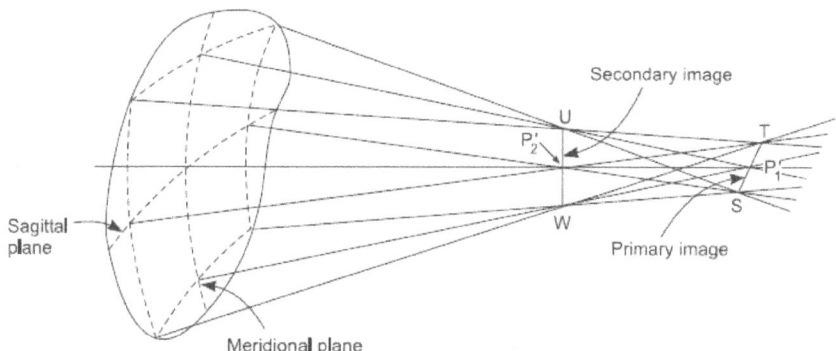

FIGURE 4.17. Divergent lens produces negative astigmatism.

Each point on the extended object, in an exactly similar manner, gives rise to its corresponding primary image, secondary image, and the circle of least confusion. Their respective loci are surfaces of revolution about the lens axis, paraboloidal in form, and are called the *primary image surface*, the *secondary image surface*, and the *surface of best focus*. These surfaces are tangential to one another at a point Q' on the lens axis, the point being the paraxial image of the conjugate axial point Q in the object space. The amount of astigmatism present corresponding to any object point is measured by the difference between the primary and the secondary image surfaces measured along the principal ray through that point, and in an uncorrected system it increases approximately as $\tan^2\theta$ where θ is the obliquity of the point object.

4.3.1 Reduction or Elimination of Astigmatism

1. Figure 4.18 illustrates the positions of the primary and the secondary images for a convergent as well as for a divergent lens.

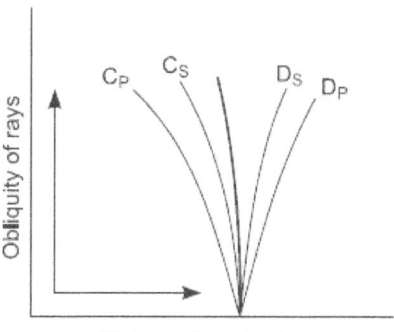

FIGURE 4.18. Positions of the primary and the secondary images for a convergent as well as for a divergent lens.

By proper spacing of convergent and divergent lenses, a combination may be designed in which the astigmatic differences compensate for one another to some extent and the images are formed on a single paraboloid surface, shown by a thick line. There is thus present another aberration called the *curvature of field*, which must be independently corrected for in the lens design. We shall discuss this aberration in detail in the next section.

2. With a single lens, astigmatism may be reduced by cutting out the oblique rays by means of coaxial apertures called stops, placed judiciously. The essential condition for complete elimination of astigmatism is that the primary and the secondary image surfaces must have the same curvature. The principal ray will now pass eccentrically through the lens, and the astigmatism is markedly reduced. Moreover, with lens shape, as shown in Figure 4.19, astigmatism may be eliminated completely.

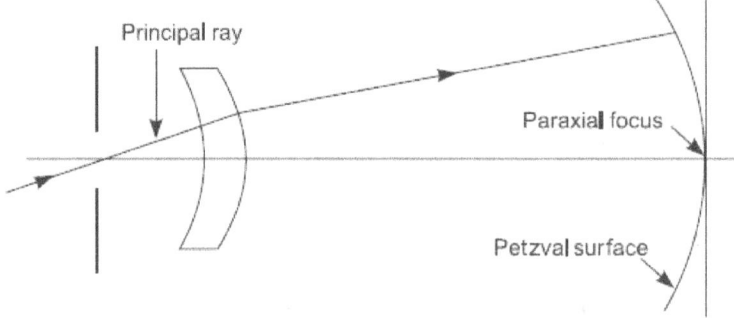

FIGURE 4.19. Correction for astigmatism.

4.4 CURVATURE OF FIELD

Even if the lens system is free from spherical aberration, coma, and astigmatism, that is, the zonal focal length of the system does not vary from its axial focal length for off-axial object points, the image of an object plane is, in general, a curved surface. This is due to the reason that since off-axial object points are farther away from the lens system than the axial point, in the case of a convergent lens forming a real image, the image of an off-axial object point is formed closer to the lens than the image of the axial object point, thus accounting for the curvature of the image shown in Figure 4.20.

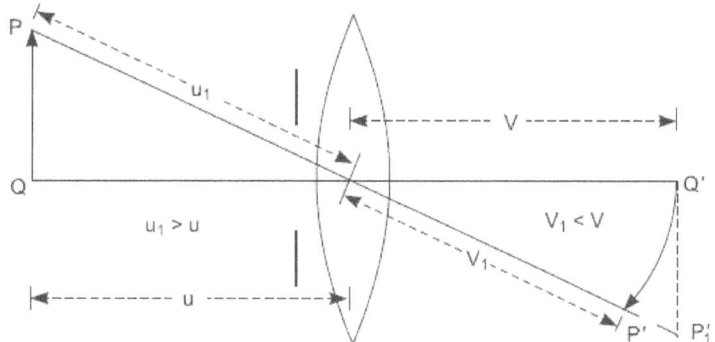

FIGURE 4.20. Curvature of field.

In a convergent system the radius of curvature of the image surface is negative, and hence the curvature is designated as negative, while the converse of this holds true in a divergent system, the concave side being toward the object in the former system while away from the object in the latter system. As a result, by their suitable combination, as explained later on, the image can be flattened. The curved image surface due to single lens refraction is called the *Petzval surface*, and the aberration is known as *curvature of field*.

4.4.1 Reduction or Elimination of Curvature of Field

1. For a system of thin lenses, the curvature in the final image is given by

$$\frac{1}{R} = \sum \left(\frac{1}{\mu f} \right) \tag{4.11}$$

Hence, for no curvature the condition is that $\sum \left(\dfrac{1}{\mu f} \right) = 0$. In the case of two lenses, this condition reduces to

$$\mu_1 f_1 + \mu_2 f_2 = 0 \tag{4.12}$$

where μ_1 and f_1 are the refractive index and focal length of one lens, and μ_2 and f_2 are the refractive index and focal length of the second lens. This condition for no curvature is known as the Petzval condition, and it holds whether the lenses are in contact or separated. Since μ_1 and μ_2 are positive, hence f_1 and f_2 must be of opposite signs. Thus, by employing a convergent and a divergent lens of different materials and satisfying the Petzval condition, the curvature of the image can be eliminated, resulting in a flat field. If f_1 refers to the convergent component, then for the combination to also be convergent, it is essential that f_2 must be greater than f_1. Therefore, in order that Equation (4.4b) may be satisfied, it is essential that μ_2 must be less than μ_1. This condition cannot be satisfied in the ordinary achromatic doublet consisting of a crown convergent lens in contact with a divergent lens of flint glass, because the flint glass has a higher refractive index and a higher dispersion, the latter condition being essential for the elimination of chromatic aberration. However, certain kinds of glasses were developed by Abbe at Jena, so that in a given pair one having a lower index had a higher dispersive power. Therefore, the achromatic doublets constructed of these glasses can be made free from curvature simply by choosing the focal lengths of the components so as to satisfy the Petzval condition. It should be emphasized that under this condition, astigmatism is also simultaneously minimized. High-quality photographic objectives are designed on the principles discussed previously and are known as an astigmat.

2. The Petzval condition is also satisfied by forming a combination of convergent and divergent lenses of equal focal lengths, made of the same glass and separated by a distance less than their individual focal lengths. Under such conditions, their combined focal length f is negative, that is, the combination is convergent free from astigmatism and curvature.

3. With a single lens, it is possible to minimize either curvature of field or astigmatism, by proper location of stops on the lens axis. To eliminate curvature, the lenses are usually of meniscus type, and the position of the front stop is adjusted so that the ray can be made to pass through the lens in such a manner that the primary and the secondary image surfaces have equal and opposite curvatures. As the surface of least confusion lies approximately midway between them, this results in the flattening of the image. Under these conditions astigmatism is, however, much more pronounced in the outer parts of the field.

4.5   DISTORTION

Even if the optical system has been corrected for spherical aberration, astigmatism, coma, and curvature of image so that the image of a transverse plane, like square, is a transverse plane in the image space, there can be present another aberration termed *distortion* in the image. This aberration refers to the departure from the strict geometrical relationship to each other of points in the image plane as their corresponding object points have in the transverse object plane. This departure arises due to a variation of magnification with the lateral distance of an object point from the lens axis. If the magnification increases with an increase in the lateral distance, then the image of a square-like object situated transverse to the principal axis, owing to the disproportionately high magnification of the corners as compared with other points, is of the form resembling a pincushion. This distortion, shown in Figure 4.21, is termed *pincushion distortion*.

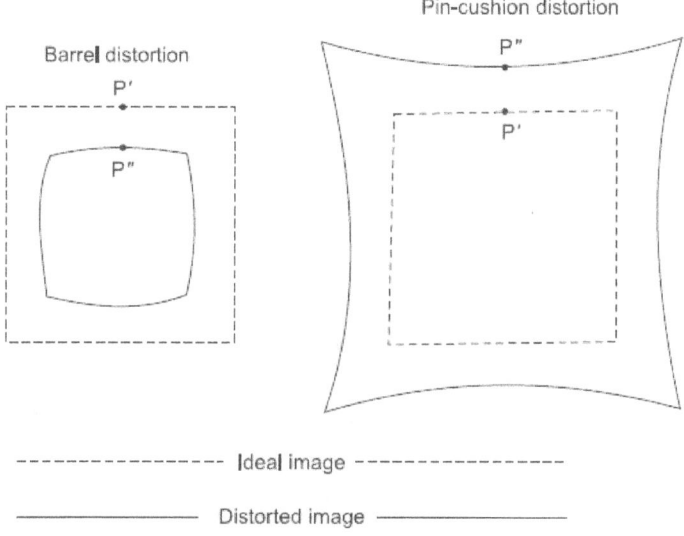

FIGURE 4.21. Illustration of two possible distortions in the images.

On the other hand, if the magnification decreases with increasing axial distance, the images of the diagonal are shortened relatively more than the images of the side of the square; that is, the opposite effect is produced, and therefore the resulting image of the square is of the form resembling a barrel. This distortion shown in Figure 4.21 is therefore termed *barrel distortion*. The ideal image in each case is shown by a dotted square. *Distortion* is measured by the distance between the actual image point P'' from the ideal image point P' of an off-axial object point P, the latter image being obtained by the paraxial condition. It would be observed from Figure 4.21 that the *pincushion distortion* and the *barrel distortion* are of opposite signs.

Before explaining the manner in which the barrel distortion or the pincushion distortion arises, we shall derive the conditions which any optical system must satisfy for freedom from distortion. A converging system with an aperture stop in front of it is shown in Figure 4.22, forming an image at Q' of a plane normal to the axis at Q, where it is assumed that the lens system is corrected for the first four aberrations with respect to the points Q and Q' so that the image at Q' is also a plane normal to the axis.

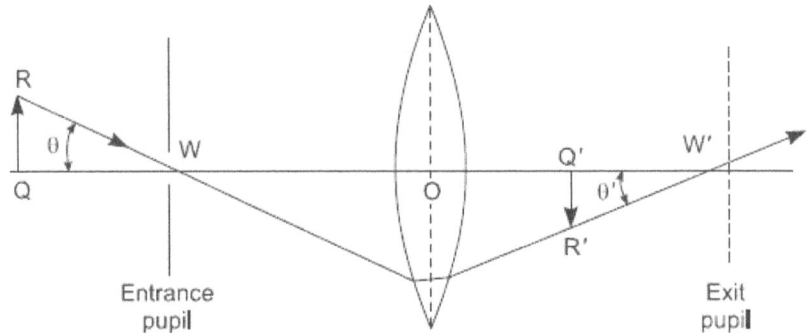

FIGURE 4.22. Condition for no distortion.

The aperture stop serves as an entrance pupil W in the object space, and in Figure 4.22 it is shown to the left of the first focal point so that the exit pupil W', which is the image of the entrance pupil W for paraxial rays, is real. Consider a ray from an object point R passing through the center of the entrance pupil W and, after refraction, through the center of the exit pupil W'. This ray is known as the *chief ray*. It cuts the image plane at R', which is consequently the paraxial image of R. The magnification produced by the optical system is given by

$$m = \frac{Q'R'}{QR} = \left(\frac{W'Q'}{WQ}\right)\left(\frac{\tan\theta'}{\tan\theta}\right)$$

From Figure 4.22 it is quite evident that the magnification of all object points will be constant, that is, the image will be free from distortion if

$$\left(\frac{W'Q'}{WQ}\right) = \text{Constant}$$

and

$$\left(\frac{\tan\theta'}{\tan\theta}\right) = \text{Constant}$$

for all values of θ. The first condition simply means that all rays through the center of entrance pupil W must also on emergence pass through the center of the exit pupil. This demands that the system be corrected for spherical aberration with respect to the entrance and exit pupils. This constancy of

$\left(\dfrac{\tan\theta'}{\tan\theta}\right)$ for all chief ray inclinations requires that the points of intersection of the entrant and the emergent chief rays must lie on a plane normal to the principal axis.

Now, to explain the manner in which barrel distortion arises, we shall consider two object points R and P, near and remote respectively from the axis. Since the point R lies in the paraxial region, the incident chief ray RW, after refraction, will pass through the center W' of the exit pupil. But the principal ray PW, being more steeply inclined to the axis, after refraction will not cross the axis at W', unless the system is corrected for spherical aberration with respect to W and W', but will cross the axis at W_1, which is closer to the lens than W', the paraxial image of W. If the system were corrected for spherical aberration with respect to W and W', that is, if the magnification were constant, the chief ray PW on emergence would have pursued the path shown by dotted line $P'W'$ in Figure 4.23 and would have crossed the image plane at P'. Hence, in such a case, we would have the relation

$$\frac{Q'P'}{QP} = \frac{Q'R'}{QR}$$

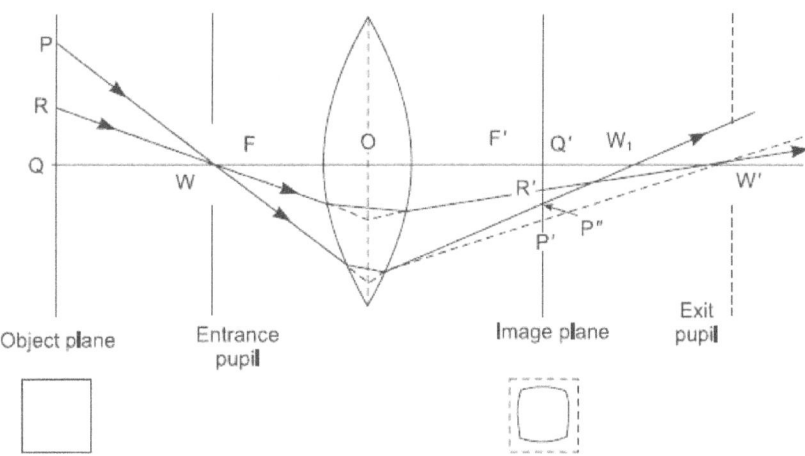

FIGURE 4.23. Illustrates the origin of barrel shaped distortion.

Now, it is quite evident that in virtue of the system not being correct for spherical aberration with respect to W and W', the relation

$$Q'P'' < Q'P'$$

holds. As a consequence, we have another relation

$$\frac{Q'P''}{QP} < \frac{Q'R'}{QR}$$

that is, the magnification decreases with increase in the axial distance of the object point, and hence barrel distortion arises in the image.

The manner in which pincushion distortion arises is shown in Figure 4.24, where the aperture stop is in the rear of the system so that it becomes the exit pupil W', the entrance pupil W being its paraxial image in the object space.

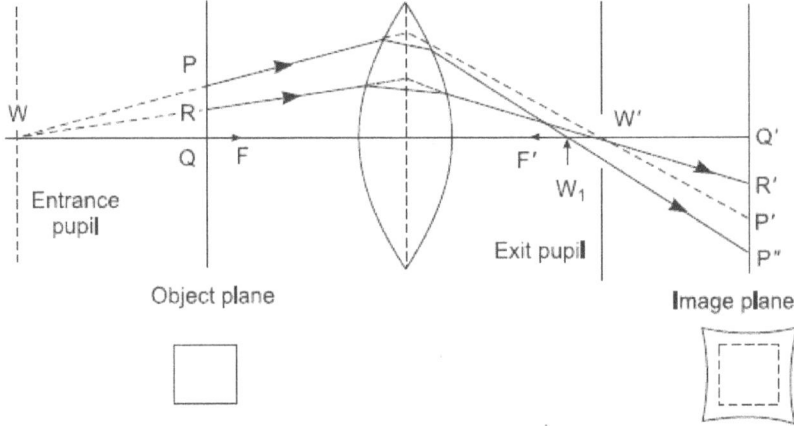

FIGURE 4.24. Illustration of the origin of pincushion distortion.

By reasoning along the lines similar to that given in the case of barrel distortion, it is quite evident that since $Q'P''$ is greater than $Q'P'$, we have the relation

$$\frac{Q'P''}{QP} > \frac{Q'R'}{QR}$$

that is, the magnification increases with increase in the axial distance, and hence pincushion distortion arises in the image. The behavior of a divergent lens is exactly opposite of the behavior of the convergent lens in both the cases discussed previously.

4.5.1 Reduction or Elimination of Distortion

A single thin lens without any stops to limit the rays is free from distortion practically for all object distances. However, it cannot be free from all other aberrations.

A method of minimizing distortion, illustrated by Figure 4.25, consists in employing two lens systems well corrected for other aberrations and placed symmetrically on opposite sides of a central aperture.

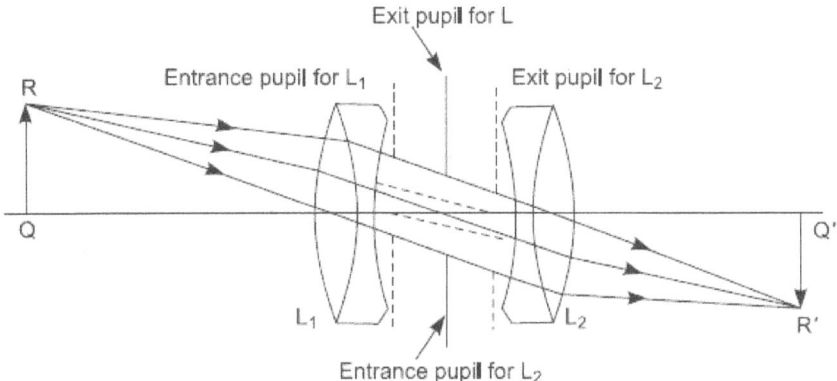

FIGURE 4.25. A symmetrical doublet is relatively free from distortion.

This system is quite distortionless at unit magnification; otherwise, it must of course be corrected for spherical aberration with respect to the center of the aperture stop. In effect, the pincushion distortion introduced by the first lens system is compensated by the barrel distortion produced by the second. Many camera and projection lenses are constructed in this way.

It should be emphasized that the combination of lenses sketched in Figure 4.25 cannot be simultaneously corrected for spherical aberration with respect to the object and image positions as well as with respect to the entrance and exit pupils. Therefore, the lens system, if free from distortion, will suffer with spherical aberration and astigmatism.

4.6 CHROMATIC ABERRATION

Even if the lens system is somehow corrected simultaneously for spherical aberration, coma, curvature, astigmatism, and distortion, there will be present another aberration called chromatic aberration in the images formed by lenses, when, instead of monochromatic light, the light diverging from an object is a heterogeneous one. Light waves of different colors have different velocities in a refracting medium. Every refracting medium, therefore, has a different refractive index for each color or wavelength of the spectrum, the index being least for the red and maximum for the violet color. Now, the focal length of a thin lens is related to the refractive index of the optical material, of which the lens is composed by the relation.

$$\frac{1}{f'} = (\mu - 1)\left(\frac{1}{R_1} - \frac{1}{R_2}\right)$$ (4.13)

As a consequence of this relationship it follows that since $\mu_v > \mu_R$, hence $f_v' < f_R'$. Thus, the focal length decreases as we pass from red to the violet end of the spectrum. A single lens, therefore, forms not merely one image of an object point but a series of colored images at varying distances from the lens, one for each of the colors constituting the incident beam. Such a series of colored images of an axial point object P are shown in Figure 4.26.

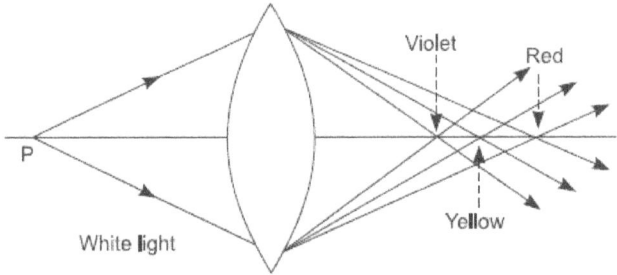

FIGURE 4.26. Longitudinal chromatic aberration.

Furthermore, as a consequence of the dependence of magnification on the focal length and, therefore, on the color of light, the size of the image of the extended object varies with the wavelength. This is illustrated in Figure 4.27, which shows for simplicity of the diagram only the red and violet images of an extended white object PQ. Since $f_v' < f_R'$, the violet rays are focused nearest the lens, and therefore the violet image is smallest, while the red rays are focused farthest away, and therefore the red image is the largest.

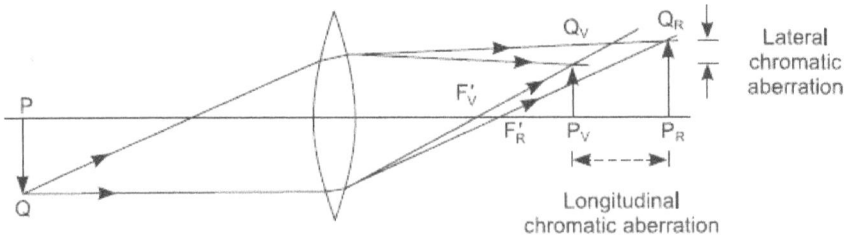

FIGURE 4.27. Two types of chromatic aberration of an image.

The images of other colors, although not shown in the diagram, are formed at the intermediate positions and are of intermediate sizes. Evidently there is no one plane in which all images are simultaneously in good focus. Therefore, on a screen we shall get a blurred image due to superposition of numerous colored images out of focus.

The variation of image distance along the axis of lens due to change in wavelength, $\left(\dfrac{dv}{d\lambda}\right)$, is called the axial or longitudinal chromatic aberration, while the variation in the image size with wavelength, $\left(\dfrac{dy}{d\lambda}\right)$, is called the *lateral chromatic aberration*. The former is measured by the linear distance along the axis between the extreme images and is said to be positive when the violet image is situated to the left of the red image, while the latter is measured by the difference in the lateral sizes of extreme images and is said to be positive when the red image is more magnified than the violet. A divergent lens, as shown in Figure 4.28, produces a negative longitudinal chromatic aberration but a positive lateral chromatic aberration.

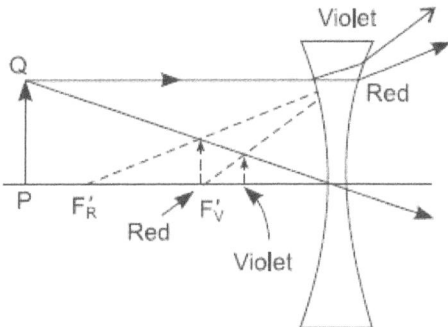

FIGURE 4.28. Divergent lens.

4.7 AXIAL CHROMATIC ABERRATION FOR OBJECT AT INFINITY

The refractive indices of glass are usually given for *Fraunhofer lines*, that is, C-Red ($\lambda_C = 6563$ A.U.), D-Yellow ($\lambda_D = 5893$ A.U.), and F-Blue ($\lambda_F = 4862$ A.U.).

Let f_C', f_D', and f_F' be the focal length of the lens for C, D, and F lines. Then we write for a thin lens

$$\frac{1}{f_D'} = \left(\mu_D - 1\right)\left(\frac{1}{R_1} - \frac{1}{R_2}\right) \tag{4.14}$$

and similar expressions for $\left(\dfrac{1}{f_C'}\right)$ and $\left(\dfrac{1}{f_F'}\right)$ may be easily written,

$$\left(\frac{1}{f_F'}\right)-\left(\frac{1}{f_C'}\right)=\left(\mu_F-\mu_C\right)\left(\frac{1}{R_1}-\frac{1}{R_2}\right)$$

or

$$\frac{f_C'-f_F'}{f_F'f_C'}=\left(\mu_F-\mu_C\right)\left(\frac{1}{R_1}-\frac{1}{R_2}\right) \tag{4.15}$$

Since f_D' is intermediate in value between f_C' and f_F', we may write as an approximation

$$\left(f_D'\right)^2=f_C'f_F'$$

in the preceding equation which, therefore, reduces to

$$\frac{f_C'-f_F'}{f_D'}=\frac{\mu_F-\mu_C}{\mu_D-1}=\omega$$

or

$$f_C'-f_F'=\omega f_D' \tag{4.16}$$

The axial separation $\left(f_C'-f_F'\right)$ between the foci for red and violet rays is called the longitudinal chromatic aberration of the lens for parallel rays, and it is equal to the product of the geometric mean of the extreme focal lengths and the dispersive power, ω, of the material of which the lens is composed.

Example 4.1

The focal lengths of a thin convex lens are 100 cm and 96.8 cm for red and blue rays respectively. Calculate the dispersive power of the material of the lens.

Solution:

The dispersive power of the material of the lens in terms of its principal focal lengths is given by the expression

$$\omega=\frac{\left(f_C'-f_F'\right)}{f_D'}$$

In the given problem, $f_C' = 100$ cm, the focal length for red rays, and $f_F' = 96.8$ cm, the focal length for blue rays. Since f_D' is intermediate in value between f_C' and f_F', we write

$$f_D' = \sqrt{f_C' \times f_F'} = \sqrt{100 \times 96.8} = 98.38 \text{ cm}$$

Hence

$$\omega = \frac{(100 - 96.8)}{98.38} = 0.0325.$$

4.8 ACHROMATISM

An optical instrument free from chromatic aberrations is called an achromatic instrument. This simply means that the lens system of the achromatic instrument should form images of the same size and at one position for all wavelengths, not only for any one specified object position but for all positions of the object. This complete achromatism demands that not only focal points must be made the same, but also the principal planes must be made to coincide for different wavelengths. However, this ideal achromatism is extremely difficult to accomplish in the case of any actual lens system. Only partial achromatism can be accomplished, namely, either longitudinal chromatic aberration or lateral chromatic aberration can be eliminated, but not both simultaneously. Moreover, this too is possible only for two or three wavelengths at one time. Therefore, in making the lens system of the optical instrument partially achromatic, it is essential to keep in view the aperture and field of view of the optical system in deciding the kind of chromatic aberration to be eliminated, because the kind of achromatism which is advantageous in one type of instrument may be entirely disadvantageous in the other. For example, a telescope objective, which has a large aperture but a small field of view, should be corrected for longitudinal chromatic aberration. In this case correction would bring rays of all wavelengths in focus in the same plane, that is, the achromatism is with respect to the position of its focal point, but as the principal points of the lens are different for different wavelengths, the focal length would not be equal, with the result that magnification would not be constant. But owing to the close proximity of images to the principal axis, the variation in their sizes is not annoying and is consequently comparatively unimportant. On the other hand, if the optical system has a small aperture and a large field of view, it should be corrected for lateral chromatic aberration. Here the correction would make the focal length the same for all wavelengths, and hence magnification would be constant, but the images would lie in different planes. It should be emphasized that no lens composed only of two kinds of glass can be achromatic for light of all colors. The achromatism in eyepieces is such that the colored virtual images subtend the same angle at the eye, no matter whether their actual sizes and positions are different or not, and thus the eye fails to detect any color effect in the image.

4.9   CONDITION OF ACHROMATISM FOR TWO LENSES IN CONTACT

The index of refraction of glass is usually specified for the following *Fraunhofer lines.*

Color	Wavelength
C-Red	6563 A.
D-Yellow	5893 A.
F-Blue	4862 A.
G-Violet	4308 A.

For accurate work, the yellow *D* line is disadvantageous because it is a doublet line. Therefore, for accurate work the yellow *d* line λ_d = 5875.6 A. of the helium spectrum is employed as an intermediate wavelength between the *C* and *F* lines.

An *achromatic doublet* consisting of two thin lenses in contact, one made of crown glass and the other of flint glass, may be designed so as to make the resultant focal length of the combination the same for any two wavelengths of the spectrum. We shall now derive the condition to be satisfied by two thin lenses in contact so as to be achromatic for two wavelengths, namely the *C* and the *F* lines, the combination being itself considered as a thin lens.

Let f_1 and f_2 be the mean focal lengths of two thin lenses placed coaxially in contact with each other. If *F* is the focal length of the combination, then

$$\frac{1}{F} = \frac{1}{f_1} + \frac{1}{f_2} \qquad (4.17)$$

which on partial differentiation leads to

$$\frac{\partial F}{F^2} = \frac{\partial f_1}{f_1^2} + \frac{\partial f_2}{f_2^2} \tag{4.18}$$

Since $\left(\dfrac{\partial f}{f}\right) = \omega$, the dispersive power of the lens, the previous equation can be written as

$$\frac{\partial F}{F^2} = \frac{\omega_1}{f_1} + \frac{\omega_2}{f_2} \tag{4.19}$$

where ω_1 and ω_2 are the dispersive powers of the two lenses. For no chromatic aberration, $F = 0$, hence

$$\frac{\omega_1}{f_1} + \frac{\omega_2}{f_2} = 0 \tag{4.20}$$

As the dispersive power ω_1 and ω_2 are always positive $\left(\omega = \dfrac{\mu_F - \mu_C}{\mu - 1}\right)$, $[\mu_F > \mu_C$, and $\mu > 1\ \mu > 1]$, in order that Equation (4.9a) is satisfied, the focal lengths f_1, f_2 must have opposite signs. In practice a convergent lens of crown glass and a divergent lens of flint glass are used. The convergent lens is more strongly convergent for the blue rays than for the red rays, while divergent lens is more strongly divergent for the blue rays than for the red rays. Hence, if a doublet is made of these lenses which satisfies Equation (4.20), the colored image formed by a single lens will fold on itself and the effects of chromatic aberration will be absent.

If $\omega_1 = \omega_2$, Equation (4.20) simplifies to $\left(\dfrac{1}{f_1} + \dfrac{1}{f_2} = 0\right)$ which means that the focal length of the combination is infinite. Thus, it will be obvious that the removal of chromatic aberration is impracticable if the lenses are made of the same material. To eliminate chromatic aberration, the two lenses must be so chosen that:

1. One is convergent and the other divergent.

2. The lenses should be made of different materials.

3. The choice of the dispersive power and focal length is governed by Equation (4.20).

It may be mentioned that if a combination is designed such that $\left(\dfrac{\omega_1}{f_1} + \dfrac{\omega_2}{f_2} = 0\right)$ for two colors, the images formed by lights of these colors coincide in size and position. Such a combination is said to be achromatic for these colors. The amount of chromatic aberration remaining after a combination has been made achromatic for two colors is called the *secondary spectrum*. This defect, however, has been considerably reduced by the introduction of Jena glass, which is nowadays generally used for this purpose.

4.9.1 Achromatic Doublet

Since the removal of chromatic aberration only restricts the choice of focal lengths and not the radii of curvature of the lenses, the latter may be chosen to make spherical aberration at a minimum. If loss of light is to be avoided, two surfaces of the lens should be cemented together by Canada balsam, and thus the radii of curvature of these surfaces should be the same. Only two surfaces are thus left free to be manipulated.

In practice a convergent lens of crown glass and a divergent lens of flint glass are connected together by Canada balsam as shown in Figure 4.29.

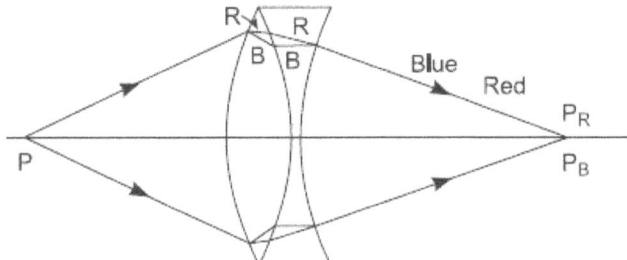

FIGURE 4.29. Achromatic doublet.

To minimize spherical aberration, the free surface of the crown glass lens is more curved than the free surface of the flint glass lens, both being convex outward. The free surface of the crown glass lens is made to face the beam. Such a combination is called an achromatic doublet.

If instead of two, three lenses are combined together, the combination will be achromatic for three colors.

The condition in that case is

$$\frac{\omega_1}{f_1} + \frac{\omega_2}{f_2} + \frac{\omega_3}{f_3} = 0$$

This device is made use of in achieving visual or optical achromatism. Such achromatism is necessary for instruments which are to be used as adjuncts to the eye. In these cases, therefore, chromatic aberration should be removed with respect to the brightest part of the spectrum. This part is the yellowish green region lying between the blue and red rays. The images formed by the red and blue rays are made to coincide with that formed by the yellowish-green region. Thus, chromatic aberration should be removed for three colors. This can be done by taking three lenses satisfying the previous equation.

Even when the combination has been made achromatic for three colors, there is still some chromatic aberration present due to the remaining colors. The remaining aberration is termed as a *tertiary spectrum*. As mentioned previously, glasses have been developed which reduce the tertiary and remaining dispersion without increasing the number of lenses. This higher kind of achromatism is called "*apochromatism*."

Example 4.2

An achromatic objective of focal length 50 cm is to be made of different kinds of glass shown as follows. Find the focal length of each lens, stating whether it is convergent or divergent.

	Glass A	Glass B
μ_{Red}	1.51	1.64
μ_{Blue}	1.52	1.66

Solution:

The mean ray in glass A will have a refractive index

$$\mu_M = \frac{(1.51 + 1.52)}{2} = 1.515$$

The dispersive power ω_1 of glass A is given by

$$\mu_M = \frac{\mu_B - \mu_R}{\mu_M - 1} = \frac{1.52 - 1.51}{1.515 - 1} = 0.0194$$

Similarly,

Now, for the combination of two lenses in contact to be achromatic for red and blue light,

$$\frac{f_1'}{f_2'} = \frac{\omega_1}{\omega_2}$$

where symbols have their usual meaning. Also, the focal length f' of the combination is given by

$$\frac{1}{f'} = \frac{1}{f_1'} + \frac{1}{f_2'}$$

or

$$\frac{f_1'}{f'} - 1 = \frac{f_1'}{f_2'} = -\frac{\omega_1}{\omega_2}$$

or

$$f_1' = \frac{\omega_2 - \omega_1}{\omega_2}(f') = \frac{(0.0307 - 0.0194)}{0.0307} \times 50 = 18.4 \text{ cm}$$

and

$$f_2' = \frac{\omega_1 - \omega_2}{\omega_1}(f') = \frac{(0.0194 - 0.0307)}{0.0194} \times 50 = -29.12 \text{ cm}$$

Thus, the lens of focal length 18.4 cm is *convergent* while that of focal length 29.12 cm is a *divergent* one.

Example 4.3

A convex lens of crown glass is perfectly cemented to a plano-concave lens of flint glass to form an achromatic combination of power +5D. Calculate the radii of curvature of the convex lens from the following data:

	Refractive Index	Dispersive Power
Crown glass	1.50	0.01
Flint glass	1.60	0.02

Solution:

The power of the achromatic lens is +5D. Hence, it is a convergent lens of focal length 20 cm formed by cementing two lenses together.

Let f_1' be the focal length of the crown glass and f_2' the focal length of the flint glass lens. Then, as explained in Example (4.2), if the combination is achromatic, we have

$$f_1' = \frac{\omega_2 - \omega_1}{\omega_2}(f') = \frac{(0.02 - 0.01)}{0.02} \times 20 = 10 \text{ cm}$$

and

$$f_2' = \frac{\omega_1 - \omega_2}{\omega_1}(f') = \frac{(0.01 - 0.02)}{0.01} \times 20 = -20 \text{ cm}$$

Let the concave surface of the plano-concave divergent lens be of radius of curvature R_1'; we write

$$-\frac{1}{20} = \frac{1.60 - 1}{R_1'}$$

or

$$R_1' = -20 \times 0.6 = -12 \text{ cm}$$

Let R_1 and R_2 be the radii of curvature of the surfaces of the crown glass convergent lens. Since the lenses are perfectly cemented together, we have

$$R_2 = R_1' = -12 \text{ cm}$$

Now,

$$\frac{1}{f_1'} = (\mu_1 - 1)\left(\frac{1}{R_1} - \frac{1}{R_2}\right)$$

Substituting the values with the proper sign, we get

$$\frac{1}{10} = (1.5 - 1)\left(\frac{1}{R_1} + \frac{1}{12}\right) \rightarrow R_1 = 8.57 \text{ cm}$$

The radii of curvature of surfaces of the convex lens are 8.57 cm and 12 cm.

4.10 CONDITION OF ACHROMATISM OF A SEPARATED DOUBLET

If two lenses of mean focal lengths f_1 and f_2 are separated by a certain distance "d," the focal length of the equivalent lens is given by

$$\frac{1}{F} = \frac{1}{f_1} + \frac{1}{f_2} - \frac{d}{f_1 f_2}$$

The condition for achromatism can then be obtained by a partial differentiation of the previous equation which gives

$$\frac{\partial F}{F^2} = \frac{\partial f_1}{f_1^2} + \frac{\partial f_2}{f_2^2} - \frac{d\left(f_1 \partial f_2 + f_2 \partial f_1\right)}{f_1^2 f_2^2}$$

$$= \frac{\omega_1}{f_1} + \frac{\omega_2}{f_2} - \frac{d}{f_1 f_2}\left(\omega_1 + \omega_2\right) \tag{4.21}$$

where ω_1 and ω_2 are the dispersive powers of the two lenses. For no chromatic aberration, $\partial F = 0$, hence

$$\frac{\omega_1}{f_1} + \frac{\omega_2}{f_2} = +\frac{d}{f_1 f_2}\left(\omega_1 + \omega_2\right)$$

Therefore,

$$d = +\frac{\omega_1 f_2 + \omega_2}{\left(\omega_1 + \omega_2\right)} \tag{4.22}$$

Thus, in order that the combination be achromatic, the distance between the two lenses must be determined by Equation (4.22). If the lenses are made of the same material, $\omega_1 = \omega_2 = \omega$, and the condition simplifies to

$$d = \frac{1}{2}\left(f_1 + f_2\right) \tag{4.23}$$

Hence, if the lenses are made of the same material, chromatic aberration can be eliminated by keeping them separated by half the sum of their focal lengths. The combination will then be achromatic for all the colors near those for which the mean focal lengths f_1 and f_2 have been calculated.

It may be noted that both the errors of chromatic aberration cannot be removed in the case of two coaxial lenses not in contact with each other.

Such a system of lenses whose distance apart is given by Equation (4.23) is generally used in the construction of eyepieces or oculars. It has been mentioned before that an eyepiece should be achromatic in the sense that different colored images should subtend equal angles at the eye. This is very easily achieved in the case of a separated doublet, since the lateral chromatic aberration is highly corrected through constancy in focal length. However, the longitudinal chromatic aberration is relatively large, because the equivalent planes are not in the same position in different colors.

4.11 THE BALANCING OF ABERRATIONS

We have seen in the preceding pages that a single lens, in general, is afflicted with seven major aberrations—five monochromatic aberrations and two chromatic aberrations. It is quite evident from the discussion of these aberrations that it is not possible to eliminate or even to minimize simultaneously all seven aberrations for a single lens. However, it is possible to design a compound lens from a number of lenses in such a way that the aberrations of one part of the field are balanced against those of the other. The greater the number of component lenses in the compound lens, the greater the degree of accuracy which may be secured. Even then, no compound lens can ever contain a sufficient number of elements permitting the simultaneous elimination of all seven aberrations, even for a single position of the object. In practice, therefore, the attention of a lens designer is always directed towars minimizing those aberrations in the lens system which are likely to be most detrimental to the work for which the lens system is designed. For example, a telescope objective, which is required to cover only a small angular field, should always be so designed as to minimize spherical aberration. On the other hand, a photographic objective, which is always required to cover large angular field, should be so designed as to minimize astigmatism, coma, and distortion while other aberrations should be partly corrected, and so on.

4.12 EXERCISES

1. What does the focal length of a lens depend on?

2. Describe and explain with the help of suitable diagrams (a) longitudinal spherical aberration and (b) lateral spherical aberration. How can the spherical aberration be minimized in the case of an ordinary lens?

3. Describe astigmatism, coma, curvature, and distortion. How may they be reduced to a minimum?

4. What is spherical aberration? How is it minimized when two thin lenses of the same medium are placed at a distance from each other? Show that a Huygens eyepiece satisfies the condition of minimum spherical aberration.

5. What is spherical aberration? How is it minimized in a microscope objective and an eyepiece?

6. Explain chromatic aberration and derive an expression for the axial chromatic error for a thin lens.

7. A combination of two thin lenses, one of crown and the other of flint glass, has the same focal length for a given pair of wavelengths. Derive expressions for the powers of the crown and flint lenses in terms of the combination and dispersive powers of glasses. What do you understand by the term achromatism? Derive and discuss the conditions of achromatism for two thin lenses of focal lengths f_1 and f_2 (a) when they are made of different materials but placed in contact and (b) when they are made of the same material but separated by a distance.

8. A thin crown glass lens is in contact with a thin flint glass lens, the radius of curvature of the common surface being 25 cm. If the combination forms an achromatic combination of 40 cm focal length, find the radii of curvature of the second face of the two lenses.

    μ for crown glass = 1.50
    μ for flint glass = 1.60
    Dispersive power of crown glass = 0.021
    Dispersive power of flint glass = 0.045

9. Show that a combination of two lenses separated by a distance cannot be achromatic in the same sense as a telescope objective is achromatic.

10. Describe and explain chromatic aberration. Deduce a condition for the achromatism of two lenses separated by a distance. Explain what is meant by an achromatic system.

11. Write a brief essay on aberrations of optical images.

12. Calculate the focal length of a lens of flint glass of dispersive power 0.45 which will render achromatic a converging lens of crown glass of focal length 75 cm and of dispersive power 0.21.

13. Calculate the focal lengths of the components of an achromatic telescope objective having a focal length of 50 cm made from components of crown and flint glasses. Dispersive power of crown glass = 0.01654. Dispersive power of flint glass = 0.02766.

14. A convex lens of focal length 35 cm, achromatic for the C and F lines, is to be made from the two glasses, data for which are as follows:

    Hard crown: $\mu_D=1.5175$; $\mu_F - \mu_C = 0.00856$
    Dense flint: $\mu_D=1.6264$; $\mu_F - \mu_C = 0.01722$

    Find the focal lengths of the components.

15. It is required to make a converging combination of focal length 100 cm for the D lines, consisting of a crown glass convex lens in contact with a

flint glass plano-concave lens, so as to be achromatic for the C and F lines. If the faces of the lenses in contact have the same radii of curvature, find the radii of curvature of the two surfaces of the convex lens, assuming the data of refractive indices given as follows:

	μ_C	μ_D	μ_F
Crown glass	1.531	1.534	1.540
Flint glass	1.583	1.587	1.597

16. An achromatic object glass is to be made with a thin crown glass equi-convex lens and a thin flint glass equi-concave lens. Calculate the focal length of the achromat.

	μ_{red}	μ_{blue}
Crown glass	1.480	1.500
Flint glass	1.610	1.670

Radius of curvature of concave lens = 30 cm.

17. An achromatic doublet is to be made of two thin lenses cemented together, and the focal length of the combination for D lines is to be 25 cm. The first lens is a symmetric convex lens of glass A, and the other lens is a concave lens of glass B. Find the radii of the surfaces on the supposition that it is achromatic for the C and F colors.

	μ_D	$\mu_F - \mu_C$
Glass A	1.6112	0.01747
Glass B	1.5205	0.01957

18. Two thin lenses of the same kind of glass, one convex and the other concave, and both of focal length 4 cm, are adjusted on the same axis until the colored images of a white object placed 12 cm in front of the convex lens are formed at the same place. Show that the interval between the lenses must be 12 cm.

19. It is required to obtain a system, achromatic for the focal length, of two lenses of the same material. The focal length of the achromatic lens is to be 30 cm. If the focal length of one lens is 20 cm, calculate the focal length of the other lens. How far apart should they be mounted?

20. A crown glass convergent lens is cemented to a flint glass plano-concave lens to form an achromatic doublet. The radius of curvature of the free surface of the convergent lens is 1 meter. Compute the radius of curvature of the spherical surface separating the two lenses, using the requirement that the doublet should have the same focal length for two C and F lines. Values of μ_C, μ_D, μ_F for crown glass and flint glass may be used from question number 15.

21. Explain chromatic aberration and achromatism. Discuss the achromatism of thin lenses (a) made of the same material and (b) made of different materials.

22. Calculate the focal length of the components of an achromatic telescope objective of 30 cm focal length made from crown and flint glass. Dispersive powers for crown and flint glasses are 0.012 and 0.02 respectively.

23. Derive the condition for the achromatism of two lenses in contact. A convergent lens of 40 cm focal length is to be made out of thin crown and flint glass lenses, the surfaces in contact having a common radius of curvature of 25 cm. Calculate the radius of curvature of the second face of each lens, given that the values of the dispersive power and mean refractive index are respectively 0.017 and 1.5 for crown glass and 0.034 and 1.7 for flint glass.

24. Explain the meaning of chromatic aberration in lenses. Derive the condition for the achromatism of two thin lenses separated by a distance.

25. What is a crossed lens? What is its utility?

26. What is meant by shape factor and position factor? How are they to be related so as to minimize spherical aberration?

27. Find the positions of the aplanatic points of a glass-sphere of $(r)(i)(n_1)$ surrounded by a medium of $(r)(i)(n_2)$. Where do you find the application of this to remove spherical aberration?

28. What is dispersion? What it is due to?

29. What is the normal dispersive medium?

30. Does the focal length of a lens depend on color of the light? In case of a glass lens, which is greater, f_v or f_r?

31. What is a normal spectrum? Is a prismatic spectrum normal?

32. In an achromatic doublet, what type of lenses are to be used?

CHAPTER 5

OPTICAL INSTRUMENTS

Chapter Outline

5.0 INTRODUCTION

Optical instruments may be conveniently divided into two general classes: (a) the image forming instruments, and (b) the analyzing instruments. Telescopes, microscopes, prismatic binoculars, and so on come under the first class, because these instruments are employed to examine objects by forming their images, much magnified in size as compared with those formed by the unaided eye. Prism-spectrometers, grating spectrometers, and interferometers belong to the second class, because these instruments are primarily employed to discover what wavelengths are present in a given beam of light, incident on the instrument, thereby analyzing the given light. The function of most of the image-forming instruments is to enable us to see better, and consequently the eye should be regarded as an essential part of every visual optical instrument. Before describing the optical system of any visual instrument, therefore, it is best to describe the essential parts constituting the eye and discuss its image-forming properties, limiting our discussion only to the physical principles involved in the process of image formation.

5.1   THE EYE

The essential parts of the human eye, considered as an optical instrument, are illustrated by the sectional Figure 5.1 of the eye.

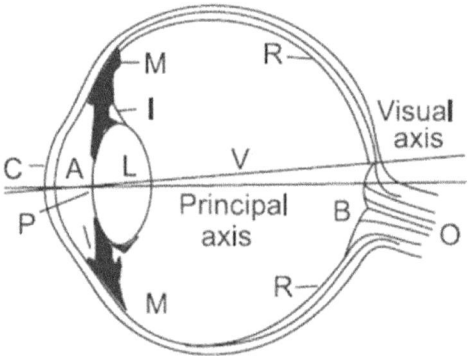

FIGURE 5.1. The eye.

The eye is roughly spherical in shape, the front part being slightly more curved, and is covered with a tough transparent membrane C, called the *cornea*. Behind the cornea is a weak salt solution known as the *aqueous humor A*, backed by a *crystalline lens L* of varying optical density. The lens is of fibrous jelly, harder at the center and then gradually becoming softer toward the surface, the refractive index varying between 1.38 and 1.41. It is held in position by the *suspensory ligaments I* through which it is connected to the *ciliary muscle M*, which is spread all around the lens. The remainder of the eyeball is filled with a thin, transparent jelly consisting largely of viscous liquid called the *vitreous humor V*. The refractive index of both the vitreous humor and the aqueous humor is 1.336. The cornea and the crystalline lens essentially constitute a compound convergent lens system which forms on sensitive nerve fibers and cells, called the *retina R*, inverted images of objects, situated near or far away in front of the eye. The nerve fibers constituting the retina are in fact the branches of the optic nerve O and terminate in minute structures called *rods* and *cones*. In and about them circulates a bluish liquid known as *visual purple*. The optic nerve is connected to the mind, and thus the image formed on the retina is transmitted via the optic nerve to the mind. The point at which optic nerve enters the eye is called the *blind spot B*, because it is not possible to see an object if its image is formed at this point.

Immediately in front of the crystalline lens is the *Iris I*, having the central hole called the *pupil P*, and this serves the purpose of a diaphragm in admitting light into the eye. The size of the pupil automatically increases if the intensity of the object is low and decreases if the intensity is high. The crystalline lens is also flexible; its shape and hence its focal length can be varied by a change in the tension of the suspensory ligaments by the ciliary muscles. As a consequence, focal length of the compound lens changes, and the eye can be made to focus on objects at different distances. This process is known as accommodation, and the extremes of the range of vision over which the eye can bring objects in focus are called the far point and the near point of the eye. The object is said to be at the far point when it can be seen by the completely relaxed eye. The second focal point of the normal eye under relaxed ciliary muscles is at the retina. The object is said to be at the near point when the lens acquires maximum curvature by the appropriate contraction of the muscles in accommodation. The focal length of the lens decreases to focus the near object. For the normal eye, the far point is at infinity while the near point is at 25 cm and is called the least distance of distinct vision. However, the far point and the near point depend upon age. The range of accommodation gradually diminishes with advancing age on account of the lens becoming less flexible. This loss of accommodation is known as presbyopia. There are three other common defects of vision, namely, (a) myopia or shortsightedness, (b) hyperopia or farsightedness, and (c) astigmatism. These defects are corrected by the use of appropriate spectacle lenses. It is needless to describe these defects in detail at this time, for it can be easily presumed that the students already possess a working knowledge of them.

5.2 THE COMPOUND MICROSCOPE

The compound microscope was invented by Galileo in 1610. The optical parts of a compound microscope, illustrated by Figure 5.2, are (a) an objective and (b) an eyepiece.

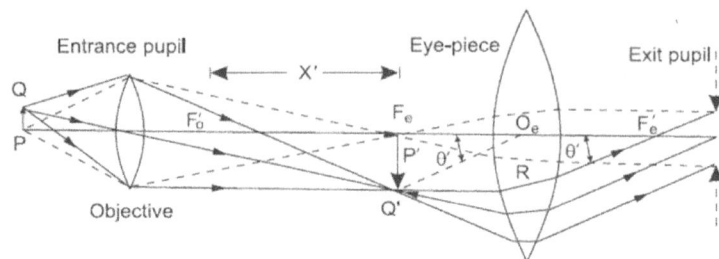

FIGURE 5.2. The compound microscope.

The focal length of the objective is very small, while that of the eyepiece is somewhat larger. Figure 5.2 shows the objective as well as the eyepiece as a single lens, but in reality they are highly corrected compound lenses; in the former spherical aberration, coma and chromatic aberrations are specially corrected for object positions at a very close distance.

The object PQ to be examined is placed just outside the first principal focal point F_o of the objective, which therefore forms a real magnified image $P'Q'$ either within the first principal focal distance of the eyepiece or exactly at the first principal focal point F_e of the eyepiece. The eyepiece is therefore positioned as a simple magnifier with respect to the image $P'Q'$. As a consequence, the eyepiece forms a large virtual image either somewhere between infinity and the least distance of distinct vision from the eye or at infinity, depending upon whether the image $P'Q'$ is within the first principal focal distance or at the first principal focal point of the eyepiece. The ray construction illustrating the formation of the final image at infinity by two thin lenses, constituting the objective and the eyepiece, is shown in Figure 5.2. When a microscope is to be used for making some measurements on the image, we must employ Ramsden's eyepiece.

The *magnifying power* of a compound microscope, like the angular magnification of a simple magnifier, is defined by

$$M = \frac{\tan \theta'}{\tan \theta} \tag{5.1}$$

where θ is the angle subtended at the unaided eye by the object situated at a distance of 25 cm, the least distance of distinct vision, and θ' is the angle subtended at the eye by the final magnified image. We can very easily evaluate the magnifying power when the image formed by the objective is at the first principal focal point F_e of the eyepiece. Let the height of the object under examination be y while that of its image formed by the objective be y'. From the triangle $O_e R F_e'$ in Figure 5.2, we get

$$\tan \theta' = \frac{O_e R}{O_e F_e'} = \frac{P'Q'}{O_e F_e'} = \frac{y'}{f_e'}$$

Also, $\tan \theta = \dfrac{y}{25}$

Where f_e' is the second principal focal length of the eyepiece. Hence,

$$M = \frac{\tan \theta'}{\tan \theta} = \frac{\left(\dfrac{y'}{f_e'}\right)}{\left(\dfrac{y}{25}\right)} = \left(\frac{y'}{y}\right) \times \left(\frac{25}{f_e'}\right) \qquad (5.2)$$

In the previous equation, $\left(\dfrac{y'}{y}\right)$ is the lateral magnification produced by the objective and $\left(\dfrac{25}{f_e'}\right)$ is the angular magnification of the eyepiece. Therefore, we have the following general relation

$$\begin{pmatrix} \textit{Magnifying power} \\ \textit{of microscope} \end{pmatrix} = \begin{pmatrix} \textit{Lateral magnification} \\ \textit{by objective} \end{pmatrix} \times \begin{pmatrix} \textit{Angular magnification} \\ \textit{by eye-piece} \end{pmatrix}$$

Lateral magnification produced by the objective is given by

$$m_o = \frac{y'}{y} = -\frac{x'}{f_o'}$$

where x' is the distance of the image y' from the second principal focal point of the objective. When the final image is formed at infinity, the angular magnification produced by the eyepiece is

$$\gamma_e = \frac{25}{f_e'}$$

Hence, in this case, the magnifying power of the compound microscope is given by

$$M = m_o \times \gamma_e = -\left(\frac{x'}{f_o'}\right) \times \left(\frac{25}{f_e'}\right) \qquad (5.3)$$

But when the final image is formed at the least distance of distinct vision, 25 cm from the eye, then the angular magnification of the eyepiece is

$$\gamma_e = \left(1 + \frac{25}{f_e'}\right)$$

Hence, in this case, the magnifying power of the compound microscope is given by

$$M = m_o \times \gamma_e = -\left(\frac{x'}{f_o'}\right) \times \left(1 + \frac{25}{f_e'}\right) \tag{5.4}$$

It is evident from Equations (5.3) and (5.4) that the compound microscope must comprise a *short focus objective* and a *short focus eyepiece* in order to have high magnifying power. The magnifying power can also be increased by increasing x', in other words, by increasing the separation of the objective and the eyepiece. The distance between the first principal focal point F_e of the eyepiece and the second principal focal point F_o' of the objective is called the *optical length* of the microscope tube, and this distance is standardized to 18 cm by the makers of this instrument. This helps in correcting the microscope objective for aberrations for only a single image distance. In the standard microscope, when the final image is formed at infinity, $x' = 18$ cm and the magnifying power by Equation (5.3) becomes

$$M = -\left(\frac{18}{f_o'}\right) \times \left(\frac{25}{f_e'}\right) \tag{5.5}$$

where f_o' and f_e' are expressed in centimeters.

In every microscope, the periphery of the objective limits the cone of rays entering the instrument, and therefore it also functions as the *entrance pupil* of the instrument. Consequently, the *exit pupil* of the instrument is simply the image of its objective formed by the eyepiece. To have a full field of view, all the rays emerging from the microscope must enter the pupil of the eye. Therefore, it is evident from Figure 5.2 that the eye must be placed at the position of the exit pupil, because all the rays entering the objective and emerging from the eyepiece also pass through the exit pupil of the instrument. Moreover, the diameter of the exit pupil should be equal to that of the pupil of the eye. Under these conditions the magnifying power of the microscope is called the *normal magnifying power*. In order to make full use of the available resolving power of the microscope objective, the overall magnification must be at least equal to the normal magnifying power.

5.3 MICROSCOPE OBJECTIVE

We have seen that the objects examined by the microscope, are to be very close to the objective, and considerations of the brightness of the image and the limit of resolution of the microscope objective demand that a wider cone of rays should enter the objective. Consequently, if a single lens is employed as an objective, the *spherical aberration* and the *coma* will be much more pronounced in the image formed by it. Moreover, as the objects are usually illuminated by white light, a single lens objective will also exhibit chromatic aberrations. The microscope objectives must therefore be carefully designed to eliminate, as far as possible, the chromatic aberrations, the *spherical aberration*, and the *coma*.

In a low-power microscope, the objective is a doublet achromatized for two wavelengths and corrected for spherical aberration and coma as shown in Figure (5.3 a). A high-power microscope objective consists of a number of achromatic doublets, and the deviation of any ray is distributed into small amounts at each surface; thus, the spherical aberration is eliminated. One such objective, called *apochromat*, sketched in Figure (5.3 b), is achromatized for three wavelengths and is corrected for spherical aberration at two wavelengths. The spherical aberration and the chromatic aberrations introduced by the front hemispherical lens must be corrected by the latter lenses in the objective.

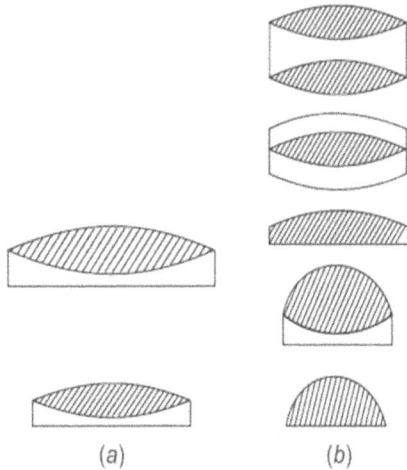

(a)   (b)

FIGURE 5.3. Microscope objectives: (a) low-power and (b) high-power apochromatic.

High-quality microscope objectives are of an oil immersion type. One such objective, designed by Amici about the middle of the last century, is illustrated by Figure 5.4.

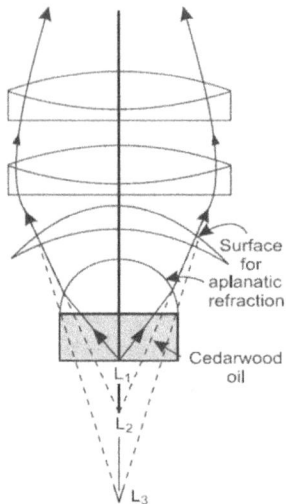

FIGURE 5.4. High-power oil immersion objective.

The space between the cover glass and the front lens is filled with a drop of cedarwood oil which has a refractive index $\mu = 1.51414$ and a dispersive power very nearly the same as for the glass. The front lens and the cover glass have the same refractive index as the oil. As a consequence, the object is virtually imbedded within the first lens, and therefore the first refraction occurs only at the spherical surface of the front lens. The movement of the objective in focusing brings the object at the nearest *aplanatic point L_1* of the spherical surface of the front lens. The rays of light after refraction by this hemispherical surface are rendered less divergent, and whatever may be the inclination of the incident rays to the axis after refraction appear to come from the conjugate *second aplanatic point L_2*. The second lens in the objective is a meniscus lens, the center of curvature of its first surface being at L_2. Therefore, the rays enter the meniscus lens without any deviation. The second surface of this lens is of such a curvature that L_2 becomes the *first aplanatic point* for it. The rays after refraction through the second surface, therefore, appear to come from the conjugate *aplanatic point L_3*. The rays are rendered still less divergent and finally convergent by the other achromatic lenses, which also serve to eliminate the chromatic aberrations introduced by the first two lenses.

The *numerical aperture* of the objective is defined as $\mu \sin \alpha$, where μ is the refractive index of the object space and α is the semivertical angle of the cone of rays entering the objective. Oil immersion-type objectives, therefore, have a large numerical aperture as compared with that of objectives having air in the object space. As a consequence, the least distance between two close points resolvable by the microscope objective, defined by

$$s = \frac{1.22\lambda}{2NA}, (NA \text{ is numerical aperture})$$

decreases by employing an oil immersion objective. For the sake of convenience, we may postpone the derivation of the previous expression and its discussion to the chapter on resolving power.

5.4   DARK FIELD ILLUMINATION

For observing *ultra-microscopic particles*, some sort of dark field illumination is used, of which one type employing a *paraboloidal mirror* as a condenser is illustrated by Figure 5.5.

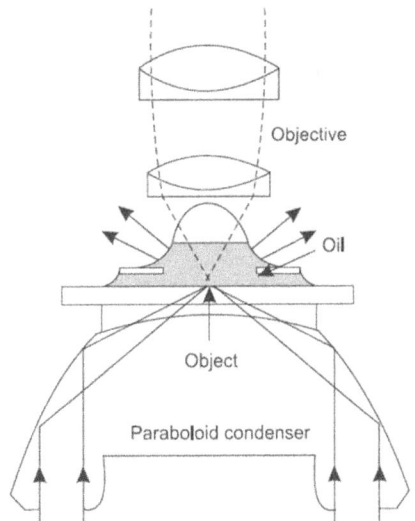

FIGURE 5.5. Illuminating system of the ultra-microscope.

Dark field illumination is accomplished by illuminating the particles at an angle which is larger than the angular aperture of the objective. Thus, no direct light can enter the microscope objective, but the particles scatter the light from the condenser in all directions, and a part of this scattered beam enters the objective. The particles can therefore be regarded as self-luminous, and they are seen as bright points against the dark background. Hence, this method of illumination is known as *dark field illumination*, and the microscope with such a condenser is called the *ultra-microscope*.

5.5 TELESCOPES

Telescopes are generally employed to form an image at infinity of an infinitely distant object like a star. They may also be employed, for example, for simple laboratory experiments, to form an image at a finite distance of an object located at a finite distance from it. Telescopes may be conveniently divided into two general classes: (a) those employing lenses as objectives, called refracting telescopes; and (b) those employing mirrors as objectives, called reflecting telescopes. These are further subdivided into the astronomical class and the terrestrial class, depending upon whether the final magnified image as seen by the eye is inverted or erect with respect to the object under examination.

5.6 THE ASTRONOMICAL REFRACTING TELESCOPE

The optical system of the astronomical refracting telescope is illustrated by Figure 5.6.

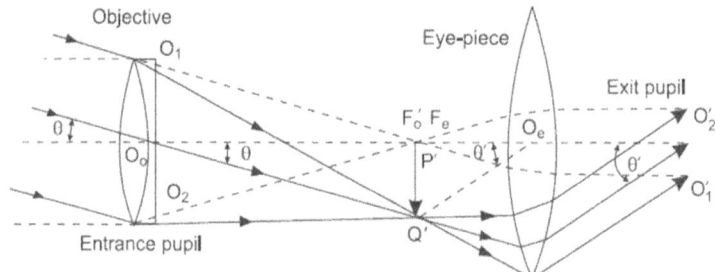

FIGURE 5.6. The astronomical telescope.

$O_1 O_2$ is the objective of a long focal length on which is incident a parallel beam of light from an off-axial distant object point, making an angle θ with the axis of the instrument. The objective refracts this beam so that the refracted beam converges at Q' in the second principal focal plane of the objective. This image, which is real and inverted with respect to the distant object, acts as a real object for the eyepiece, and if the first principal focal plane of the eyepiece coincides with the second principal focal plane of the objective, the so-called normal adjustment, then after refraction by the eyepiece, rays emerge as a parallel beam. The direction of the emergent beam is parallel to the line joining the point Q' to O_e, the optical center of the eyepiece, because if we imagine the ray $Q'Q_e$ to exist, it would emerge from the eyepiece without any deviation from its path. Thus, a virtual magnified image is formed at infinity of the real image $P'Q'$, the latter being the image of an object and formed due to the refraction through the objective.

The objective is at one end of a large tube, while the eyepiece is usually fixed in a smaller tube which is inserted in the larger one at its other end. Thus, the focus can be adjusted to suit any eye. The rim of the objective limits the width of the beam incident on the telescope. As a consequence, the objective itself serves as an entrance pupil of the instrument. A real image of the objective formed by the eyepiece is therefore the exit pupil of the instrument. The ray entering the telescope through the optical center of the objective, which is also the center of the entrance pupil, is called the *chief ray*, and this ray must finally pass through the center of the exit pupil. Rays entering at the rim of the objective must pass through the conjugate points on the rim of the exit pupil, as illustrated by Figure 5.6. The pupil of the eye should be placed at the exit pupil position of the instrument to receive all the emergent rays, and this incidentally leads us to the conclusion that in order that all rays emerging from the telescope may enter the eye, the diameter of the exit pupil should not be greater than that of the eye pupil.

The *angular magnifying power* of a telescope is defined as the ratio of the tangent of the angle subtended at the eye by the final image to the tangent of the angle subtended at the unaided eye by the object itself. A distant object subtends an angle θ at the objective and would subtend essentially the same angle at the unaided eye. Also, since the pupil of the eye is at the exit pupil, the angle subtended at the eye by the final magnified image is θ'. Accordingly, the angular magnifying power is

$$\gamma = \frac{\tan \theta'}{\tan \theta} \qquad (5.6)$$

Let the height of the real image $P'Q'$ at F_o' be denoted by y', which according to our sign convention is negative. From the right triangle $O_oF_o'Q'$, since θ and y' are negative and $O_oF_o' = f_o'$ is positive, we get

$$\tan \theta' = \left(\frac{y'}{f_o'} \right)$$

From the right triangle $O_e F_e' Q'$, since y' and f_e are negative and θ' is positive, we get

$$\tan \theta' = \left(\frac{y'}{f_e} \right) = -\frac{y'}{f_e'}$$

Since $f_e = -f_e'$, the medium is air on both sides of the eyepiece. Substitution in Equation (5.6) yields

$$\gamma = \frac{\tan \theta'}{\tan \theta} = \frac{\left(\dfrac{-y'}{f_e'} \right)}{\left(\dfrac{-y'}{f_e} \right)} = -\frac{f_o'}{f_e'} \tag{5.7}$$

where it will be recalled that f_o' and f_e' are the second principal focal lengths of the objective and the eyepiece respectively. From Equation (5.7) it follows that a *short focal length eyepiece* and a *long focal length objective* are essential to impart a large angular magnifying power to the instrument. Since f_o' and f_e' are positive, hence γ is negative. Therefore, the final image is inverted with respect to the object under examination.

We may also express the angular magnifying power γ in terms of the diameter of the entrance pupil (objective) and that of the exit pupil of the instrument. Applying the lens equation,

$$\frac{1}{v} - \frac{1}{u} = \frac{1}{f'}$$

to the image formation of the objective (the entrance pupil) by the eyepiece, we have

$$\frac{u}{v} = 1 + \frac{u}{f_e'} \tag{5.8}$$

Also, if D and d represent the diameters of the objective (entrance pupil) and its image (exit pupil), we have from similar triangles $O_1'O_e O_2'$ and $O_1 O_e O_2$,

$$\frac{u}{v} = \frac{D}{d} \tag{5.9}$$

From Equations (5.8) and (5.9), it follows that

$$\frac{D}{d} = 1 + \frac{u}{f_e'}$$

According to sign convention, u is negative, and its magnitude is

$$u = f_o' - f_e = f_o' + f_e'$$

Substituting the value of u with a negative sign in the preceding equation, we have

$$\frac{D}{d} = 1 - \frac{f_o' + f_e'}{f_e'} = -\frac{f_o'}{f_e'} \tag{5.10}$$

As a consequence of Equations (5.7) and (5.10), we get

$$\gamma = \frac{D}{d} \tag{5.11}$$

as an alternative expression for the angular magnifying power of a telescope. Equation (5.11) offers a convenient method of determining the angular magnifying power simply by measuring the diameters of the entrance pupil (objective) and of the exit pupil, the image of the objective formed by the eyepiece. When the diameter, d, of the exit pupil becomes equal to the diameter, d_E, of the eye pupil, the magnification is called normal magnification and is denoted by γ_N.

$$\gamma_N = \frac{D}{d_E} \tag{5.12}$$

In order to make full use of the available resolving power of the objective, the overall magnification must be the normal angular magnification.

5.7 OCULARS OR EYEPIECES

An ocular or eyepiece is employed to magnify an image formed by a lens or lenses preceding it in any optical instrument. For example, a microscope or telescope objective forms a real image of an object under examination, and an ocular functions as a simple magnifier in enlarging this image. This final magnified image, as seen by the eye placed behind the ocular, must be free from the various aberrations. It is impossible for a single lens magnifier to satisfy this requirement. Moreover, there is another disadvantage in employing a single lens as an eyepiece, arising from an entirely different cause. A real object diverges rays in all directions, whereas when an image is used as an object, rays from any point of it are simply confined to a narrow cone. As a consequence, each point of the image is seen through a single lens, only by a limited cone of rays. The full image can therefore be seen only when all such cones simultaneously enter the pupil of the eye in any given position for it. It is quite evident from Figure 5.7 that the whole image can be visible only when the eye is at a point A, which of course is at a considerable distance from the lens.

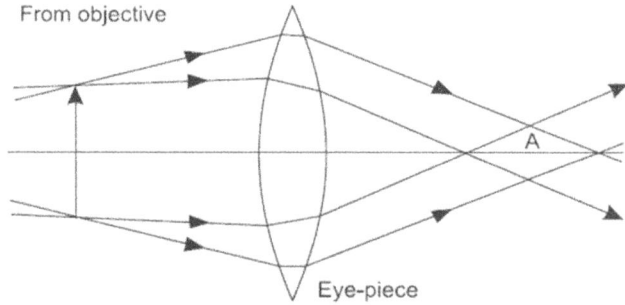

FIGURE 5.7. Single lens eyepiece.

However, to shield the eye from the light not coming through the telescope, it is desirable to place it immediately behind the magnifier. In this position of the eye the field of view becomes limited, simply due to the reason that rays from every point of the real image cannot enter the pupil of the eye.

To overcome these defects, it is usual to employ in every optical instrument a compound system of lenses, instead of a single lens, as an eyepiece. The most popular eyepieces consists of two convergent lenses, constructed from the same glass and separated by a distance so chosen that the spherical aberration and the chromatic aberrations are minimized. For spherical aberration to be at a minimum, the separation between the component lenses should be

$$d = f_1' - f_2'$$

and for chromatic aberration to be at a minimum, the separation should be

$$d = \frac{1}{2}\left(f_1' + f_2'\right)$$

The spherical aberration can also be further reduced by adjusting the four radii of curvature of the surfaces of the lenses constituting the eyepiece.

In the eyepiece, the lens nearest to the eye is called the eye-lens, and its principal function is to magnify the image formed by the objective. The lens which faces the objective is called the field-lens, and its principal function is to enlarge the field of view, simply by deviating toward the axis the cone of rays from each point of the real image so that these cones pass through the central region of the eye-lens. Thus, the pupil of the eye placed close to the eye-lens receives every cone of rays simultaneously, thereby making the whole field of view visible simultaneously.

The function of the field lens in enlarging the field of view, which can be seen by an eye-lens of a given diameter, may be further explained with reference to Figure 5.8.

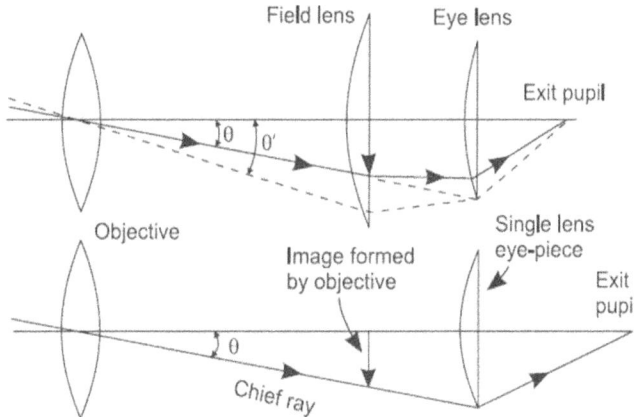

FIGURE 5.8. Field lens enlarges the field of view.

The image formed by the objective coincides with the field lens in the upper figure while the field lens is absent in the lower figure. Only the chief ray of the pencil transmitted by the objective is shown in each figure. In the lower figure, it would be observed that the chief ray passing through the boundary of the eye-lens limits the object field to a narrow cone of semivertical angle θ. In the upper figure, owing to the deviation of the same chief ray toward the axis by the field lens, it is incident, nearer to the axis on the eye-lens. As a consequence, the semi-angular field can be increased to θ' by the field lens because the chief ray inclined at θ' to the axis and shown by dotted line in the upper figure now passes through the periphery of the eye-lens.

In addition to increasing the field of view, the eyepiece should be so designed as to make full use of resolution due to the telescope or microscope objective. This simply means that the power of the eyepiece should be so chosen that the resultant magnifying power of the instrument satisfies the relation.

Note that (Limit of resolution of objective) × (Normal magnifying power) = (Limit of resolution of eye). This statement will be fully discussed and derived in the chapter on resolving power.

5.7.1 The Kellner Eyepiece

In the Kellner eyepiece, as shown in Figure 5.9, the field lens and the eye-lens are plano-convex lenses of equal focal lengths f separated by a distance numerically equal to the focal length of either.

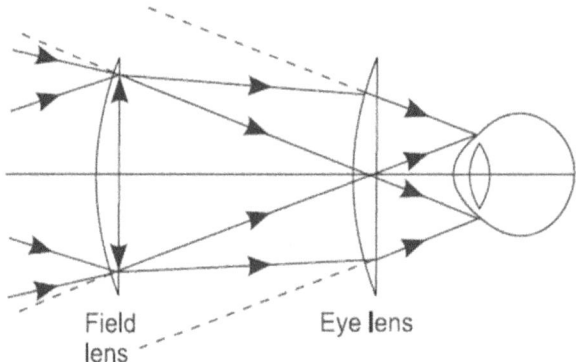

Field lens   Eye lens

FIGURE 5.9. The Kellner eyepiece.

The condition of achromatism, in other words,

$$d = \frac{1}{2}\left(f_1' + f_2'\right) = f'$$

is therefore satisfied in the case of this eyepiece. This condition simply means that the red and the blue images, although of different sizes and formed at different positions, subtend an equal angle at the eye. Thus, the color effect in the retinal image almost vanishes. This achromatism may be improved by making the eye-lens an achromatic combination of a convergent lens of crown glass in contact with a divergent lens of flint glass.

In order to locate the point where a telescope or microscope objective must form a real image of the object under examination if the final image is to be at infinity, we have to trace backward a set of parallel rays emerging from the eye-lens. This set of rays must have come from a point in the first principal focal plane of the eye-lens. But the separation of the field lens from the eye-lens is numerically f, the focal length of either lens. Hence, the principal plane of the field-lens is coincident with the first principal focal plane of the eye-lens. Therefore, the objective must form a real image in the principal plane of the field lens. As a consequence, the field lens produces no magnification but only deflects toward the axis the incoming rays from the objective, every point of the real image being the origin about which deflection occurs. Thus, Kellner's eyepiece has a very wide field of view, but since the magnification is entirely produced by the eye-lens, the final magnified image will suffer from spherical aberration.

A serious disadvantage in Kellner's eyepiece lies in the circumstance that any dust particles or scratches present on the field-lens would be also magnified by the eye-lens along with the real image. To remedy this drawback, two other main classes of eyepieces, the Huygens and the Ramsden types, have been designed, and they are in wide use.

5.7.2 The Ramsden Eyepiece

Ramsden's eyepiece is illustrated by Figure 5.10.

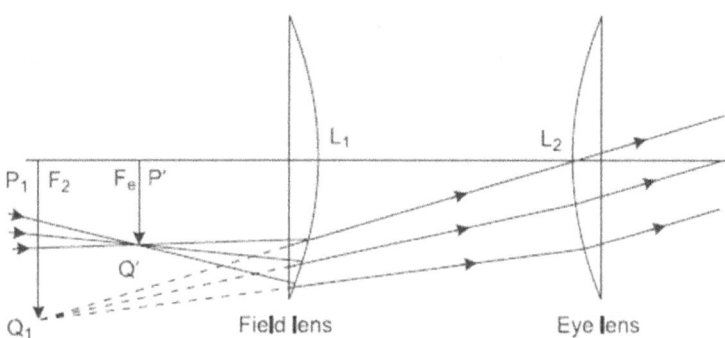

FIGURE 5.10. The Ramsden eyepiece.

It consists of two plano-convex lenses of equal focal length, separated by a distance equal to two-thirds of their common focal length f, with their curved surfaces turned toward each other. The *cardinal points* of this eyepiece may be located as shown in Figure 5.11.

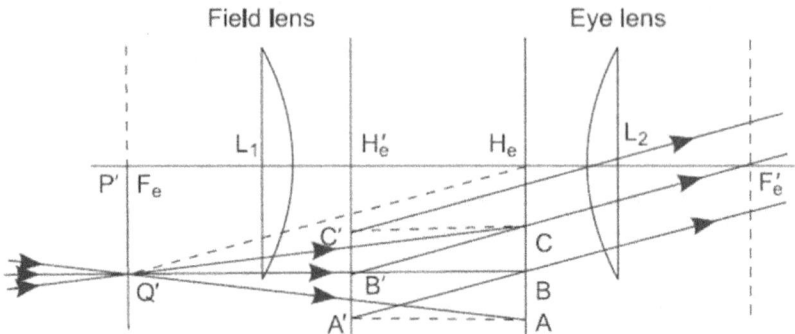

FIGURE 5.11. Cardinal points of Ramsden's eyepiece.

In Chapter 3, it is shown that the *optical separation* Δ between the two lenses constituting a compound lens, having the same medium on either side of the component lenses, is expressed by Equation (3.65), which is

$$\Delta = d - f_1' - f_2'$$

where f_1' and f_2' are the second principal focal lengths of the lenses, and d is the actual separation between them. In Ramsden's eyepiece,

$$d = \left(\frac{2}{3}\right) f; f_1' = f \text{ and } f_2' = f$$

since the second principal focal lengths f_1' and f_2', according to our sign convention, are positive. f is the numerical value of the focal length.

Hence, the *optical separation* Δ of lenses in Ramsden's eyepiece is given by

$$\Delta = \left(\frac{2}{3}\right) f - f - f = -\left(\frac{4}{3}\right) f$$

The *second principal focal length* f_e' of the eyepiece, according to Equation (3.64), is

$$f_e' = -\frac{f_1' f_2'}{\Delta} = -\frac{f \times f}{\left(-\frac{4}{3} f\right)} = \left(\frac{3}{4}\right) f$$

Since the medium is the same (air) on either side of the component lenses, the *first principal focal length* f_e of the eyepiece is

$$f_e' = -f_e' = -\left(\frac{3}{4}\right)f$$

The distance of the first principal plane from the field lens L_1 according to Equation (3.77) is

$$L_1 H_e = \frac{f_1 d}{\Delta} = \frac{(-f)\left(\frac{2}{3}f\right)}{\left(\frac{4}{3}-f\right)} = \frac{f}{2}$$

where f_1, the first principal focal length of L_1, according to our sign convention, is negative. The first principal point H_e is, therefore, at $\left(\frac{1}{2}\right)f$ from the field lens in the rear of it. The first principal focal length is measured from the first principal point. Therefore, the first principal focal point F_e is situated at a distance $\left(\frac{3}{4}\right)f$ from the first principal point and in front of it.

Also, the distance of the first principal focal point from the field lens is

$$-L_1 F_e = F_e H_e - L_1 H_e = \left(\frac{3}{4}\right)f - \left(\frac{1}{2}\right)f = \left(\frac{1}{4}\right)f \rightarrow L_1 F_e = -\left(\frac{1}{4}\right)f$$

i.e., F_e is numerically at $\left(\frac{1}{4}\right)f$ in front of the field lens.

The distance between the eye-lens and the second principal plane of the eyepiece, according to Equation (3.74), is

$$L_1 H_e' = \frac{f_2' d}{\Delta} = \frac{(f)\left(\frac{2}{3}f\right)}{\left(-\frac{4}{3}f\right)} = -\frac{f}{2}$$

Thus, the second principal point H_e' is at $\left(\frac{1}{2}\right)f$ in front of the eye-lens. The second principal focal length of the eyepiece is measured from the second principal plane. Therefore, the second principal focal point F_e' is situated at a distance $\left(\frac{3}{4}\right)f$ from the second principal point H_e' and in the rear of it. It can be easily shown that the distance of the second principal focal point from the eye-lens is $\left(\dfrac{1}{4}\right)f$ in the rear of it.

The nodal points of the eyepiece obviously coincide with its principal points, since the medium is the same on both sides of the eyepiece. The focal points and its principal points are shown in Figure 5.11.

We may now explain image formation by the Ramsden eyepiece on the basis of the properties of cardinal points. In order that the final magnified image may be formed at infinity, it follows from the property of the principal focal points that the objective of any instrument must form a real image in the first principal focal plane of the eyepiece. The real image $P'Q'$ is, therefore, at a distance $\left(\dfrac{1}{4}\right)f$ in front of the field lens. Three rays, $Q'A$, $Q'B$, and $Q'C$, through the tip Q' of the real image are shown in Figure 5.10, intersecting the first principal plane at points A, B, C respectively. The path of the ray conjugate to $Q'A$ must be drawn through the point A' in the second principal plane at the same distance from the principal axis as A, because of the unit positive lateral magnification characterizing the principal planes of the optical system; similar consideration also applies to rays conjugate to $Q'B$ and $Q'C$. Rays conjugate to $Q'A$, $Q'B$, and $Q'C$ must form a parallel beam, since the real image $P'Q'$ which serves as an object for the eyepiece is situated in its first principal focal plane. Moreover, the direction of this parallel beam must be parallel to the line joining Q' to H_e, the first principal point, because if we imagine the ray $Q'H_e$ to exist, its conjugate ray would emerge parallel to it from H_e' by virtue of H_e and H_e' also being the nodal points of the eyepiece. This explains the ray construction in Figure 5.11.

In reality the final image is obtained by two steps. The real image $P'Q'$ formed in the first principal focal plane of the eyepiece by the objective serves as a real object for the field lens and, as obtained above, $L_1F_e = u = -\left(\dfrac{1}{4}\right)f$. In the lens equation

$$\frac{1}{v} - \frac{1}{u} = \frac{1}{f_1'}$$

if we substitute $u = -\dfrac{f}{4}$ and $f_e' = f$, it is easy to deduce $v = -\dfrac{f}{3}$. In other words, the field lens forms a virtual image P_1Q_1 of $P'Q'$, and the former is situated at $\left(\dfrac{f}{3}\right)$ in front of the field lens or at $\left(\dfrac{2f}{3} + \dfrac{f}{3} = f\right)$ in front of the eye-lens. The virtual image P_1Q_1 which lies in the first principal focal plane of the eye-lens serves as an object for it. Hence, P_1Q_1 is imaged at infinity by the eye-lens.

The first principal focal point F_e of the Ramsden eyepiece, as explained previously, lies in front of the field lens. As a consequence, it is more convenient to introduce a micrometer scale or cross wires, movable by a fine calibrated screw, in the plane in which the objective forms a real image $P'Q'$ of an object under examination. Accordingly, the cross wires or micrometer scale will be magnified by the eyepiece to the same extent as the real image $P'Q'$. Hence, it becomes practicable to take micrometer measurements of the image with sufficient accuracy with the help of this eyepiece. This is the main reason which led to the designing of Ramsden's eyepiece.

Various aberrations in the final image are minimized to a great extent as follows: since the final image is seen by refraction at four surfaces, the spherical aberration and the coma are reduced by providing total deviation, as far as possible, in four equal increments at the four refracting surfaces. Spherical aberration is further reduced by employing plano-convex lenses as the field lens and the eye-lens. The circle of least confusion due to spherical aberration and astigmatism is also very small, owing to the narrow pencil of rays transmitted through the eyepiece. Finally, in order to correct for lateral chromatic aberration, the eye-lens and the field lens must be separated numerically by the distance

$$d = \frac{1}{2}\left(f_1' + f_2'\right) = f' = f$$

This condition of achromatism would, however, bring the field lens in the plane of the real image formed by the objective. Thus, the combination would simply be Kellner's eyepiece associated with the undesirable feature of bringing into focus dust particles and scratches, if any, on the field lens. As a consequence, in Ramsden's eyepiece, lenses are separated by $\left(\frac{2}{3}\right)f$ instead of f, thus sacrificing some lateral achromatism in order to eliminate the undesirable features of Kellner's eyepiece. Moreover, this sacrifice has the additional advantage of making the first principal focal plane of the eye-piece real, associated with the convenience of introducing cross wires in this focal plane.

The first principal focal point F_e lies on the negative side of field lens, and hence this eyepiece is often called a *negative eyepiece*.

5.7.3 The Huygens Eyepiece

The kind of eyepiece which enjoys the widest use is the one designed by Huygens in order to improve the corrections for the spherical aberration as well as for the chromatic aberrations present in the Ramsden eyepiece. This eyepiece is constructed of two plano-convex lenses, made of the same material, the focal length of the field-lens being three times that of the eye-lens. The two lenses are coaxially mounted with their curved faces turned towards the incident light and separated by a distance of 2f, where f is the focal length of the eye-lens. The combination of lenses in this eyepiece satisfies the condition of achromatism as follows

$$d = \frac{1}{2}\left(f_1' + f_2'\right) = \frac{1}{2}(3f + f) = 2f$$

as well as it satisfies the condition of minimum spherical aberration as follows

$$d = \left(f_1' - f_2'\right) = (3f - f) = 2f$$

The *cardinal points* of the Huygens eyepiece may be located as follows:

The *optical separation* between the field lens and the eye-lens can be obtained by Equation (3.65), which is

$$\Delta = d - f_1' - f_2'$$

and by substituting in it $d = 2f$, $f_1' = 3f$, and $f_2' = f$, according to our sign convention the second principal focal length of the convergent lens is positive. Accordingly, the *optical separation* Δ of lenses in Huygens eyepiece is

$$\Delta = 2f - 3f - f = -2f$$

The *second principal focal length* of the eyepiece, according to Equation (3.64), is

$$f_e' = -\frac{f_1' f_2'}{\Delta} = -\frac{(3f)(f)}{-2f} = \left(\frac{3}{2}\right)f$$

Since the medium is the same (air) on either side of the component lenses, the *first principal focal length* f_e of the eyepiece is

$$f_e = -f_e' = -\left(\frac{3}{2}\right)f$$

The distance of the first principal plane from the field lens L_1, according to Equation (3.77), is

$$L_1 H_e = \frac{f_1 d}{\Delta} = \frac{(-3f)(2f)}{(-2f)} = 3f$$

Thus, the first principal point H_e is at $3f$ from the field-lens and in the rear of it. The separation between the field-lens and the eye-lens is $2f$; hence, H_e lies at a distance f from the eye-lens in the rear of it. The first principal focal length is measured from the first principal point. Therefore, the first principal focal point F_e is situated at a distance $\left(\frac{3}{2}\right)f$ from H_e and in front of it or at a distance $\left(\frac{1}{2}\right)f$ in front of the eye-lens. Thus, the first principal focal point F_e lies in between the two lenses.

The distance between the eye-lens and the second principal plane of the eyepiece, according to Equation (3.74), is

$$L_2 H_e' = \frac{f_2' d}{\Delta} = \frac{(f)(2f)}{(-2f)} = -f$$

Thus, the second principal point H_e' is at f in front of the eye-lens. Therefore, the second principal focal point F_e' lies at a distance

$$L_2 F_e' = H_e' F_e' - H_e' L_2 = \frac{3}{2}f - f = \frac{1}{2}f$$

from the eye-lens; that is, F_e' is at $\left(\frac{1}{2}\right)f$ in the rear of the eye-lens.

The nodal points of the eyepiece obviously coincide with the corresponding principal points, since the medium is the same on both sides of the eyepiece. The principal points and the focal points of this eyepiece are shown in Figure 5.12.

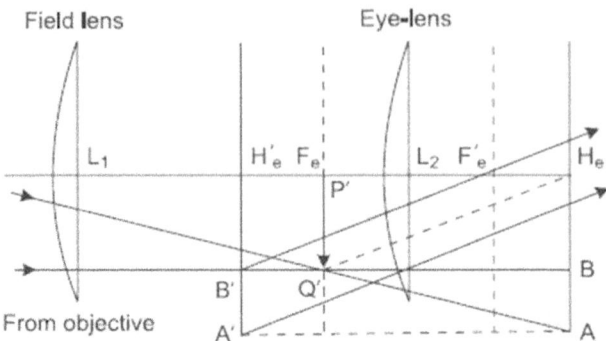

FIGURE 5.12. Cardinal points of Huygens's eyepiece.

We may now explain the image formation by this eyepiece utilizing the properties of the cardinal points of the optical system. The first principal focal point F_e as explained previously lies in between the two lenses at $\left(\dfrac{f}{2}\right)$ in front of the eye-lens. In order that the final magnified image may be formed at infinity by the eyepiece, it follows from the property of the principal focal planes that the objective of an instrument must form an image $P'Q'$ in the first principal focal plane of the eyepiece. Two rays $Q'B$ and $Q'A$ are shown through the tip Q' of the image $P'Q'$; $Q'B$ is parallel to the principal axis of the lens system and intersects the first principal plane at B while the second ray intersects this plane at A. By virtue of the positive unit lateral magnification, the property characterizing the principal planes, the ray conjugate to $Q'A$ is drawn through a point A' in the second principal plane so that $\left(H_e A = H_e{}'A'\right)$, and a similar consideration applies to the ray conjugate to $Q'B$. Rays emerging from the eyepiece must form a parallel beam; since $P'Q'$ is in the first principal focal plane, it must be imaged at infinity by the eyepiece. The direction of this beam must be parallel to the line joining Q' to H_e, the first principal point of the eyepiece, because if we imagine the ray $Q'H_e$ to exist, its conjugate ray would emerge parallel to it through $H_e{}'$ by virtue of H_e and $H_e{}'$ also being the nodal points of the eyepiece. This explains the ray construction shown in Figure 5.12.

In reality the final image is obtained by two steps. The image $P'Q'$, which the objective would form if the field lens were absent in the first principal focal plane of the eyepiece, serves as a virtual object for the field lens, and according to our sign convention, $\left(L_1 P' = u = \dfrac{3}{2}f\right)$, since F_e is in the rear of the field-lens. In the lens equation $\left(\dfrac{1}{v}-\dfrac{1}{u}=\dfrac{1}{f_1'}\right)$, if we substitute $u=\left(\dfrac{3}{2}\right)f$,

$f_1' = +3f$, it is easy to deduce $v = f$. In other words, the field lens forms, at f in the rear of it, a real image P_1Q_1 of $P'Q'$. Since the separation between the field-lens and eye-lens is $2f$, therefore the real image P_1Q_1 is situated in the first principal focal plane of the eye-lens. Consequently, P_1Q_1 is imaged at infinity by the eye-lens. A field stop should be placed in the plane of the image P_1Q_1 to bring into effect a well-defined final image.

In order to carry out the measurements of the image, a micrometer scale or cross wires should be employed between the eye-lens and the field-lens, and it must be mounted in the plane of the real image P_1Q_1 of the object under examination. The eye-lens alone magnifies the scale, so there would be present some distortion and other aberrations in the image of the scale or cross wires. The reason for this is that although the eyepiece is corrected for lateral chromatic aberration and it satisfies the condition for minimum spherical aberration, the eye-lens alone is not corrected. The accurate measurements of any object cannot be made by the help of this eyepiece.

The Huygens eyepiece, as illustrated in Figure 5.13, exhibits some spherical aberration, astigmatism, and curvature of field but is free from lateral chromatic aberration, although longitudinal chromatic aberration is not completely absent. The image is of a pincushion distortion type.

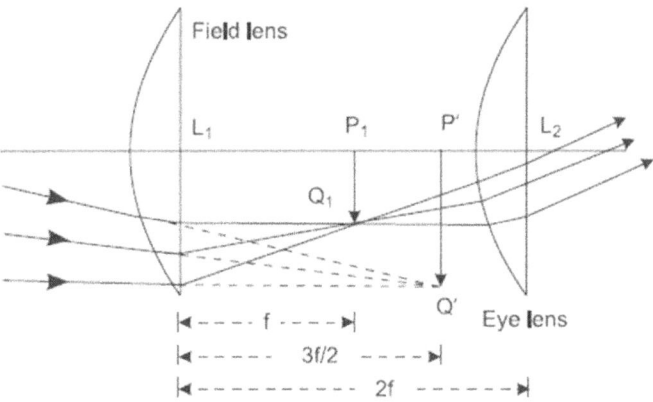

FIGURE 5.13. The Huygens eyepiece.

The first principal focal point F_e lies on the positive side of the field lens, and hence this eyepiece is often called a positive eyepiece.

5.7.4 Comparison of Ramsden's Eyepiece and Huygens's Eyepiece

The chief advantage of Ramsden's eyepiece over Huygens's type lies in the fact that in the former, the first principal focal plane lies in front of the field lens, while in the latter, the principal focal plane lies between the two lenses. As a consequence of this, Ramsden's eyepiece can be used to carry out accurate measurements of an image with the help of a micrometer scale or movable cross wires. Moreover, it can also be employed as a simple magnifier to examine real objects, while Huygens's eyepiece can only examine images. On the other hand, the chief advantage of Huygens's eyepiece is the elimination of the lateral chromatic aberration, but other aberrations are more satisfactorily reduced in Ramsden's eyepiece. Of the two types, Huygens's eyepiece is chosen where the object to be examined is illuminated by white light and the residual lateral chromatic aberration, if Ramsden's eyepiece were used, would be objectionable, which is, however, specially the case while examining biological slides. Thus, Huygens's eyepiece is invariably employed in microscopes used in biological work. In telescopes fitted in spectrometers and in other kinds of spectroscopes, Ramsden's eyepiece is employed. Since only one wavelength region is viewed at a time in such instruments, the residual lateral chromatic aberration becomes unimportant. Huygens's eyepiece also gives a slightly higher field than that given by Ramsden's type.

In the case of Ramsden's eyepiece, the final magnified image is almost flat, but so far as Huygens's eyepiece is concerned, it is convex toward the eye.

5.7.5 The Gauss Eyepiece

In a telescope used in a spectrometer, it is sometimes convenient to employ the Gauss eyepiece. This eyepiece is illustrated by Figure 5.14.

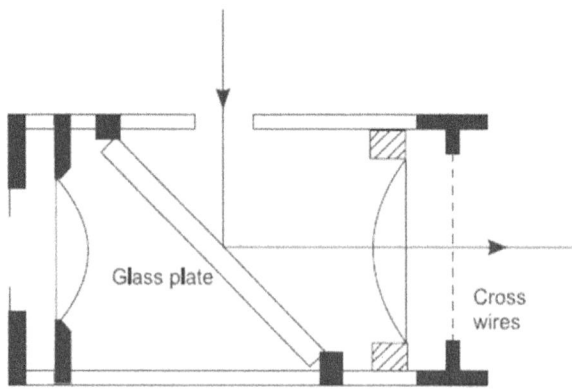

FIGURE 5.14. The Gauss eyepiece.

It is merely a Ramsden eyepiece, except that between the two lenses a thin transparent glass plate is mounted at 45° to the axis of the tube, and there is an opening in the side of the tube carrying the lenses through which light can be admitted within the eyepiece. This eyepiece is used to set the optical axis of the spectrometer telescope at right angles to its axis of rotation in the spectrometer.

5.8   THE SPECTROMETER

The spectrometer is one of the most important optical instruments. It is usually employed for the study of a spectrum produced by the transmission of light through dispersion-producing devices like a prism or a grating. Quantitatively, it is always employed for the measurement of an angle, for example, the angle of a prism, the angle of minimum deviation, and the angle of the diffraction of light due to transmission of light through a plane diffraction grating.

The essential parts of the spectrometer, illustrated by Figure 5.15, are

1.   A collimating device or collimator, which serves the purpose of rendering parallel the rays of light to be examined by the spectrometer.

2.   A turntable on which a dispersion-producing device, like a prism or a grating, can be conveniently mounted.

3.   A telescope for examining the spectrum.

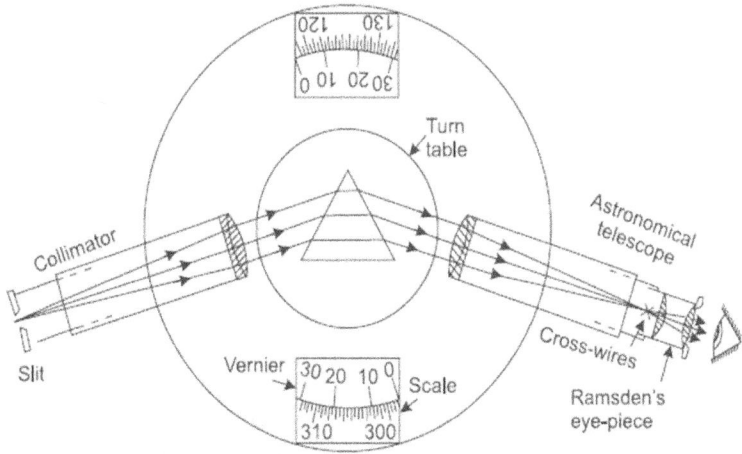

FIGURE 5.15. The spectrometer.

The collimator simply employs a well-constructed achromatic lens at one end of a tube, and sliding within it at its other end is another tube carrying a vertical slit of adjustable width. The relative position of the two tubes can be adjusted by a rack and pinion arrangement, and in this way the slit can be adjusted in the first principal focal plane of the collimating lens. In this setting of the collimator, the rays of light under examination, after passing through the slit, are rendered parallel by the collimator lens. The collimator tube is usually fixed in the instrument with its axis horizontal and intersecting the vertical axis, which passes through the center of the circular scale, graduated in half-degree, fixed at the base of the instrument.

The parallel beam of light emerging from the collimator falls on the prism resting on a table which can be rotated, independently of the collimator and the telescope, about a vertical axis which passes through its center as well as through the center of the circular scale. The rotation of the table can be read by the help of the diametrically positioned verniers sliding over the circular scale. The turntable as shown in Figure 5.16 can also be adjusted to any desired height and clamped.

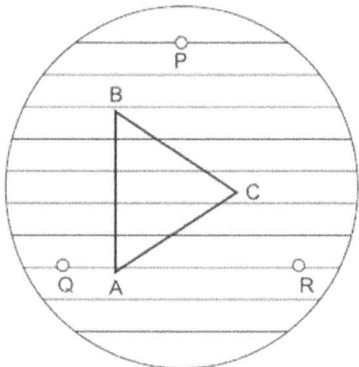

FIGURE 5.16. Turntable.

The slow rotation can be now given with a tangent screw provided at the base of the instrument. The table is provided with three screws P, Q, and R, which form the three corners of an equilateral triangle. By the help of these screws, the plane of the table can be made as nearly horizontal as possible. The surface of the table is usually ruled with straight lines parallel to the line joining two of the leveling screws. These lines assist in setting the prism with one of its faces normal to them.

The refracted beam from the prism is examined by an astronomical telescope described in Section 5.6 and provided with a Ramsden eyepiece. The tube carrying the lenses which constitute the Ramsden eyepiece can slide within the tube carrying the cross wires. The two tubes are fixed to the third tube, which slides within the tube carrying the objective. Thus, the distance between the plane of cross wires and the objective can be varied by the help of a rack and pinion arrangement. The telescope tube with its axis horizontal and pointing toward the axis of the spectrometer is fixed in an arm which can be rotated about the vertical axis of the instrument. This rotation can also be read with the help of another pair of diametrically positioned verniers which slide, with the telescope arm, on the circular scale. The telescope arm can be clamped, and slow rotation can be accomplished with the help of a tangent screw. The collimator and the telescope are provided with leveling as well as locking screws.

5.8.1 Adjustment of the Spectrometer

Before using the spectrometer for any experiment, the following adjustments must be made in the instrument:

1. The axis of the telescope and that of the collimator must intersect the principal vertical axis of rotation of the telescope.

2. The telescope must be focused for parallel rays and the collimator must be adjusted for rendering the rays of light from the illuminated slit parallel.

3. The optical axis of the telescope and that of the collimator must be perpendicular to the principal axis of the instrument.

4. The refracting edge of the prism, if one is employed, must be made parallel to the vertical axis of the rotation of the prism table and the telescope.

5.9   EXERCISES

1. What is an eyepiece? What are its advantages over a single lens? Describe Huygens's eyepiece and show why it is free from both spherical and chromatic aberrations.

2. An ocular consists of two identical positive thin lenses, each of focal length 2 in. separated by a distance of one inch. Determine the positions of the focal points and principal points of the combination.

3. The lenses in a Huygens eyepiece have focal lengths of 2 and 4 cm. Find the distance between them and locate the cardinal points.

4. Describe the construction of a higher-power microscope and discuss the relation between magnifying and resolving powers.

5. The focal length of the more convergent lens of a Huygens eyepiece is 0.5 cm. Calculate the focal length of the eyepiece and locate on a diagram the positions of its focal points.

6. Prove that when two lenses are separated by the algebraic sum of their focal lengths, the linear magnification is independent of the position of the object and is equal to the reciprocal of the angular magnification for the case of an object at infinity.

7. A telescope with an objective of focal length F and an eyepiece of focal length f is focused on an object at a finite distance d from the objective. Calculate its magnifying power and show that it varies up to the limiting value.

8. A distant star is seen with relaxed accommodation through an astronomical telescope fitted with Huygens's eyepiece. Draw a neat diagram of the path of the rays from the star to the eye. Use the diagram to explain that the eyepiece is negative. (Hint: *According to our sign convention, it is positive.*)

9. Mention the characteristic features of the Ramsden's eyepiece. Draw a neat diagram of the path of rays through a telescope fitted with this eyepiece when the telescope is focused with relaxed accommodation on a distant point source of light on its axis.

10. Write notes on:
    a. Oil immersion objectives
    b. Eyepieces
    c. Huygens's eyepiece and its achromatism
    d. Ultra-microscope
    e. Ramsden's eyepiece
    f. Numerical aperture

11. Explain the function of an eyepiece in an optical instrument. Explain the construction and theory of Ramsden's and Huygens's eyepieces. Discuss their respective merits.

12. Describe fully the construction and workings of Huygens's eyepiece. Show that this eyepiece is achromatic in the same sense as an ordinary magnifier. Why is Huygens's eyepiece also called a negative eyepiece? (Hint: *According to our sign convention, it is positive.*)

CHAPTER 6

NATURE OF LIGHT AND SIMPLE HARMONIC MOTION

Chapter Outline

6.0 INTRODUCTION

In the preceding chapters we have studied the image-forming property of lenses and mirrors, the aberrations of optical images, and the action of many optical instruments. We have based our arguments on the theory that light is propagated accurately in straight lines without, however, concerning ourselves with the actual nature of light. After this long study, the question would naturally arise to an inquisitive reader as to what is light. The answer to this question and experimental evidence in support of it constitute what is called *physical optics.*

Speculations about the true nature of light have gone on through the centuries. The *Pythagoreans* believed light to consist of minute particles shot out at high speed from luminous objects. *Plato* and his peers regarded light as some kind of emanation from the eye by means of which objects are scanned and made visible when struck by it. *Aristotle* thought of light as something nonmaterial occurring in the space intervening the eye and the object examined.

Light is really a form of energy, and so it is regarded as the transfer of energy from the luminous source to the eye, either directly or through the object seen. Energy can be transferred from one point to the other either by means of a wave disturbance traveling through the intervening medium or by the motion of material particles between the two points. Accordingly, two different theories as to the nature of light were brought forward in the seventeenth century, that is, *wave* (or *undulatory*) *theory* and *emission* (or *corpuscular*) *theory*. Huygens, Hooke, and Descartes championed the wave theory. The *emission theory* was upheld by Huygens's great contemporary Newton, although Young, who in 1801 developed further wave theory, gave credit of his ideas to Newton's writings, which contained some basic notions pointing toward the wave picture of light.

This chapter shows that the phenomena of interference, diffraction, and polarization of light can be satisfactorily explained only by postulating that light is a form of wave motion. The physicists concerned with the earlier development of such a concept indeed thought that light was a form of mechanical energy transported through space and matter very much like the flow of acoustic energy. Although this concept turned out to be erroneous, inasmuch as the light waves were subsequently shown to be associated with the transport of electromagnetic energy, it still went a long way in our understanding of the previous phenomena and many more. For a subsequent understanding of the transport of electromagnetic energy, it is therefore worthwhile to study wave motion in general terms and see how far the parallelism mentioned previously takes us along in this chapter.

6.1 THE CORPUSCULAR THEORY OF LIGHT

In 1675, Newton sent to the Royal Society a paper in which he presented the fundamental postulates of his corpuscular (emission) theory of the nature of light. According to this theory, light consists of a stream of minute invisible particles, called corpuscles, moving at a great speed. These corpuscles were emitted in straight lines from a luminous source, and their mechanical impact on the retina stimulated the sensation of vision. Different colors were ascribed to different sized corpuscles. Newton also postulated "an ethereal medium much of the same constitution with air but far rarer, subtler and more strongly elastic." Ether "pervades the pores of all material bodies but with a greater degree of rarity in these pores than in the free ethereal spaces." When corpuscles "impinge on any refracting or reflecting superficies (a surface layer), must as necessarily excite vibrations in the ether, as stones do in water when thrown into it." Nevertheless, Newton rejected the theory that vibrations of the ether should be regarded in themselves to constitute light, since it appeared to him to conflict with the fact of the rectilinear propagation of light. He arrived at these conclusions from considerations which are best expressed in his own words:

> If it (light) consisted in pression (pressure in ether) or motion, propagated either in an instant or in time, it would bend into the shadow. But light is never known to follow crooked passages nor to bend into the shadows. For the fixed stars by the interposition of any of the planets cease to be seen. The rays of light are very small bodies emitted every way from shining substances. For such bodies will pass through uniform mediums in right lines without bending into the shadows, which is the nature of rays of light. Pellucid (transparent) substances act upon the rays of light at a distance in refracting, reflecting, and inflecting them, and the rays mutually agitate the parts of those substances at a distance for heating them; and this action and reaction at a distance very much resemble an attractive force between bodies. If refraction be performed by attraction of the rays, the sines of incidence must be to the sines of refraction in a given proportion.

By effecting the prismatic decomposition of white light into its constituent colors, Newton recognized that color is an important characteristic of light and it must be associated with the definite quality of the corpuscles or the ether vibrations. He writes "the most refrangible rays excite the shortest vibrations for making a sensation of deep violet, the least refrangible the largest for making a sensation of deep red, and the several intermediate sorts of rays vibrations of intermediate bigness to make sensation of the several intermediate colors." This is the first statement of the principle that monochromatic light is essentially periodic in its nature and that different periods correspond to different colors. It appears that Newton had also in his mind some form of wave theory regarding the nature of light, but he was not committed to it, as he never computed the wavelength of ether vibrations.

In the latter part of his life, Newton enunciated the law of gravitation, which involves no reference to ether hypothesis, and he also developed the mechanics of the motion of heavenly bodies and concluded that the ether hypothesis was totally superfluous, for he writes: "Against filling the Heaven with fluid medium, unless they be exceedingly rare, a great objection arises from the regular and very lasting motion of the Planets and Comets. For thence it is manifest, that Heavens are void of all sensible resistance, and by consequence of all sensible matter." Thus, in the latter part of his life, he was definitely inclined to a purely corpuscular conception as to the nature of light.

6.2   WAVE THEORY OF LIGHT

In 1690, Dutch Physicist Christian Huygens published a *Treatise on Light* in which he presented a theory, which was first developed in a definite form in 1678, bringing out clearly the necessity of wave conception, instead of corpuscular conception, regarding the nature of light. According to his ideas, a beam of light consists of a large number of longitudinal pulses which are propagated as condensations and rarefactions, like sound waves, from the point of origin to the point of observation of light. He totally rejected the theory that light consists of the motion of small corpuscles emitted from luminous sources.

Huygens arrived at these conclusions from the following considerations, which are best expressed in his own words:

> When we consider the very great speed with which light is propagated in all directions, and the fact that when rays come from different directions, even those directly opposite, they cross without disturbing each other, it must be evident that we do not see luminous objects by means of matter translated from the object to us, as a shot or an arrow travels through the air. For certainly this would be in contradiction to the two properties of light which we have just mentioned and especially to the latter. Light is then propagated in some other manner, like that in which sound travels through the air. Now there can be no doubt that light also comes from the luminous body to us by means of some motion impressed upon the matter

which lies in the intervening space: for we have already seen that this cannot occur through the translation of matter from one point to another. If, in addition, light requires time for its passage—a point we shall presently consider—it will then follow that this motion is impressed upon the matter gradually, and hence is propagated, as that of sound, by surfaces and spherical waves.

For the propagation of every wave, we require some medium as a *vehicle*. As, however, the propagation of light also occurs in a vacuum, Huygens postulated as a vehicle for these waves an omnipresent, all pervading, universal, and continuous medium called *luminiferous ether*, which according to him was composed structurally of tiny elastic particles. According to Huygens, every luminous body is a source of disturbance in ether particles whose mutual impacts transmitted from one particle to the next in continuous succession are propagated with tremendous velocity to the point of observation. Thus, he thought of light as consisting of longitudinal waves in ether. We must emphasize here that the conception of a transverse wave motion, which is the true nature of light waves, had not then occurred to Huygens. Different colors of light are attributed to light waves of different wavelengths.

6.3 CORPUSCULAR THEORY VERSUS WAVE THEORY

Obviously, the test of the adequacy of any theory consists in its ability to account for the known experimental facts with a minimum of postulates. In this light we must admit that the corpuscular theory triumphs above all prejudices, as it can explain many experimental facts like rectilinear propagation, reflection, and refraction at the plane boundary separating two media, of course, with a minimum of hypothesis.

Twelve years after the presentation of his wave theory, Huygens was also successful in explaining reflection and refraction, and he arrived at the conclusion theoretically that the speed of light should be greater in an optically rarer medium, whereas the theory of Newton gave the opposite result. Obviously, a measurement of the speed of light in water and air would decide the nature of light. Huygens also gave a suitable explanation of the phenomenon of double refraction, discovered in 1669 by Erasmus Bertholinus in uniaxial crystals like calcite, by attributing to ether present in the surface layer of uniaxial crystals the property of sending out two waves propagated with different speeds, within the crystal, under the influence of incident light waves. The corpuscular theory was found to be totally inadequate to explain double refraction. But Huygens's discovery in 1690 of the phenomenon of the polarization of light, namely that the two beams of light emerging from calcite crystal have sides, that is, they are not perfectly symmetrical about the direction of propagation, presented an unsurmountable difficulty in respect of the then-prevalent nature of light waves. Of course, it is inconceivable for a longitudinal wave, which is perfectly symmetrical about the direction of propagation (due to vibration of the medium along the direction of the propagation), to acquire one-sidedness or a symmetry on transmission through a calcite crystal. Besides this, Huygens could not give any satisfactory explanation of the rectilinear propagation of light, as another phenomenon called interference of light had not yet been discovered. Polarization was first interpreted by Newton on the supposition that the corpuscles were not perfectly spherical. Accordingly, they would present different sides depending on their orientation to the direction of propagation. As a consequence, the new wave theory could not receive immediate acceptance. In fact, Newton argued that we should compare the properties of light with those of sound, which is definitely a wave motion and has the property of bending around corners and other obstacles in its path. This sideway spreading of the waves or their bending around the corners is called diffraction. But up to 1821 no significant diffraction effects were observed with light, although it should be pointed out that the phenomenon of diffraction of light was first observed by Grimaldi as early as 1665. But the true significance of his observations was not realized at that time, presumably due to our inability to repeat his observations. It should be emphasized that diffraction effects arise only when the dimensions of the obstacle become comparable with the wavelength of the wave, but in Huygens time the smallness of the wavelength of light was not known. Newton, therefore, argued: "I know sound is waves, how then can light be waves if it is so different from sound in this matter of diffraction." Later, Newton also succeeded in observing diffraction fringes outside the geometrical shadow of the obstacle in the path of light. He argued that it arises due to the deflection of light corpuscles from their straight path in passing the region of variable ether density close to the edges of an opaque body. Had he not fatefully overlooked the diffraction fringes within the geometrical shadow, first noticed by Grimaldi, they would have set him on the right path. Furthermore, Newton considered rectilinear propagation as an essential property of light, and since he was successful in explaining it by the emission theory, he criticized the wave theory on these grounds also. His views were, unfortunately, interpreted by his followers with a much narrower significance than he intended to attach to them, and they rejected the wave theory in favor of the corpuscular theory.

Nevertheless, by the end of eighteenth century, the theories of light propounded by Newton and Huygens had acquired a following in the scientific world, although there was still a controversy about whether a corpuscular or a wave picture best fit in the experimental facts. Then came the discovery of the phenomenon of the interference of light in 1801 by Thomas Young (1773–1829). He championed the wave theory to explain the observed uneven distribution of light on a screen placed in front of another screen with two extremely close pinholes, when the sunlight was allowed to fall on these holes after traversing through another pinhole. The corpuscular theory was found to be totally inadequate to explain the observed uneven distribution of light because, on this conception, the intersection of two rays at a point could only produce an increased brightness equal to twice the intensity of each one separately but never a partial or complete darkening. In fact, the observed intensity in bright regions was found to be four times the intensity caused by each beam acting separately. According to wave theory, since the intensity at any point is proportional to the square of the amplitude of the light wave, the addition of two light waves of equal amplitude, originating from the same source as found in Young's experiment, could produce under suitable conditions (arriving in equal phase) a wave of twice the amplitude, and hence the intensity would be four times that due to component waves alone. In Young's own words, "When two undulations from different origins coincide their joint effect is a combination of motion belonging to each." Young's experiment was thus a crucial one at that time, because it added further evidence to the growing belief in the wave nature of light. Besides, he also explained colors of thin films, a phenomenon discovered by Newton himself, by attributing it to the interference between light waves reflected from the upper and lower surfaces of thin films.

Owing to Newton's great authority, even Young's experiment and his explanation of the interference of light could not lead to the triumph of the wave theory. It was, however, left to the brilliant contribution of young French engineer Augustin Fresnel (1788–1827) to bring about the general conversion of the scientific world to the wave nature of light. He employed Huygens's wave theory together with the principles of the interference of light waves and succeeded not only in giving a satisfactory quantitative explanation of the uneven distribution of light in all diffraction phenomenon then known to scientists, but above all he also explained the rectilinear propagation (approximate) of light, thereby refuting one of the chief arguments against the wave nature of light.

Although Fresnel's explanation of interference and diffraction effects paved the way for the success of the wave theory, still there were scientists who were hesitating to accept this theory, chiefly on account of its inability to account for the phenomenon of the polarization of light waves on the prevalent nature (viz. longitudinal) of light waves. Credit again goes to Fresnel, who in 1816, after studying the interference of polarized light, put forward the highly revolutionary hypothesis that light must be regarded as a transverse wave motion in ether analogous to those in an elastic solid, in contradiction to Huygens's conception of the longitudinal nature of light waves. In the case of transverse waves, vibrations of the medium are executed at right angles to the direction of propagation, but if the vibrations are confined to one place, then this kind of wave would definitely exhibit one-sidedness. Thus, the phenomenon of the polarization of light could be easily explained by Fresnel's hypothesis. There is no doubt that this revolutionary hypothesis proved to be a great triumph for wave theory over the corpuscular theory.

The death blow to the corpuscular theory and the universal acceptance of wave theory were, however, achieved simultaneously when Leon Foucault, in 1850, showed experimentally that the speed of light is less in water than in the outside atmosphere. The wave theory of Huygens, although radically modified by Fresnel, finally triumphed over the rival emission theory of Newton.

6.4 ELECTROMAGNETIC THEORY OF LIGHT

Huygens and Fresnel regarded light as simply the propagation of mechanical energy. A completely different conception regarding the nature of light waves developed in 1873 chiefly due to the theoretical researches of a Scottish scientist, James Clerk Maxwell, Professor of Physics in Cambridge University. He put forward the hypothesis that light must be thought essentially as an electromagnetic wave propagation, so that a beam of light propagates not mechanical but electromagnetic energy.

Maxwell explained by theoretical reasoning that the oscillatory electric circuit must produce, in the medium occupying the space, the periodic variations of electric and magnetic intensities having all the characteristics of transverse waves. The vectors representing electric and magnetic field strengths are mutually perpendicular and also at right angles to the direction of the propagation of the waves.

Furthermore, he deduced that the velocity of propagation of electromagnetic waves through this medium (electric ether) is numerically equal to the number of electrostatic units of electric charge, which are equivalent to one electromagnetic unit of charge. Light travels as a wave motion in luminiferous ether. To the inquisitive mind of Maxwell, the question naturally arose: were light waves and electromagnetic waves identical? The problem and method of solution were stated by Maxwell himself in his *Treatise on Electricity and Magnetism*, as follows:

> In several parts of this treatise an attempt has been made to explain electromagnetic phenomenon by means of mechanical action transmitted from one body to another by means of medium occupying the space between them. The undulatory theory of light also assumes the existence of a medium. We have now to show that the properties of the electromagnetic medium are identical with those of the luminiferous ether.

> But the properties of bodies are capable of quantitative measurement. We therefore obtain the numerical value of some property propagated through it which can be calculated from electromagnetic experiments, and also directly in the case of light. If it should be found that the velocity of propagation of electromagnetic disturbance is the same as the velocity of light, we shall have strong reasons for believing that light is an electromagnetic phenomenon.

Maxwell's predictions were remarkably confirmed in 1888 by Helmholtz's brilliant pupil Heinrich Hertz. He experimentally detected electromagnetic waves in space, generated by the rapidly accelerated charges in an oscillatory spark discharge. At the same time, he also demonstrated that these waves could be reflected, refracted focused by a lens, polarized, and so on. In short, they exhibited properties similar to those of light. In addition to this, it was found experimentally that electromagnetic waves are propagated precisely with the same velocity as those of light in a vacuum. The evidence went far to show that the light consisted of electromagnetic waves. Fresnel's ideas did not meet any opposition, but they were interpreted not in terms of mechanics but in terms of electrodynamics.

By electromagnetic theory, light is regarded as the propagation of a vibration of an electric vector **E** coupled with the propagation of a vibration of a magnetic vector **H**. Figure 6.1 shows the fields **E** and **H** in a plane transverse electromagnetic wave.

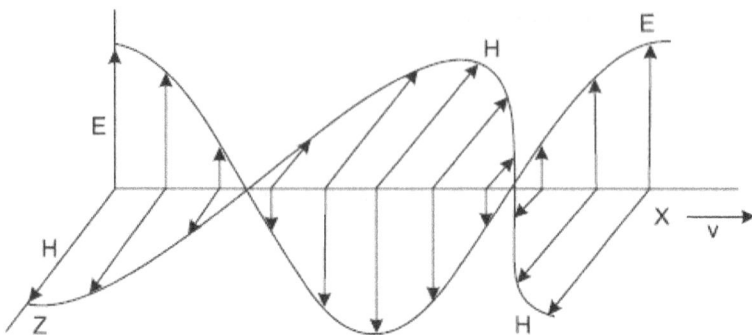

FIGURE 6.1. Fields *E* and *H* in a plane transverse electromagnetic wave.

Both vectors in synchronism are to be always in equal phase, mutually at right angles, and at right angles to the direction of the velocity vector **V** of the wave. Wiener's experiment on stationary light waves has shown that the electric vector plays the role of the light vector. The experimental confirmation of the electromagnetic nature of light waves has since then accumulated, for example, the magnetic splitting of spectral lines (Zeeman effect) and the rotation of the plane of polarization under the influence of the magnetic field (Faraday effect). The electromagnetic theory completely obliterated the corpuscular conception of light, and the former reigned supreme in the domain of optics, of course, only up to the end of the nineteenth century, when the corpuscular conception was revived to explain new discoveries, for example, the photoelectric effect, Compton's scattering of X-rays, and the emission of light.

6.5 QUANTUM THEORY OF LIGHT

By a strange irony of fate, Hertz, who provided an experimental confirmation of Maxwell's electromagnetic theory of light, discovered another phenomenon called the photoelectric effect, which was destined to provide the greatest setback to the established supremacy of wave theory and which, ultimately again, brought into light the corpuscular picture of radiant energy but in slightly modified form. This phenomenon consists of the instantaneous ejection of loosely bound electrons—valence electrons—from alkali atoms (Na, K, Li, etc.) under the stimulus of light, X-rays, and γ-rays. An expulsion of an electron requires absorption of energy by the atom, equal to the binding energy of the electron. According to wave conception, it would take several days for the atom to absorb the requisite amount of energy from a beam of light of moderate intensity to bring about its expulsion. Actually, the expulsion of electrons starts the moment light strikes the metal surface. Einstein in 1905 suggested that in order to give an adequate explanation of these experimental observations, it is essential to postulate that energy in a beam of light is not uniformly distributed over the whole wave-front, but is made up of localized centers of energy traveling through free space, and each of them is capable of being absorbed by the atoms as a whole. The localized concentrations of energy, which he called photons, are propagated like particles with the speed of light. The prestige of the wave picture was retained inasmuch as the energy of the photon was assumed to be proportional to the frequency, v, of the associated light wave, that is,

$$E = hv \tag{6.1}$$

where h is the universal constant known as Planck's constant, and its value is 6.624×10^{-27} ergs-sec. In reality, Einstein took his cue from the epoch-making hypothesis put forward on December 11, 1900 by Max Planck, Professor of Mathematical Physics at Berlin, in connection with the explanation of the energy distribution in black body radiation. Planck's hypothesis, which was destined to represent one of the outstanding achievements in physics and to win for him the Nobel Prize in 1918, may be stated as follows: Every atom, usually called the oscillator, absorbs or gives off energy in intermittent and discontinuous amounts equal to an integral multiple of a certain energy unit, hv, which he called quanta (or energy), where v is the oscillation frequency of the oscillator. According to wave theory, the process of absorption or radiation of energy is a continuous one.

Planck's quantum hypothesis, which constitutes a clear break with the doctrines of classical physics, along with Einstein's ideas of the photon character of light, constitutes what is now called the quantum theory of light. Another striking confirmation of the photon character of light is provided by the interpretation of emission and absorption line spectra in 1913 by Neils Bohr, and Compton's experimental observation and interpretation in 1921 of the scattering of X-rays.

Our present view is that light has a dual character. In experiments on propagation, reflection, refraction, interference, diffraction, and polarization, light exhibits those properties which are definitely attributable to waves, while in experiments on interaction with matter (photoelectric effect), in the process of emission, absorption, and scattering, it exhibits those properties which are definitely of a particle or photon character. The far-reaching consequences of this dual character of radiant energy were, however, not fully recognized until de-Broglie announced his new view on the subject, which is best expressed in his own words as follows:

> A consideration of these problems led me, in 1923, to the conviction that in the theory of matter as in the theory of radiation it was essential to consider corpuscles and waves simultaneously, if it were desired to reach a single theory, permitting of the simultaneous interpretation of the properties of light and those of matter. It then became clear at once that, in order to predict the motion of a corpuscle it is necessary to construct a new mechanics, a theory closely related to that dealing with wave phenomenon, and one in which motion of corpuscle is inferred from the motion in space of a wave. In this way, there will be, for example, light corpuscles, photons but their motion will be connected with that of Fresnel's waves, and will provide an explanation of the phenomena of interference and diffraction, Meanwhile, it will no longer be possible to consider the material corpuscles or electrons as discrete isolated entities, it will on the contrary, have to be assumed in each case that they are accompanied by a wave which is bound up with their own motion.

Thus, in this unified theory known as *quantum mechanics*, a photon is not to be thought of as a mere concentration of energy but is to be associated with a periodic phenomenon extended over the space. We can conclude that the photon and wave ideas are in reality complementary rather than rival conceptions.

6.6 MOTION ABOUT A CENTRAL POINT

We start with one of the simplest and most useful concepts of physics, namely the linear to and fro motion of a particle about a central point. So long as the maximum displacement of the vibrating particle as measured from its mean position is small, its motion is termed as simple harmonic. The motion of a simple pendulum and the vibrations of a flat spiral spring and a stretched string are a few examples of simple harmonic motion encountered in everyday life.

Consider a particle P moving along the circumference of a circle with a constant velocity v as in Figure 6.2.

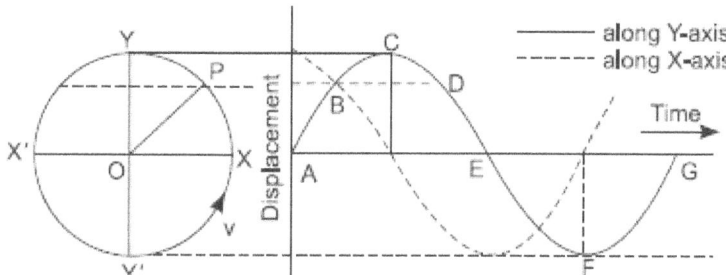

FIGURE 6.2. A particle *P* moving along the circumference of a circle with a constant velocity *v*.

Suppose we get interested in studying its dynamics the moment it crosses the position x $(t = 0)$. At this instant its displacement from that "center O" measured along the horizontal direction is equal to the radius "a" of the circle, while that measured along the vertical direction is zero. When it arrives at P, after a lapse of a few seconds, its respective displacements along the X and Y axes, measured from the central point O, are $a\cos\omega t$ and $a\sin\omega t$, where ω is the angular frequency of rotation and is defined as the angle traced by the radius vector in unit time. On arrival at "Y" the horizontal displacement vanishes, but the vertical displacement attains its maximum value, namely "a." In the second quadrant of the circle, the vertical displacement gradually decreases, vanishing completely when the particle arrives at X', but the horizontal displacement simultaneously increases, becoming a maximum at X'. This maximum is nevertheless in the negative X-direction. Thus, we find that as the particle completes one half-cycle from X to X', to the displacement measured along the Y-axis, it starts from zero, rises to a maximum, and falls back to zero. In the next half-cycle (from X' to X) the vertical displacement again increases (though in the negative direction), becoming maximum when the particle reaches Y', and starts decreasing till it becomes zero when the particle arrives at X. At each instant the displacement in the Y-direction as measured from the central point O is given by

$$Y = a\sin\omega t \tag{6.2a}$$

where "t" is measured from the instant we become interested in following this motion. A plot of Y against "t" gives the sine curve expressed in Equation (6.2a) and has been shown to the right in Figure 6.2.

Similarly, the displacement measured along the X-axis is given by

$$X = a\cos\omega t \tag{6.2b}$$

and has been shown by the dotted curve to the right of Figure 6.2 and can be obtained by displacing the continuous curve to the left by an interval $\left(\dfrac{T}{4}\right)$, where "$T$" is the time required by the particle to complete one full cycle along the circumference.

Thus, we find that as P moves along the circumference of the circle, its coordinates along the X and Y directions execute a vibratory motion along the two axes about the central point O. Moreover, it can be seen from a double differentiation of Equations (6.2a) and (6.2b) with respect to time that the accelerations along the two axes are directly proportional to the instantaneous displacements of the coordinates along those directions. Thus, from Equation (6.2a)

$$y' = \frac{dy}{dt} = a\omega \cos \omega t \ \text{ and } \ y'' = \frac{d^2 y}{dt^2} = -a\omega^2 \sin \omega t \qquad (6.3)$$

or

$$y'' = \frac{d^2 y}{dt^2} = -\omega^2 y \qquad (6.4)$$

Similarly,

$$x'' = \frac{d^2 x}{dt^2} = -\omega^2 x \qquad (6.5)$$

The negative sign in Equations (6.3) and (6.5) indicates that the instantaneous accelerations are directed along directions opposite to that of the increasing displacements from the central point O.

These then are the necessary and sufficient conditions for a motion to be called *simple harmonic*. The motion of a particle along the circumference of a circle can thus be resolved into two mutually perpendicular simple harmonic motions of the same amplitude obeying certain conditions. So long as these conditions are satisfied, the converse of this statement is also true.

6.7 VARIABLES OF A SIMPLE HARMONIC MOTION

The most general equation of a simple harmonic motion is given by

$$y = a\sin(\omega t - \delta) \qquad (6.6)$$

where a is the amplitude of vibration, ω the angular frequency, and $(\omega t - \delta)$ the phase of the vibration. It may be mentioned that δ is a constant depending upon the particular motion under study and may be evaluated with the help of certain known conditions satisfied by it.

The maximum value of displacement y [obtained by putting $\sin(\omega t - \delta) = 1$] is called *amplitude*. The quantity

$$v = \left(\frac{\omega}{2\pi}\right) = \left(\frac{1}{T}\right) \tag{6.7}$$

is called the *frequency* and is defined as the number of vibrations executed by the particle in one second. The angular frequency ω is the number of vibrations completed by the particle in 2π seconds. If t in Equation (6.6) is replaced by $t + T$, we find that

$$y = a\sin\left[\omega(t+T)-\delta\right]$$
$$= a\sin\left[\omega\left(t+\frac{2\pi}{\omega}\right)-\delta\right]$$
$$= a\sin\left[2\pi+(\omega t-\delta)\right]$$
$$y = a\sin\left[\omega t-\delta\right]$$

The nature of the function thus remains unaltered and, consequently, T is called the *periodic time* of the simple harmonic motion. The phase of the vibration, which is the argument of the sine (or cosine) function in Equation (6.6), can be expressed in terms of

1. the time (fraction of T) elapsed since the vibrating particle passed through its mean position,

   or

2. the angle through which the particle has turned after passing through its mean position.

   Thus, when the vibrating particle is at its extreme position in the positive direction, it has turned through an angle of $\left(\dfrac{\pi}{2}\right)$ radians and the time expressed as a function of T is $\left(\dfrac{T}{4}\right)$ seconds, both being reckoned with reference to its mean position. If there are two particles vibrating along the same straight line with the same frequency, the following possibilities may arise:

1. They pass through their mean positions simultaneously moving in the same direction. They are then said to be in the same phase. Their phase difference may be 0, 2π, 4π, etc., since the function expressed in Equation (6.6) repeats itself after every 2π radians.

2. They pass through their mean positions simultaneously moving in opposite directions. They are then said to be in opposite phase, and their phase difference may be π, 3π, etc.

3. They pass through their mean positions at different times, either in the same or opposite directions. Their phase difference will then be determined according to the conditions of the problem and may have a value lying between 0 and 2π radians.

6.8 PARTICLE VELOCITY, ACCELERATION, AND ENERGY IN A SIMPLE HARMONIC MOTION

We have seen that the instantaneous displacement of a particle executing simple harmonic motion is given by

$$y = a \sin(\omega t - \delta) \tag{6.8}$$

A differentiation of this equation with respect to time gives the particle velocity at any instant. Thus

$$\text{Velocity} = y' = a\omega \cos(\omega t - \delta) \tag{6.9}$$

The particle velocity therefore undergoes a sinusoidal variation with respect to time. The frequency of this variation, however, remains unchanged, but the amplitude, that is, the maximum value of velocity, is (ω) times the amplitude of displacement. Moreover, there is a phase difference of $\left(\dfrac{\pi}{2}\right)$ between the two functions represented by Equations (6.8) and (6.9). Thus, when one becomes a maximum, the other is a minimum and vice versa.

A second differentiation of Equation (6.9) with respect to time gives particle acceleration. Thus

$$y'' = -a\omega^2 \sin(\omega t - \delta) = -\omega^2 y \tag{6.10}$$

The acceleration of the particle also undergoes periodic variations of the same frequency as its displacement, but its amplitude is (ω^2) times the amplitude of displacement. Moreover, the negative sign in Equation (6.10) indicates that the two variations are in opposite phase. The graphs of these variations are shown in Figure 6.3.

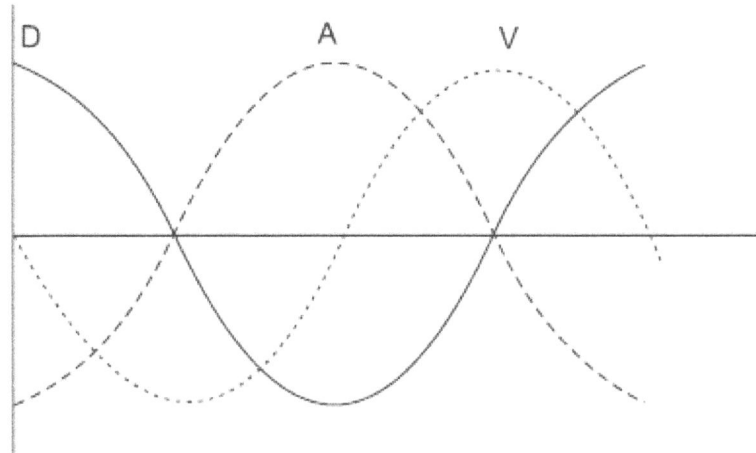

FIGURE 6.3. Particle velocity, displacement, and acceleration in a simple harmonic motion.

Thus, we find from Equation (6.9) that as the particle passes through its mean position $\left(\dfrac{\delta}{\omega}, T+\dfrac{\delta}{\omega}, 2T+\dfrac{\delta}{\omega}, \text{ etc.}\right)$, its velocity has a maximum value $(a\omega)$. Its kinetic energy $\left(\dfrac{1}{2}mv^2\right)$ will therefore also have a maximum value. Its acceleration while passing through the mean position is zero, Equation (6.10), and will go on increasing as it moves farther and farther away but is directed toward the central point (hence deceleration). On reaching the extreme position, the particle loses all its kinetic energy, the deceleration attains its maximum value, and the particle starts its journey back toward its mean position, experiencing all the time an ever-decreasing acceleration till it arrives at its mean position once again, where its kinetic energy becomes maximum. It overshoots this position and starts experiencing an ever-increasing deceleration till its other extreme position, and so on.

At any instant the energy of the vibrating particle is made up of

1. its kinetic energy

2. its potential energy

If m is the mass of the vibrating particle, its kinetic energy at any instant (assume $\delta = 0$ for simplicity) is

$$E_{kin} = \left(\frac{1}{2}\right)mv^2 = \left(\frac{1}{2}\right)my'^2$$

$$E_{kin} = \left(\frac{1}{2}\right) ma^2 \omega^2 \cos^2 \omega t \qquad (6.11)$$

It will be always experiencing a force *my″* trying to bring it back to its mean position. If the displacement of the particle at any instant is y, it will have a certain energy by virtue of its position in the field of this restoring force. The local potential energy can thus be found out in terms of the work done in displacing the particle through a further distance dy against this restoring force. Hence,

$$dW = \text{Force} \times \text{distance}$$

$$dW = m\omega^2 y dy$$

Therefore, the total work done in displacing the body through a distance y from its mean position is

$$W = E_{pot} = \int_0^y m\omega^2 y dy = \left(\frac{1}{2}\right) m\omega^2 y^2$$

$$W = \left(\frac{1}{2}\right) ma^2 \omega^2 \sin^2 \omega t \qquad (6.12)$$

Hence, the total energy of the particle is

$$E = E_{kin} + E_{pot}$$

$$E = \left(\frac{1}{2}\right) ma^2 \omega^2 \left(\sin^2 \omega t + \cos^2 \omega t\right)$$

$$E = \left(\frac{1}{2}\right) ma^2 \omega^2 \qquad (6.13)$$

Thus, the total energy of the particle executing a simple harmonic motion remains independent of its coordinates and depends only on its angular frequency and amplitude of vibration. It also follows from Equations (6.11) and (6.12) that when the kinetic energy is maximum, the potential energy is zero and vice versa. But the individual energies, kinetic and potential, are seen to be functions of time and hence the position of the vibrating particle. Their average values however are $\left(\frac{1}{4}\right) ma^2 \omega^2$ each (since the average values of $\sin^2 \theta$ and $\cos^2 \theta$ are each equal to half).

6.9 DIFFERENTIAL EQUATION OF SIMPLE HARMONIC MOTION AND ITS SOLUTION

Equation (6.10) gives the differential equation of simple harmonic motion e obtained from the definition f of such a motion. We know that the restoring force acting on the vibrating particle is directly proportional to the instantaneous displacement from its mean position and is always directed toward this central point. Hence, according to Newton's second law for *linear motion*

$$my'' = -\mu y$$

where μ is the constant of proportionality and gives the restoring force for unit displacement. On putting $\left(\dfrac{\mu}{m} \right) = \omega^2$ $\mu/m = -2$, this equation can be reduced to Equation (6.14), that is,

$$y'' = -\omega^2 y \tag{6.14}$$

The easiest method of solving such equations is by substitution. We suppose that the solution is of the form

$$y = Ae^{\alpha t}$$

Hence,

$$y' = A\alpha e^{\alpha t} \quad \text{and} \quad y'' = A\alpha^2 e^{\alpha t} \tag{6.15}$$

Substituting the value of y'' in Equation (6.14), we get

$$A\alpha^2 e^{\alpha t} + \omega^2 A e^{\alpha t} = 0$$

Since $y = Ae^{\alpha t}$ is not equal to zero, we get

$$\alpha^2 + \omega^2 = 0$$

So that, $\alpha = \pm i\omega$, where $i = \sqrt{-1}$

The possible solution of Equation (6.14) is therefore

$$y = A_1 e^{i\omega t} \text{ or } y = A_2 e^{-i\omega t}$$

It is the property of linear differential equations that the algebraic sum of all admissible solutions is also one of its solutions. Hence, the most general solution of Equation (6.14) is of the form

$$y = A_1 e^{i\omega t} + A_2 e^{-i\omega t} \tag{6.16}$$

It can be shown that

$$e^{\pm i\omega t} = \cos\omega t \pm i\sin\omega t$$

so that Equation (6.5c) can be written as

$$y = (A_1 + A_2)\cos\omega t + i(A_1 - A_2)\sin\omega t$$

$$y = A\cos\omega t + B\sin\omega t \tag{6.17}$$

Where $A = A_1 + A_2$ and $B = i(A_1 - A_2)$. Equation (6.17) can be written in another form by putting $A = R\sin\theta$ and $B = R\cos\theta$, where $R = \sqrt{A^2 + B^2}$ and $\tan\theta = \left(\dfrac{A}{B}\right)$.

With these values Equation (6.17) becomes

$$y = R\sin(\omega t + \theta) \tag{6.18}$$

Equations (6.17) and (6.18) are the most general solutions of Equation (6.16). The pairs of constants A and B or R and θ can be evaluated from the conditions of the problem under study. For instance, if for a motion

$y = y_o$ and $y' = 0$ at $t = 0$, i.e., initially

then it can be shown that $B = 0$, $\theta = \left(\dfrac{\pi}{2}\right)$, and $R = A = y_o$. This can be easily

recognized as the case of a simple pendulum.

6.10 COMPOSITION OF SIMPLE HARMONIC MOTION

Very often problems are encountered in physics where the same particle is under the simultaneous action of two or more simple harmonic motions acting either along the same straight line or at right angles to each other. There are two general methods for solving such problems based on (a) a graphical treatment and (b) analytical treatment of the dynamics of the particle. The graphical treatment is based on the rectilinear projection of uniform circular motion, while the analytical treatment is based on finding the vector sum of the individual motions either with the help of trigonometric functions or writing them as complex quantities. The last of these methods is usually the easiest to handle and is at the same time more informative. It will nevertheless be worthwhile to understand the general outline of the other two.

1. *Composition of simple harmonic Motions (Graphical Method).* Let the vector x_1 as shown in Figure 6.4 represent the amplitude of a simple harmonic motion.

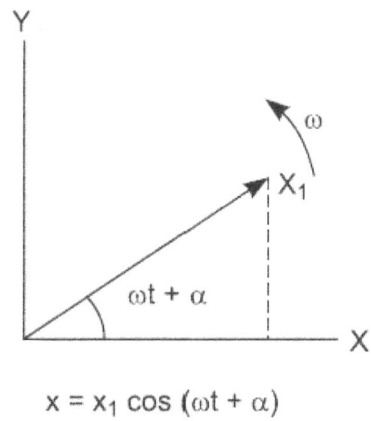

$$x = x_1 \cos (\omega t + \alpha)$$

FIGURE 6.4. Amplitude of a simple harmonic motion.

By convention of such representation, the vector is assumed to rotate counterclockwise with an angular frequency ω. Its projection along the X-axis will give the displacement equation of the simple harmonic motion.

$$x = x_1 \cos(\omega t + \alpha)$$

Where α is the starting angle at $t = 0$. If there are several collinear simple harmonic motions of different amplitude, frequency, and phase acting simultaneously on a particle, each of these can be represented by an appropriate rotating vector like x_1. The projections of all these vectors will give the individual displacements as previously. Suppose the amplitude vectors of two simple harmonic motions at say $t = 0$ are represented by x_1 and x_2 as shown in Figure 6.5.

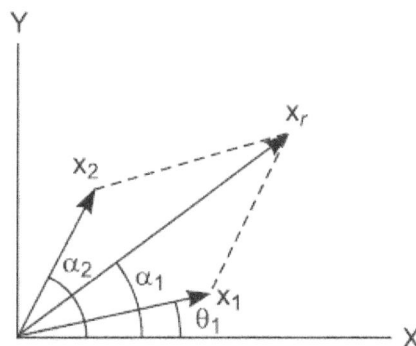

FIGURE 6.5. Amplitude vectors of two simple harmonic motions.

Resolving them along two mutually perpendicular directions X and Y, we get

$$X_x = x_1 \cos\alpha_1 + x_2 \cos\alpha_2$$
$$\text{and}$$
$$X_y = x_1 \sin\alpha_1 + x_2 \sin\alpha_2$$
(6.19)

Hence, if the amplitude vector of the resultant is represented by x_r, we have

$$X_r = \sqrt{X_x^2 + X_y^2} = \sqrt{x_1^2 + x_2^2 + 2x_1 x_2 \cos(\alpha_2 - \alpha_1)}$$
(6.20)

The phase of the resultant at $t = 0$ will therefore be

$$\alpha_r = \tan^{-1}\left(\frac{X_y}{X_x}\right)$$
(6.21)

In general, the magnitude of the resultant vector can be obtained from the *parallelogram law* as in Equation (6.21), which is given by

$$X_r = \sqrt{\left(\sum X_x\right)^2 + \left(\sum X_y\right)^2}$$
(6.22)

and

$$\tan\alpha_r = \frac{\left(\sum X_y\right)}{\left(\sum X_x\right)}$$
(6.23)

where X_x and X_y are respectively the sums of the x and y components of the amplitude vectors at $t = 0$. The resultant motion is then given by

$$x = x_r \cos(\omega t + \alpha_r)$$
(6.24)

Thus, we find that the resultant of a number of simple harmonic motions is always another simple harmonic motion. If the two motions are at right angles to each other, the resultant motion of the particle executing them simultaneously is no longer simple but depends upon their difference of phase and frequency. Let us consider two simple harmonic motions, one along the direction AB and the other along $A'B'$ as in Figure 6.6 and suppose that by the time the rotating amplitude vector completes one revolution along the lower circle, that of the upper circle completes two rotations so that their time periods are in the ratio $2:1$. Let the two motions start in the same phase (positions marked 0, 0). We can divide the circumference of the two circles into parts such that the two vectors reach the points 0, 1, 2, 3, ... 8 at the same time.

The points of intersection of the perpendiculars drawn from corresponding points (1, 1), (2, 2), . . . (8, 8) will then give the locus of the vibrating particle if it were under the simultaneous action of the two simple harmonic motions. This is seen to be a figure of eight in this particular case. If the initial conditions of the two vibrations are changed, the resulting curve may have a different shape. Such curves are called *Lissajou's* figures.

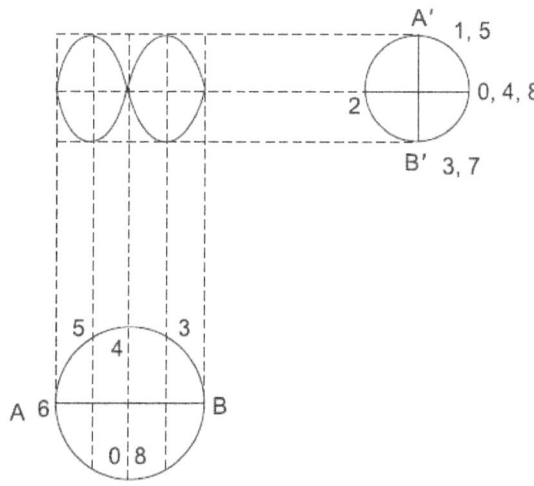

FIGURE 6.6. Composition of simple harmonic motions at right angles.

2. *Analytical Method.* Let a particle execute two collinear simple harmonic motions of the same frequency

$$x_1 = A\cos(\omega t + \alpha_1) \text{ and } x_2 = B\cos(\omega t + \alpha_2)$$

The resultant displacement of the particle will be given by the vector sum of the individual motions, so that

$$x = x_1 + x_2$$

Therefore,

$$X = A\cos(\omega t + \alpha_1) + B\cos(\omega t + \alpha_2)$$

$$X = A\left[\cos\omega t \cos\alpha_1 - \sin\omega t \sin\alpha_1\right] + B\left[\cos\omega t \cos\alpha_2 - \sin\omega t \sin\alpha_2\right]$$

$$X = \left(A\cos\alpha_1 + B\cos\alpha_2\right)\cos\omega t - \left(A\sin\alpha_1 + B\sin\alpha_2\right)\sin\omega t$$

If we put

$$R\cos\theta = A\cos\alpha_1 + B\cos\alpha_2 \text{ and } R\sin\theta = A\sin\alpha_1 + B\sin\alpha_2$$

the resultant motion can be written as

$$X = R(\cos\omega t\cos\theta - \sin\omega t\sin\theta)$$

$$X = R\cos(\omega t + \theta) \tag{6.25}$$

This is the equation of a simple harmonic motion of amplitude

$$\tan\theta = \sqrt{A^2 + B^2 + 2AB\cos(\alpha_1 - \alpha_2)} \tag{6.26}$$

and initial phase angle at $t = 0$ given by

$$\tan\theta = \frac{A\sin\alpha_1 + B\sin\alpha_2}{A\cos\alpha_1 + B\cos\alpha_2} \tag{6.27}$$

whose period is the same as that of the two individual motions. If the two simple harmonic motions superposed on a particle are mutually perpendicular, they can be represented by equations of the form

$$x = a\cos\omega t \tag{6.28}$$

$$y = b\cos(n\omega t + \alpha) \tag{6.29}$$

whose periods are in the ratio $n:1$ and initial phase difference is α. For finding the resultant motion, we proceed as follows: from Equation (6.28), we get

$$\left(\frac{x}{a}\right) = \cos\omega t$$

and hence

$$\sqrt{1 - \left(\frac{x^2}{a^2}\right)} = \sin\omega t$$

These values of the sine and cosine functions can be substituted in the expansion of $\cos(n\omega t + \alpha)$ on the right-hand side of Equation (6.6k) to get the equation of the resultant motion. For instance, if $n = 2$, we have from Equation (6.29)

$$\left(\frac{y}{b}\right) = \cos 2\omega t\cos\alpha - \sin 2\omega t\sin\alpha$$

$$\left(\frac{y}{b}\right) = \left(2\cos^2\omega t - 1\right)\cos\alpha - 2\sin\omega t\cos\omega t\sin\alpha$$

or

$$\left(\frac{y}{b}\right) = \left(2\left(\frac{x^2}{a^2}\right) - 1\right)\cos\alpha - 2\left(\frac{x}{a}\right)\sqrt{\left(1 - \left(\frac{x^2}{a^2}\right)\right)}\sin\alpha$$

or

$$\left[\left(\frac{y}{b}\right) - \left(2\left(\frac{x^2}{a^2}\right) - 1\right)\cos\alpha\right]^2 = 4\left(\frac{x^2}{a^2}\right)\left(1 - \left(\frac{x^2}{a^2}\right)\right)\sin^2\alpha$$

which on simplification gives

$$\left(\frac{y^2}{b^2}\right) + 4\left(\frac{x^4}{a^4}\right) - 4\left(\frac{x^2}{a^2}\right)\left(1 + \left(\frac{y}{b}\right)\cos\alpha\right) + 2\left(\frac{y}{b}\right)\cos\alpha + \cos^2\alpha = 0 \qquad (6.30)$$

This is the general equation of the resultant motion of a particle under the simultaneous action of two mutually perpendicular simple harmonic motions whose periods are in the ratio 2 : 1. Depending upon the value of the phase difference α between the two motions, the curve will be a figure of eight, parabola, and so on.

We can obtain the locus of the particle vibrating under the simultaneous action of two mutually perpendicular simple harmonic motions of any other frequency ratio in a similar manner.

3. *Use of Complex Quantities.* Calculations with harmonic motions are simplified by the use of complex quantities instead of the usual trigonometric functions. Thus, if a simple harmonic motion is represented by the displacement equation

$$y = b\cos(\omega t + \delta) \qquad (6.31)$$

we can also represent it in the form

$$y = R\left(be^{-i(\omega t - \delta)}\right) \qquad (6.32)$$

where $i = \sqrt{-1}$, R is the real part of the exponential function, and $\cos x - i\sin x = e^{-ix}$. If the operations on "$y$" are linear, we may drop the symbol R in Equation (6.31) and work with the exponential function, only providing we extract out the real part from the final expression, which will give the result under study. But if the operations on "y" are nonlinear, such as squaring, and so on, we should be careful to take the real part first and then operate with these alone. It may be mentioned that such representations of harmonic motions cause considerable simplicity in calculations and have other advantages which shall be discussed in a different context.

(*i*) *Composition of Collinear Simple Harmonic Motions*: Let two simple harmonic motions be represented by

$$x_1 = a\cos(\omega t + \alpha_1) \tag{6.33}$$

$$x_2 = b\cos(\omega t + \alpha_2) \tag{6.34}$$

These motions have the same period but different amplitudes and phase; $(\alpha_1 \sim \alpha_2)$ is called the phase difference between the two motions.

As outlined previously, we can write

$$x_1 = R\left(ae^{-i(\omega t + \alpha_1)}\right)$$

$$x_2 = R\left(be^{-i(\omega t + \alpha_2)}\right)$$

It may be mentioned that the amplitudes "*a*" and "*b*" here are not complex quantities. Since the resultant motion is the vector sum of individual motions, we can write

$$x = x_1 + x_2 = e^{-i\omega t}\left(ae^{-i\alpha_1} + be^{-i\alpha_2}\right) \tag{6.35}$$

Since the operation performed here has been linear, we can afford to drop the symbol R on the understanding that only the real part of the final expression will give the desired result. Equation (6.35) can now be written as

$$x = e^{-i\omega t}\left[(a\cos\alpha_1 + b\cos\alpha_2) - i(a\sin\alpha_1 + b\sin\alpha_2)\right]$$

$$x = e^{-i\omega t}\left[A(\cos\phi - i\sin\phi + b\cos\alpha_2)\right] \tag{6.36}$$

where

$$A = \sqrt{(a\cos\alpha_1 + b\cos\alpha_2)^2 + (a\sin\alpha_1 + b\sin\alpha_2)^2}$$

$$A = \sqrt{a^2 + b^2 + 2ab\cos(\alpha_1 - \alpha_2)} \tag{6.37}$$

and

$$\tan\phi = \frac{a\sin\alpha_1 + b\sin\alpha_2}{a\cos\alpha_1 + b\cos\alpha_2} \tag{6.38}$$

Hence, the resultant motion can be derived from

$$x = Ae^{-i\omega t}e^{-i\phi} = Ae^{-i(\omega t + \phi)}$$

Taking the real part of this expression, we get

$$x = A\cos(\omega t + \phi) \tag{6.39}$$

This is in accordance with Equation (6.25) and the remarks following it.

(*ii*) *Composition of Rectangular Simple Harmonic Motions*: Let the two motions be represented by

$$x = a\cos\omega t \tag{6.40a}$$

$$x = b\cos(n\omega t + \alpha) \tag{6.40b}$$

Equation (6.40*a*) represents a vibratory motion of amplitude "*a*" along the X-axis while Equation (6.40*b*) is another similar motion of amplitude "*b*" along the Y-axis. The periods of the two motions are in the ratio $n:1$. In terms of complex quantities, we can write these motions as

$$\left(\frac{x}{a}\right) = R\left(e^{-i\omega t}\right)$$

$$\left(\frac{y}{b}\right) = R\left(e^{-i(n\omega t + \alpha)}\right)$$

$$\left(\frac{y}{b}\right) = R\left(e^{-in\omega t}e^{-i\alpha}\right)$$

$$\left(\frac{y}{b}\right) = R\left[(\cos\alpha - i\sin\alpha)e^{-in\omega t}\right] \tag{6.41}$$

According to *De Moivre's theorem*, if "*n*" is real, one value of $(\cos\phi \pm i\sin\phi)^n$ is $\cos n\phi \pm i\sin n\phi$. Hence, the real part of Equation (6.41) will be

$$\cos\alpha\cos n\omega t - \sin\alpha\sin n\omega t$$

So that

$$\left(\frac{y}{b}\right) = \cos\alpha\cos n\omega t - \sin\alpha\sin n\omega t$$

or

$$\left(\frac{y}{b}\right) - \cos\alpha\cos n\omega t = -\sin\alpha\sin n\omega t \tag{6.42}$$

We had to extract the real part, because the next operation is going to be nonlinear. Squaring both sides of Equation (6.42) and simplifying, we get

$$\left(\frac{y^2}{b^2}\right) - 2\left(\frac{y}{b}\right)\cos\alpha\cos n\omega t + \cos^2 n\omega t = \sin^2\alpha \tag{6.43}$$

where the value of $\cos n\omega t$ has to be found with the help of Equation (6.43). This is the general equation of the locus of the particle under the simultaneous action of two rectangular simple harmonic motions.

Special Cases: (*i*) *Frequency ratio* $1:1$

If the time periods of the two simple harmonic motions are equal, $n = 1$ and Equation (6.43) can be written as

$$\left(\frac{y^2}{b^2}\right) - 2\left(\frac{y}{b}\right)\cos\alpha \cos\omega t + \cos^2\omega t = \sin^2\alpha$$

Since

$$\cos\omega t = \left(\frac{x}{a}\right),\ \cos^2\omega t = \left(\frac{x}{a}\right)^2,\ \text{and we get}$$

$$\left(\frac{y^2}{b^2}\right) - 2\left(\frac{y}{ab}\right)\cos\alpha + \left(\frac{x^2}{a^2}\right) = \sin^2\alpha \tag{6.44}$$

as the equation of the locus in this case. This is the general equation of a conic whose shape will depend upon the value of the phase difference α between the two motions.

a. If $\alpha = 0$ the equation of *Lissajou*'s figure is seen to be

$$\left(\frac{y}{b} - \frac{x}{a}\right)^2 = 0$$

which is a pair of coincident straight lines inclined at an angle $\tan^{-1}\left(\frac{b}{a}\right)$ to the X-axis.

b. If $\alpha = \left(\frac{\pi}{4}\right)$, $\sin\alpha = \cos\alpha = \frac{1}{\sqrt{2}}$, the equation of the curve becomes

$$2b\left(\frac{y^2}{b^2}\right) - \sqrt{2}\left(\frac{xy}{ab}\right) + \left(\frac{x^2}{a^2}\right) = \frac{1}{2}$$

This is the equation of an ellipse inscribed in a rectangle whose length parallel to the X-axis is $2a$ and breadth. The ellipse touches this rectangle at points $\left(\pm a, \pm\frac{b}{\sqrt{2}}\right)$ and $\left(\pm\frac{a}{\sqrt{2}}, \pm b\right)$

c. If $\alpha = \left(\frac{\pi}{2}\right)$, Equation (6.44) reduces to

$$\left(\frac{y^2}{b^2}\right) + \left(\frac{x^2}{a^2}\right) = 1$$

which is the familiar equation of an ellipse whose center coincides with the origin and whose axes are parallel to the coordinate axes X and Y. If the two simple harmonic motions have the same amplitude, the locus of the particle becomes a circle ($a = b$, hence, $x^2 + y^2 = a^2$).

d. If $\alpha = \left(\dfrac{3\pi}{4}\right)$, *Lissajou's* figure is represented by the equation

$$\left(\frac{y^2}{b^2}\right) - \sqrt{2}\left(\frac{xy}{ab}\right) + \left(\frac{x^2}{a^2}\right) = \frac{1}{2}$$

which is again an ellipse with its axes rotated by $\left(\dfrac{\pi}{2}\right)$ with respect to that of case (b).

e. If $\alpha = \pi$, we again get a pair of coincident straight lines given by

$$\left(\frac{y}{b} + \frac{x}{a}\right)^2 = 0$$

But this time they are inclined at an angle, $\tan^{-1}\left(-\dfrac{b}{a}\right)$, to the axis of X.

All these cases have been summarized in Figure 6.7.

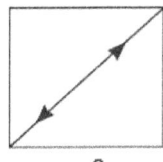 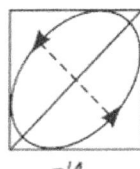 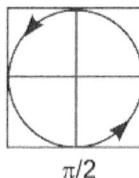 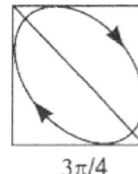 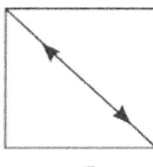

| $\alpha = 0$ | $\pi/4$ | $\pi/2$ | $3\pi/4$ | π |

FIGURE 6.7. Lissajou's figures (period ratio 1:1).

(ii) Frequency Ratio 1:2

The general equation of Lissajou's figure when the time periods are in the ratio 2 : 1 can be obtained by putting $n = 2$ in Equation (6.43). Thus, we get

$$\left(\frac{y^2}{b^2}\right) - 2\left(\frac{y}{b}\right)\cos\alpha\cos 2\omega t + \cos^2 2\omega t = \sin^2\alpha \qquad (6.45)$$

where

$\cos\omega t = \dfrac{x}{a}$. Now,

$\cos 2\omega t = 2\cos^2\omega t - 1$

$\cos 2\omega t = \left(2\dfrac{x^2}{a^2} - 1\right)$

and

$$\cos^2 2\omega t = 4\left(\frac{x}{a}\right)^4 - 4\left(\frac{x}{a}\right)^2 + 1$$

Substituting these values in Equation (6.45), we get

$$\left(\frac{y^2}{b^2}\right) - 2\left(\frac{y}{b}\right)\left(2\frac{x^2}{a^2} - 1\right)\cos\alpha + \left(2\frac{x^2}{a^2} - 1\right)^2 = \sin^2\alpha$$

which on rearranging gives

$$\left(\frac{y^2}{b^2}\right) - 4\left(\frac{x^4}{a^4}\right)\left(1 + \frac{y}{b\cos\alpha}\right) + \left(\frac{2y}{b\cos\alpha} + \cos^2\alpha\right)^2 = 0 \qquad (6.46)$$

An inspection of this equation shows that the general conic is symmetrical about the X-axis only; the actual shape of the curve will of course depend upon the phase difference "α" between the two motions. The resulting *Lissajou* figures for different values of _ are shown in Figure. 6.8.

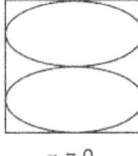 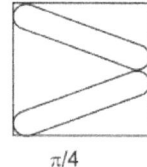 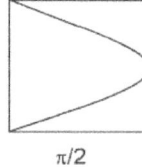 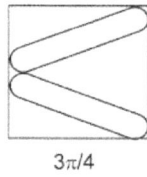 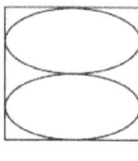

| $\alpha = 0$ | $\pi/4$ | $\pi/2$ | $3\pi/4$ | π |

FIGURE 6.8. Lissajou's figures (period ratio 2:1).

It may be mentioned that these figures can be demonstrated on the screen of a cathode ray oscillograph by applying audio-frequency voltage signals on the X and Y plates. Stable figures can be obtained only if the two frequencies bear a whole number ratio. As this ratio goes on increasing, the figures get more and more crowded but always remain closed curves. These figures provide a very sensitive test for determining any unknown frequency by combining it with a known frequency in a perpendicular direction. It may be mentioned that the composition of simple harmonic motions is an important problem in physics and is encountered very frequently in wave optics and acoustics.

6.11 DAMPED VIBRATIONS OF A PARTICLE

Until now we have considered the linear vibratory motions of a particle under the long action of a restoring force trying perpetually to bring it back to its mean position. A little reflection will show that this situation is an ideal case of the general phenomenon. In actual practice the vibratory motion is also opposed by frictional or dissipative forces generated in the fluid in which the motion is taking place. Such dissipative forces are usually proportional to some power of the velocity of the particle; for small values of velocity, the power can be taken as unity without loss of any significant accuracy. With this restriction in view, the dissipative forces can be written as

$$F = -ry'$$

where the constant of proportionality r is the dissipative force per unit velocity of the particle, and the negative sign signifies a restraining influence on the vibration of the particle.

According to Newton's second law, the equation of motion in this case will be

$$my'' = -r\,y' - \mu y$$

or

$$y'' + \left(\frac{r}{m}\right) y' + \left(\frac{\mu}{m}\right) y = 0 \qquad (6.47)$$

The motion of the particle defined by Equation (6.47) is called damped harmonic motion. Such equations arise many times in different branches of physics, the most important being that describing the transient behavior of an LCR circuit. It may be added that the solution to Equation (6.47) in the present context is of a similar nature as that encountered for an LCR circuit, but in order to obtain any meaningful results from its application to the circuit theory, the similarity conditions or analogy between the two cases must be closely studied. Moreover, there are very many equations in electrodynamics which have a close parallel in mechanics, and a solution of the former leads to solution of the latter provided due attention is paid to the analogy between the two fields. In fact, this has led to the development of analog computers.

Coming back to Equation (6.47), we find that it can be reduced to a simpler form by proper substitutions. Let

$$y = xe^{-bt} \qquad (6.48)$$

where "b" is an arbitrary constant, then

$$y' = x'e^{-bt} - bxe^{-bt} \text{ and } y'' = x''e^{-bt} - bx'e^{-bt} - bx'e^{-bt} + b^2xe^{-bt}$$

Substituting these values in Equation (6.47), we get

$$x'' + \left(\frac{r}{m} - 2b\right)x' + \left(\frac{\mu}{m} + b^2 - \frac{br}{m}\right)x = 0 \tag{6.49}$$

Since b is an arbitrary constant, we can take it equal to $\left(\dfrac{r}{2m}\right)$ so that the coefficient of y' may vanish. Hence, Equation (6.49) becomes

$$x'' + \left(\omega^2 - b^2\right)x = 0 \tag{6.50}$$

an equation of the type we have solved before. Knowing the solution to Equation (6.50), we can write down the value of y with the help of Equation (6.48). There are, however, three important cases which are discussed as follows.

Case I: Suppose $\omega^2 < b^2$, i.e., $\left(\dfrac{\mu}{m}\right) < \left(\dfrac{r^2}{4m^2}\right)$

In the case of large frictional losses, the difference $\left(\omega^2 - b^2\right)$ is negative, say $\left(-K^2\right)$. Then Equation (6.50) can be written

$$\left(D^2 - K^2\right)x = 0 \ \text{ or } \ \left(D - K\right)\left(D + K\right)x = 0$$

where

$$D = \frac{d}{dt} \ \text{ and } D^2 = \frac{d^2}{dt^2}$$

The solution of this differential equation is of the form

$$x = A_1 e^{+Kt} + A_2 e^{-Kt}$$

Hence,

$$y = A_1 e^{\left(-b + \sqrt{b^2 - \omega^2}\right)t} + A_2 e^{\left(-b - \sqrt{b^2 - \omega^2}\right)t} \tag{6.51}$$

where A_1 and A_2 are the constants of integration to be determined from the conditions of each problem.

It will be noticed that $b > \sqrt{b^2 - \omega^2}$, and hence the powers of the exponential will be negative in both terms of Equation (6.51). Consequently, the displacement of the particle will fall exponentially to zero from its initial value of $(A_1 + A_2)$ at the start of its damped motion as in Figure 6.9. How soon the motion dies out will depend upon the magnitude of difference between "b" and "ω."

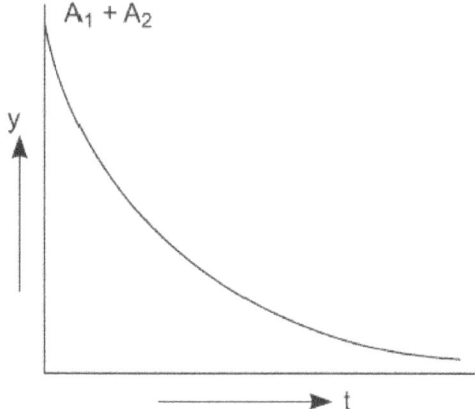

FIGURE 6.9. Damped motion of a particle.

Case II: Suppose $\omega^2 = b^2$ or $\left(\dfrac{\mu}{m}\right) = \left(\dfrac{r^2}{4m^2}\right)$

This is the condition of a critically damped motion. In this case Equation (7.50) reduces to $x'' = 0$, which is the equation of the straight line $x = A_1 t + A_2$, where A_1 and A_2 are again the constants of integration, both assumed positive as before.

Hence,

$$y = e^{-bt}\left(A_1 t + A_2\right)$$

A plot of this equation has been shown in Figure 6.10.

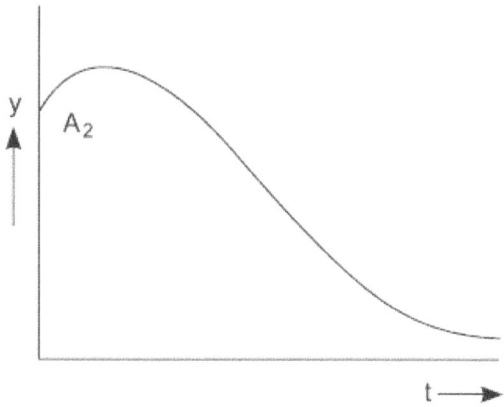

FIGURE 6.10. Critically damped motion.

But for an initial increase (provided $A_1 > A_2$), the amplitude of motion goes on decreasing as time progresses. This case is of significance in the design of scientific equipment.

Case III: Suppose $\omega^2 > b^2$
Or

$$\left(\frac{\mu}{m}\right) > \left(\frac{r^2}{4m^2}\right)$$

In this case where the dissipative forces are small, Equation (7.50) reduces to the form of a simple harmonic motion.

$$x'' = -\left(\omega^2 - b^2\right)x$$

The solution of such an equation has been seen to be of the form

$$x = A\cos\left(\sqrt{\omega^2 - b^2}\; t + \theta\right)$$

where A and θ are the two constants of integration. Therefore, the displacement of the damped vibration will be given by

$$y = Ae^{-bt}\cos\left(\sqrt{\omega^2 - b^2}\; t + \theta\right) \tag{6.52}$$

From Equation (6.52) we find that the effective amplitude (Ae^{-bt}) of a damped motion dies out exponentially with time. The rate at which this happens is of course governed by "b." If the dissipative forces increase, the exponential decay becomes much more rapid. In addition to this the dissipative forces also affect the time period of the vibrations of the particle. Whereas in the case of free vibrations the periodic time is $\left(\dfrac{2\pi}{\omega}\right)$, in the case of damped vibrations this is increased to $\left(\dfrac{2\pi}{\sqrt{\omega^2 - b^2}}\right)$. The frequency of the oscillations is therefore lowered. As "b" increases, this oscillation frequency approaches zero as b tends to "ω."

The damped oscillations are shown in Figure 6.11. The dotted curve represents the exponential decay of the amplitude. These damped vibrations are of importance in musical instruments, where it is necessary that they die out quickly. This is accomplished by arranging for a large value of "b."

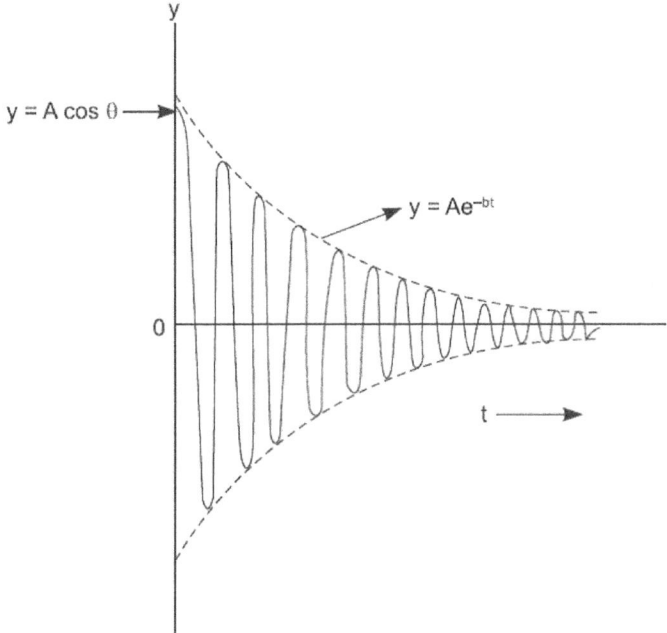

FIGURE 6.11. Damped oscillations of a particle.

6.12 FORCED VIBRATIONS

It has been discussed that in practical cases, the natural vibrations of a particle die out with the lapse of time. If it is desired to maintain its vibrations, extra force has to be supplied from outside. If this force is periodic, the particle begins to vibrate with a frequency equal to that of the applied force. Such oscillations are called *forced oscillations*. Let the externally applied force be

$$F \cos pt = R\left(Fe^{-ipt}\right)$$

The equation of motion of the particle then becomes

$$my'' + ry' + \mu y = F\cos pt \tag{6.53}$$

There are several methods of solving this equation, but we shall try to solve it with the help of complex quantities. The solution to such equations consists of two parts:

1. *The complementary function*: This is obtained by putting the right-hand side equal to zero and solving. This solution is of course exactly the same as that of Equation (6.47) and has two arbitrary constants. This part of the solution is called the *transient term*.

2. *The particular integral*: This is obtained by solving Equation (6.53) as it is and does not contain any arbitrary constants. This is called the *steady state solution*.

To obtain the steady state solution, we shall replace the right-hand side of Equation (6.53) by the complex quantity Fe^{-ipt} and solve the equation with the understanding that the real part of the solution will be of relevance to us. Thus, we write

$$my'' + ry' + \mu y = Fe^{-ipt} \tag{6.54}$$

Let us assume that the solution of this equation is of the form

$$y = Ae^{-ipt}$$

where A is in general complex. Then

$$y' = -iApe^{-ipt}$$

and

$$y'' = -Ap^2 e^{-ipt}$$

Substituting these values in Equation (6.54), we get

$$\left(Ap^2 m - iApr + A\mu\right)e^{-ipt} = Fe^{-ipt}$$

For this equation to be satisfied at all values of t, A must be given by

$$A = \frac{F}{\left(\mu - mp^2\right) - ipr}$$

The complex displacement will then be

$$y = \frac{F}{\left(\mu - mp^2\right) - ipr} e^{-ipt}$$

$$y = \frac{iF}{p\left[r + i\left(mp - \dfrac{\mu}{p}\right) - ipr\right]} e^{-ipt} \tag{6.55}$$

Suppose we put

$$Z_M = r + i\left(mp - \frac{\mu}{p}\right) = r + iS = Z_o e^{i\phi} \tag{6.56}$$

where

$$Z_o = \sqrt{r^2 + S^2}$$

and

$$\tan\phi = \left(\frac{S}{r}\right)$$

then Equation (6.56) becomes

$$y = \frac{iF}{pZ_o e^{-i(pt-\phi)}}$$

The solution to Equation (6.53) is the real part of the previous solution. Thus, we find that

$$y = \frac{F}{pZ_o \sin(pt-\phi)}$$

$$y = \frac{F}{p\sqrt{r^2 + \left(mp - \frac{\mu}{p}\right)^2}} \sin(pt-\phi) \tag{6.57}$$

is the steady state solution of the problem under study.

An examination of Equation (6.57) reveals that the steady state displacement of the particle executing forced oscillations is not in step with the driving force; there is a phase difference of between the two. The amplitude of these forced vibrations is less than that of the driving force but depends upon F. The frequency of forced vibrations is equal to that of the externally applied force, which is different from either that of its free vibrations or that of its damped vibrations. These are given as follows:

$$\upsilon = \frac{\omega}{2\pi} \qquad\qquad \text{(Free vibrations)}$$

$$\upsilon = \sqrt{\frac{\omega^2 - b^2}{2\pi}} \qquad\qquad \text{(Damped vibrations)}$$

$$\upsilon = \frac{p}{2\pi} \qquad\qquad \text{(Forced vibrations)}$$

The complete solution of Equation (6.53) is therefore

$$y = Ae^{-bt} \cos\left(\left(\sqrt{\omega^2 - b^2}\right)t + \theta\right) + \frac{F}{pZ_o}\sin(pt - \phi) \tag{6.58}$$

where

$$Z_o = \sqrt{r^2 + \left(mp - \frac{\mu}{p}\right)^2}$$

The constants of integration A and θ in the transient term of Equation (6.58) are determined from the initial conditions of the problem.

When a particle is subjected to "forced vibrations," its displacement in the beginning is made up of the transient as well as the steady state components. After a certain time "t" such that $bt \gg 1$, the transient term rapidly fades away, leaving only the steady state components in action. Before proceeding with a discussion of forced vibrations, it is interesting to note the striking analogy of Equation (6.53) with that for an LCR circuit under an impressed alternating voltage. If a voltage $E_o \cos pt$ is impressed on a circuit containing an inductance L, capacity C, and resistance R, the current i satisfies the following equation:

$$L\frac{di}{dt} + Ri + \frac{1}{C}\int i\,dt = E_o \cos pt$$

And the charge $q = \int i\,dt$. The equation then can be reduced to

$$Lq'' + Rq' + \frac{1}{Cq} = E_o \cos pt \tag{6.59}$$

Comparison of Equations (6.53) and (6.59) leads to the solution

$$q = \frac{E_o}{p\left(R^2 + \left(Lp - \frac{1}{Cp}\right)^2\right)}\sin(pt - \phi)$$

Hence

$$i = \frac{E_o}{p\left(R^2 + \left(Lp - \frac{1}{Cp}\right)^2\right)}\cos(pt - \phi)$$

where

$$\tan\phi = \frac{Lp - \left(\dfrac{1}{Cp}\right)}{R}$$

By analogy with the electrical impedance $\left(\sqrt{R^2 + \left(Lp - \dfrac{1}{Cp}\right)^2}\right)$, the

quantity under the square root sign in Equation (6.57) is called the mechanical impedance. The analogy can be carried still further: the mechanical resistance r is analogous to electrical resistance R, the mass m to the electrical inductance L, and the restoring force per unit displacement μ is called the *stiffness factor* to the reciprocal of capacitance. Hence, we can call the quantity $\left(mp - \dfrac{\mu}{p}\right)$ as mechanical reactance.

6.13 RESONANCE

We have seen that the steady state displacement of a particle undergoing forced vibrations is given by

$$y = \frac{F}{p\sqrt{r^2 + \left(mp - \dfrac{\mu}{p}\right)^2}}\sin(pt - \phi)$$

Hence, the particle velocity under this condition is

$$y' = \frac{F}{p\sqrt{r^2 + \left(mp - \dfrac{\mu}{p}\right)^2}}\cos(pt - \phi)$$

The power dissipated in working against dissipative forces will be

$$force \times velocity = ry'^2 = \frac{rF^2}{\left(r^2 + \left(mp - \dfrac{\mu}{p}\right)^2\right)}\cos^2(pt - \phi)$$

The average value of this power dissipation over one complete cycle is $\left(\text{since the average value } \cos^2 \theta = \dfrac{1}{2}\right)$

$$P_{av} = ry'^2 = \frac{rF^2}{2\left[r^2 + \left(mp - \dfrac{\mu}{p}\right)^2\right]}\cos^2\left(pt - \phi\right) \qquad (6.60)$$

We can say that the amplitude and phase of the particle adjust themselves so that the average power being supplied by the driving force is just equal to that being dissipated in overcoming the frictional forces. The maximum power transfer will therefore take place when the denominator of Equation (6.60) has a minimum value or when

$$mp = \frac{\mu}{p} \text{ or } p = \sqrt{\left(\frac{\mu}{m}\right)} = \omega \qquad (6.61)$$

Hence, we find that when the natural frequency of the vibrating particle is equal to that of the externally applied force, the former will pick up maximum power from the latter. Under this condition the mechanical reactance vanishes and the equation for displacement becomes

$$y = \left(\frac{F}{\omega r}\right)\sin \omega t \qquad (6.62)$$

since

$$\tan \phi = \frac{mp - \left(\dfrac{\mu}{p}\right)}{r} = 0 \text{ or } \phi = 0$$

This phenomenon is known as resonance. This resonance displacement amplitude is therefore $\left(\dfrac{F}{\omega r}\right)$, and the resonance velocity amplitude is simply $\left(\dfrac{F}{r}\right)$. But the resonance displacement amplitude should not be mixed up with the maximum displacement amplitude of the forced vibrations. This can be obtained by finding the condition under which the denominator of Equation (6.57) has a minimum value. For this

$$\frac{d}{dp}\left(p\sqrt{r^2 + \left(mp - \frac{\mu}{p}\right)^2}\right) = 0$$

which gives

$$p = \sqrt{\omega^2 - 2b^2} \ \text{ where } \ b = \frac{r}{2m}$$

It may be verified that for this value of p, the second differential of the quantity inside the curved brackets previously leads to a positive value.

The maximum displacement amplitude of the forced vibrations will therefore be

$$y_{\max} = \frac{F}{r\sqrt{p^2 + b^2}} = \frac{F}{r\sqrt{\omega^2 - b^2}} \tag{6.63}$$

Sharpness of Resonance. It has been seen previously that the average power supplied to the vibrating particle by the external force is given by

$$P_{av} = \frac{rF^2}{2\left(r^2 + \left(mp - \dfrac{\mu}{p} \right)^2 \right)}$$

At resonance, maximum power transfer takes place, that is, at $\omega = p = \dfrac{\mu}{m}$, the average power taken up by the particle is $\left(\dfrac{F^2}{2r} \right)$. Thus, it is seen that the power transfer is primarily determined by the value of the mechanical resistance "r." This has been shown in Figure 6.12, which gives the resonance curves for two values of "r."

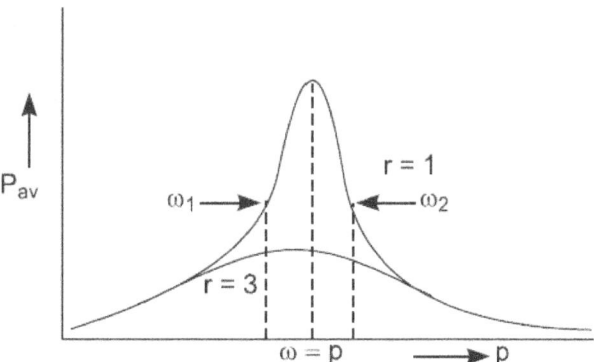

FIGURE 6.12. Resonance curves.

It is seen that for low values of "r," the resonance curve has a sharp peak. Increasing "r" by a factor of three considerably broadens the resonance peak. In the former case resonance is said to be sharp, while in the latter case it is said to be flat.

Example 6.1

A particle has a mass of 2gms. It is free to vibrate under the action of an elastic force of 128 dyne-cm^{-1} and a damping force of 8 dyne-cm^{-1} sec. A periodically varying outside force of maximum value 256 dynes is applied to the particle. Find the frequency for displacement resonance, the frequency for velocity resonance, and the approximate amplitude at resonance.

Solution:

For displacement resonance,

$$\text{Frequency} = \frac{p}{2\pi} = \frac{1}{2\pi}\sqrt{\omega^2 - 2b^2}$$

where

$$\omega^2 = \frac{\mu}{m} = 64\,\text{sec}^{-2} = \frac{128 \text{ dyne-cm}^{-1}}{2\text{gm}}$$

and

$$b = \frac{r}{2m} = \frac{8 \text{ dyne-cm}^{-1} \text{ sec}}{2 \times 2\text{gm}} = 2\,\text{sec}^{-2}$$

Therefore,

$$\text{Frequency} = \frac{1}{2\pi}\sqrt{64 - 8} = 1.19\,\text{sec}^{-1}$$

For the velocity resonance,

$$\text{Frequency} = \frac{1}{2\pi}\sqrt{\left(\frac{\mu}{m}\right)} = 1.27\,\text{sec}^{-1}$$

$$\text{Amplitude at resonance} = \frac{F}{\omega r} = \frac{256}{8 \times 8} = 4 \text{ cm}$$

Example 6.2

A particle of mass 3gm is subject to an elastic force of 48 dyne-cm^{-1} and a damping force of 12 dyne-cm^{-1} sec. If the motion is oscillatory, find its period.

Solution:

$$\omega = \sqrt{\left(\frac{\mu}{m}\right)} = \sqrt{\left(\frac{48 \text{ dyne-cm}^{-1}}{3 \text{ gm}}\right)} = 4 \text{ sec}^{-1}$$

and

$$b = \frac{r}{2m} = \frac{12}{2 \times 3} = 2 \sec^{-1}$$

Since $\omega > b$, the motion is oscillatory.

$$\text{Period time} = \frac{2\pi}{\sqrt{\left(\omega^2 - b^2\right)}} = \frac{2\pi}{\sqrt{\left(16-4\right)}} = \frac{2\pi}{\sqrt{\left(12\right)}} \cong 1.18 \text{ sec}$$

6.14 EXERCISES

1. Discuss: "Photon and wave ideas are in reality complementary rather than rival conceptions."

2. Write a short essay on the nature of light.

3. Two collinear harmonic motions of the same frequency have amplitudes of 2 cm and 3 cm respectively and corresponding phase angles of +10° and +30°. Find (a) the amplitude and (b) the phase angle of the resultant vibration.

4. Obtain an expression for the resultant motion of a particle under the simultaneous action of two rectangular motions of the same period.

5. What is sharpness of resonance? Obtain an expression for the sharpness of resonance for a mechanical system and discuss its variation with mechanical resistance.

6. Distinguish between forced vibrations and resonance. Discuss the oscillatory motion of a particle subjected to an external periodic force.

7. A certain mass is acted upon by an elastic force.

   (i) Find its period of motion.
   (ii) How will this period be modified if the vibrating particle also experiences a dissipative force?

8. A particle has a mass of 3gms. It is free to vibrate under the action of an elastic force of 135 dyne-cm^{-1} and a damping force of 10 dyne-cm^{-1} sec. A periodically varying outside force of maximum value 262 dynes is applied to the particle. Find the frequency for displacement resonance, the frequency for velocity resonance, and the approximate amplitude at resonance.

9. A particle of mass 4gm. is subject to an elastic force of 52 dyne-cm^{-1} and a damping force of 14 dyne-cm^{-1} sec. If the motion is oscillatory, find its period.

CHAPTER 7

WAVE MOTION AND LIGHT WAVES

Chapter Outline

7.0 INTRODUCTION

When a particle executes simple harmonic motion in an elastic fluid, the energy and phase changes are not confined to its immediate vicinity but are progressively communicated to distant parts of the medium. Such disturbances propagating in the medium are called waves. We are familiar with many types of waves, for example, the water waves traveling outward when a pebble is dropped into a pond, the sound waves spreading through concert or cinema halls, or the electromagnetic waves comprising the entire solar spectrum traveling through space. Then there are waves of larger amplitudes like sonic bangs or shock waves, detonations, and earthquake waves. All the waves of small amplitude have certain properties in common irrespective of their nature: they travel with finite velocities, are reflected and refracted, show interference and diffraction effects, and exhibit the phenomenon of beats and Doppler shift; only the scale of these properties is different for different type of waves. It is not surprising, therefore, that they can be represented by the same general type of equation subject to conditions that differ in each field. The importance of wave motion, therefore, cannot be overemphasized by pointing out other common properties like dispersion and the existence of phase and group velocities, as for example in water and light waves, or the significance of longitudinal and transverse waves in determining the specific heats of solids. The study of any wave motion involves an understanding of the phenomena of its emission, propagation, and absorption. This becomes a simple matter if the nature of wave motion is known.

7.1 DIFFERENTIAL EQUATION OF WAVE MOTION

Consider the propagation of *pressure waves* in a homogeneous isotropic fluid medium which is perfectly elastic (i.e., there are no dissipative forces such as those due to viscosity). Suppose the mean density ρ and mean pressure "p" have uniform values in the medium. In the presence of a wave propagating from left to right, the instantaneous values of pressure and density at any point will, in general, be different from their mean values; let them be "p'" and ρ' respectively. These variables, instantaneous pressure and density, will have reasonably constant values over an elementary volume of the fluid. We can call such volume elements *particles*. On account of the changes in pressure and density, these particles will have instantaneous velocities. Let u, v, and w be the components of the velocity of any such particle in the x, y, and z directions respectively, as illustrated in Figure 7.1.

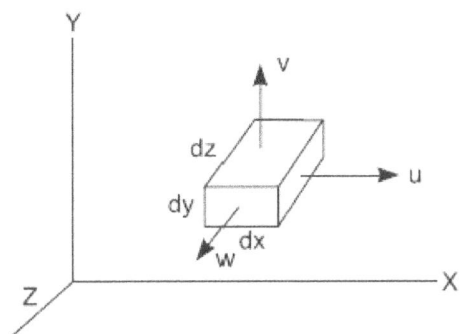

FIGURE 7.1. The components of velocity of a particle.

Consider a volume element of the medium $dxdydz$. The rate of change of the mass inside $dxdydz$ will be given by the difference between rates of influx and efflux of fluid from this element. For the X-direction, this flow is given by

$$\left(\rho'u - \left(\rho'u + \frac{\partial}{\partial x}(\rho'u)dx \right) \right)dydz = -\frac{\partial}{\partial x}(\rho'u)dxdydz \qquad (7.1)$$

Similar expressions can be written for flow along the other two directions so that the net influx in the volume element will be

$$-\left(\frac{\partial}{\partial x}(\rho'u) + \frac{\partial}{\partial y}(\rho'v) + \frac{\partial}{\partial z}(\rho'w) \right)dxdydz \qquad (7.2)$$

According to the law of conservation of mass, this expression must be equal to the rate of change of mass of fluid inside the volume $dxdydz$.
Hence

$$\frac{\partial \rho'}{\partial t} + \frac{\partial}{\partial x}(\rho'u) + \frac{\partial}{\partial y}(\rho'v) + \frac{\partial}{\partial z}(\rho'w) = 0 \qquad (7.3)$$

This is usually known as the equation of continuity.

Since ρ' is changing, we like to express Equation (7.3) in terms of the mean density ρ. Suppose we write

$$\left(\frac{\rho'}{\rho} \right) = 1 + s \qquad (7.4)$$

where s is called the condensation, then

$$\left(\frac{\partial \rho'}{\partial t} \right) = \rho \left(\frac{\partial s}{\partial t} \right) \qquad (7.5)$$

Substituting Equation (7.5) into Equation (7.3) we therefore get

$$\rho\left(\frac{\partial s}{\partial t}\right) + \rho(1+s)\left(\frac{\partial u}{\partial x} + \frac{\partial v}{\partial y} + \frac{\partial w}{\partial z}\right) + \rho\left(u\frac{\partial s}{\partial x} + v\frac{\partial s}{\partial y} + w\frac{\partial s}{\partial z}\right) = 0 \qquad (7.6)$$

If we consider waves of small amplitude, for example, acoustic waves, the condensation "s" has a negligible value as compared to unity (for speech it is of the order of 10^{-7}). Moreover, the acoustic wavelengths for the audible range are so long that all the quantities u, v, and w along with their derivatives change very slowly with x, y, and z. Hence, all the terms in Equation (7.6) involving the products of these small quantities can also be neglected so that we have

$$\left(\frac{\partial s}{\partial t} + \frac{\partial u}{\partial x} + \frac{\partial v}{\partial y} + \frac{\partial w}{\partial z}\right) = 0 \qquad (7.7)$$

in place of Equation (7.6) on taking these factors into account.

We shall now write u, v, and w in terms of the derivatives of a scalar quantity "ϕ"called the *velocity potential*. Suppose $\phi = \phi(x,y,z,t)$

$$u = \frac{\partial \phi}{\partial x}, v = \frac{\partial \phi}{\partial y}, w = \frac{\partial \phi}{\partial z}$$

then the gradients of "ϕ" along the three axes are all vector quantities, although "ϕ" itself is scalar. The introduction of velocity potential at this stage will considerably simplify matters as we proceed. From Equation (7.7), we therefore get

$$\frac{\partial s}{\partial t} + \nabla^2\phi = 0 \qquad (7.8)$$

where the symbol ∇^2 is called the *Laplacian operator* (∇ is pronounced as del and ∇^2 as del square) and has been written in place of

$$\nabla^2(x,y,z) = \frac{\partial^2}{\partial x^2} + \frac{\partial^2}{\partial y^2} + \frac{\partial^2}{\partial z^2}.$$

There is an unbalanced force acting along each of the three directions x, y, and z. From the definition of force (pressure x area), the net force acting in the positive X-direction is given by

$$\left(p' - \left(p' + \frac{\partial p'}{\partial x}dx\right)\right)dydz = -\frac{\partial p'}{\partial x}dxdydz$$

This must be equal to the rate of change of momentum.

$$\frac{\partial}{\partial t}(\rho'u)dxdydz$$

so that the equation of motion becomes

$$\frac{\partial p'}{\partial x} + \frac{\partial(\rho'u)}{\partial t} = 0 \tag{7.9}$$

under the conditions mentioned in connection with the continuity equation

$$\frac{\partial(\rho'u)}{\partial t} = \rho\left(\frac{\partial u}{\partial t}\right) \tag{7.10}$$

Hence, we have from Equations (7.9) and (7.10)

$$\frac{\partial p'}{\partial x} + \rho\left(\frac{\partial u}{\partial t}\right) = 0, \frac{\partial p'}{\partial y} + \rho\left(\frac{\partial v}{\partial t}\right) = 0, \frac{\partial p'}{\partial z} + \rho\left(\frac{\partial w}{\partial t}\right) = 0 \tag{7.11}$$

We can then write

$$\frac{\partial p'}{\partial x}dx + \frac{\partial p'}{\partial y}dy + \frac{\partial p'}{\partial z}dz + \rho\frac{\partial}{\partial t}(udx + vdy + wdz) = 0 \tag{7.12}$$

or

$$dp' + \rho\frac{\partial}{\partial t}(d\phi) = 0$$

where

$$dp' = \frac{\partial p'}{\partial x}dx + \frac{\partial p'}{\partial y}dy + \frac{\partial p'}{\partial z}dz \text{ and } d\phi = \frac{\partial\phi}{\partial x}dx + \frac{\partial\phi}{\partial y}dy + \frac{\partial\phi}{\partial z}dz$$

An integration of Equation (7.12) gives

$$p' + \rho\frac{\partial\phi}{\partial t} = C \tag{7.13}$$

where the constant of integration C can be evaluated from the conditions of the problem. Thus, when no waves are propagating through the medium $\left(\frac{\partial\phi'}{\partial t} = 0\right)$ and $(p' = p)$ so that we get

$$-\rho\left(\frac{\partial\phi}{\partial t}\right) = p' - p \tag{7.14}$$

where the right-hand side of Equation (7.14) is called the *excess pressure* (p_o).

On account of the excess pressure acting upon the volume element, the medium will be under volume strain given by

$$\frac{dV}{V} = -\frac{d\rho'}{\rho} = -\frac{\rho' - \rho}{\rho} = -s \tag{7.15}$$

Hence, the right-hand side of Equation (8.14) can be expressed in terms of the negative condensation and bulk modulus of elasticity K of the medium.

$$p' - p = dp = -\frac{dV}{KV} = Ks \tag{7.16}$$

So that

$$\frac{\partial \phi}{\partial t} = -\frac{Ks}{\rho} = -C^2 s \tag{7.17}$$

where

$$C^2 = \frac{K}{\rho} \tag{7.18}$$

Combining Equations (7.17) and (7.8), we get

$$\frac{\partial^2 \phi}{\partial t^2} = C^2 \nabla^2 \phi \tag{7.19}$$

and the access pressure

$$p_o = \rho C^2 s \tag{7.20}$$

Equation (7.19) is the three-dimensional acoustic wave equation applicable to fluids. Its solutions give the propagation of the velocity potential ϕ, the velocity of propagation C being given by

$$C = \sqrt{\frac{\text{Bulk modulus}}{\text{Density}}} \tag{7.21}$$

7.2 PLANE WAVE EQUATION

If the deformations in the medium are a function of one *Cartesian* coordinate, the wave is said to be plane. In such waves, conditions are uniform over a plane specified by the previous space coordinate. Thus, the differential equation of a plane wave of infinite width propagating in the X-direction can be written as

$$\frac{\partial^2 \phi}{\partial t^2} = C^2 \frac{\partial^2 \phi}{\partial x^2} \tag{7.22}$$

The general solution of this *wave equation* may be written in the form

$$\phi = f_1(ct - x) + f_2(ct + x) \tag{7.23}$$

It can be seen by substitution that irrespective of the nature of functions f_1 and f_2, Equation (8.23) is always a solution of Equation (8.22). Thus, a single scalar quantity ϕ is sufficient to represent the propagation of a longitudinal disturbance in any fluid medium. If this disturbance is harmonic, the solution is written in the form

$$\phi = Ae^{-i(\omega t - kx)} + Be^{-i(\omega t + kx)} \tag{7.24}$$

where A is the complex amplitude of a plane wave traveling in the positive X-direction and B that of a wave traveling in the negative X-direction, both propagating with the velocity $C = \left(\dfrac{\omega}{k}\right)$.

In the case of acoustic waves where A and B are real, we take the real part of Equation (7.24). Thus, we have

$$\phi = A\cos(\omega t - kx) + B\cos(\omega t + kx) \tag{7.25}$$

$$p_o = -\rho \frac{\partial \phi}{\partial t} = \rho A\omega \sin(\omega t - kx) + \rho B\omega \sin(\omega t + kx) \tag{7.26}$$

$$u = \frac{\partial \phi}{\partial x} = KA\sin(\omega t - kx) + KB\sin(\omega t + kx) \tag{7.27}$$

The particle displacement ξ defined by the equation $\dfrac{d\xi}{dt} = u$ is thus seen to be

$$\xi = \int u \, dt = \int KA\sin(\omega t - kx) \, dt - KB\sin(\omega t + kx) \, dt \tag{7.28}$$

or

$$\xi = -\frac{A}{C}\cos(\omega t - kx) + \frac{B}{C}\cos(\omega t + kx) \tag{7.29}$$

A wave traveling in the positive X-direction has been seen to be represented by the equation

$$\phi = A\cos(\omega t - kx) \tag{7.30}$$

The time taken by one complete wave to pass any point is called its periodic time T. Thus

$$T = \frac{2\pi}{\omega} \tag{7.31}$$

The frequency v is the number of waves passing a stationary observer in one second, so that

$$v = \frac{1}{T} = \frac{\omega}{2\pi} \tag{7.32}$$

The wave profile repeats itself at regular intervals λ given by

$$\lambda = \frac{2\pi}{k} = \frac{2\pi C}{\omega} \tag{7.33}$$

which on combination with the expression for frequency gives

$$C = v\lambda \tag{7.34}$$

All these relations can be used in writing down alternative forms of the wave equation (7.30). These are summarized as follows

$$\phi = A\cos 2\pi\left(\frac{t}{T} - \frac{x}{\lambda}\right) \tag{7.35a}$$

$$\phi = A\cos\left(\left(\frac{2\pi}{\lambda}\right)(Ct - x)\right) \tag{7.35b}$$

$$\phi = A\cos 2\pi\left(vt - \frac{x}{\lambda}\right) \tag{7.35c}$$

An inspection of Equations (7.30) and (7.35 a, b, c) reveals that at any time t different particles of the medium (different x's) are in different states of disturbance. As time passes every single particle of the medium goes through the same cycle of change of state. Such waves are called progressive waves. A few features of plane progressive waves have been shown in Figure 7.2. The analytical expressions for these variations have already been derived. Figure 7.2 indicates that the particle velocity, the excess pressure, and the condensation are all in phase. But all these quantities are $\left(\dfrac{\pi}{2}\right)$ out of phase with the displacement of the particle. Accordingly, particle velocity, condensation, and excess pressure shall attain their maximum values only when the local cross-section of the medium is passing through its central position.

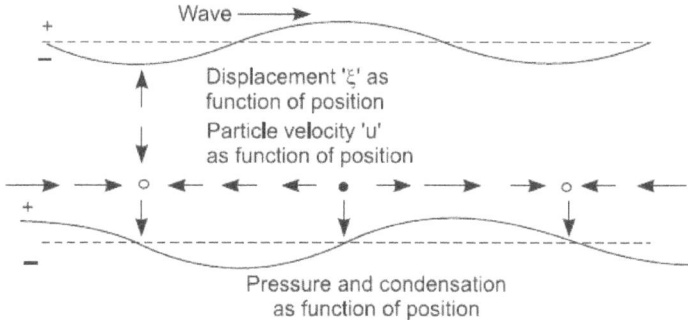

Wave ⟶

Displacement 'ξ' as function of position

Particle velocity 'u' as function of position

Pressure and condensation as function of position

FIGURE 7.2. Features of plane progressive waves.

7.3 ENERGY DENSITY OF PLANE WAVES

When a progressive wave travels through a fluid, successive particles of the medium possess local velocities at any instant. In addition, the medium is in a state of compression or expansion due to the passage of the acoustic wave. The energy involved in wave propagation is therefore of two forms—kinetic and potential. The kinetic energy of the moving particles is

$$\Delta E_{kin} = \frac{1}{2}\rho u^2 \Delta V \tag{7.36}$$

where ΔV is the volume element which can be treated effectively as a particle. The potential energy of this volume element is

$$\Delta E_{pot} = -\int p_o dV \tag{7.37}$$

where the negative sign indicates an increase in potential energy due to compression and decrease due to expansion. In this expression the excess pressure p_o has been seen to be

$$p_o = \rho C^2 s \tag{7.38}$$

and the condensation s is defined as

$$s = -\frac{\text{Change in volume}}{\text{Original volume}} = -\frac{\int dV}{\Delta V} \tag{7.39}$$

Hence, the expression for potential energy becomes

$$\Delta E_{pot} = +\int \rho C^2 s ds (\Delta V) = \frac{1}{2}\rho C^2 s^2 (\Delta V) \tag{7.40}$$

where dV has been written as
$$dV = ds(\Delta V)$$

The total acoustic energy of the volume element is therefore

$$\Delta E = \Delta E_{kin} + \Delta E_{pot} = \Delta E_{pot} = \frac{1}{2}\rho \Delta V\left(u^2 + C^2 s^2\right) \qquad (7.41)$$

and the energy density per unit volume is

$$e = \frac{\Delta E}{\Delta V} = \frac{1}{2}\rho\left(u^2 + C^2 s^2\right) \text{ ergs./c.c.} \qquad (7.42)$$

It can be shown that
$u = Cs$ for positive traveling waves, and $u = -Cs$ for negative traveling waves.

The instantaneous energy density corresponding to the presence of both waves is accordingly

$$e = e_+ + e_- = \rho\left(u_+^2 + u_-^2\right) \qquad (7.43)$$

where the subscripts + and – refer to the two waves traveling in opposite directions.

According to Equation (7.43), the particle velocity of either wave is a function of both space and time. The energy density of either wave is therefore not constant throughout the medium. We can nevertheless calculate either the time average of energy density at any point or the space average of energy density at any time due to the motion of either wave. Thus, for example, the time average density at a point is given by

$$e_{tad} = \frac{1}{T}\int_0^T e\, dt \qquad (7.44)$$

where the integration on the right-hand side is to be performed over one complete cycle. Substituting the value of e, we get

$$e_{tad} = \frac{1}{T}\int_0^T \left(kA\sin(\omega t - kx)\right)^2 dt = \frac{1}{2}\left(\rho k^2 A^2\right) \qquad (7.45)$$

Similarly, it can be shown that the time average of energy density due to a negatively traveling wave is given by

$$e_{tad} = \frac{1}{2}\left(\rho k^2 B^2\right) \qquad (7.46)$$

If we obtain the space average densities at any time due to individual waves, we find that these averages are the same as the respective time averages.

There are alternative expressions for space and time averages of energy densities at a point in terms of the velocity amplitude (kA), pressure amplitude $(P = kA\rho C)$, and displacement amplitude $\left(\varepsilon = \dfrac{kA}{\omega} \right)$, which readily follow from Equations (7.45) and (8.46).

7.4   ACOUSTIC INTENSITY

The acoustic intensity of a pressure wave is defined as the average amount of energy flowing per second through a unit area held normal to the direction of propagation. It is evident that all the energy contained in a length C of the medium will pass through a unit area in one second and, consequently, sound intensity can be written as

$$I = Ce_{tad} = \frac{1}{2}\left(\rho Ck^2 A^2 \right) \text{ or } I = \frac{1}{2}\left(\rho Ck^2 B^2 \right) \tag{7.47}$$

as the case may be. This intensity can be measured in ergs per second per square cm or watts per sq. cm. But it is usual to measure it with reference to some acoustic intensity standard. Since the range of sound intensities encountered in practice is rather large, it is more convenient to use a logarithmic scale. An added reason for this is the fact that the response of human ears to sound intensity follows an approximately logarithmic scale.

Thus, the intensity level of a sound of intensity I is defined as

$$n = 10\log\left(\frac{I}{I_o} \right) \tag{7.48}$$

where I_o is the reference intensity and n is measured decibels (*db*). Intensity levels are also sometimes expressed in terms of pressure amplitude of the sound wave

$$n = 20\log\left(\frac{P}{P_o} \right) \tag{7.49}$$

where P_o is some reference pressure. This definition is usually followed in recording and reproduction of sound.

The intensity of a 1000 cps pure note just audible to human ears is sometimes taken as reference intensity I_o for plane waves propagating in air. This works out to be 10^{-16} watt/cm². But the commonly used standard for underwater measurements is a root mean square pressure of 1 dyne/cm². This has also been adopted for airborne sound. A reference pressure of 1 dyne/cm² corresponds to a reference intensity of 6.75×10^{-6} watts/sec. in fresh water and 2.41×10^{-2} watts/sec. in air.

7.5 SPECIFIC ACOUSTIC IMPEDANCE

The ratio of acoustic pressure in a medium to the associated particle velocity is called the specific acoustic impedance Z of the medium. Thus

$$Z = \frac{\left(-\rho \dfrac{\partial \phi}{\partial t} \right)}{\left(\dfrac{\partial \phi}{\partial x} \right)} \tag{7.50}$$

These derivatives can be calculated for the positively or negatively traveling plane waves with the help of Equation (7.24). For plane waves traveling in the positive X-direction

$$Z_+ = \rho C \tag{7.51}$$

and for those traveling in the opposite direction

$$Z_- = -\rho C \tag{7.52}$$

while the specific acoustic impedance is seen to be a real quantity. In the case of plane waves, it is in general a complex quantity of the form

$$Z = r + is \tag{7.53}$$

where r is the specific acoustic resistance and s is the specific acoustic reactance of the medium for the wave motion under study. Thus, for plane waves, specific acoustic reactance is seen to be zero.

7.6 SPHERICAL ACOUSTIC WAVES, TELEGRAPHY, AND COMPLEX QUANTITIES

The wave Equation (7.19) can be expressed in coordinate systems other than Cartesian. Each of these systems is suitable under different situations. We have already seen how a rectangular coordinate system can be used for describing plane progressive waves. If there is a long cylindrical source of sound, it is usually more advantageous to express Equation (7.19) in cylindrical coordinates. On the other hand, if the source approximates a point, we should choose spherical coordinates. All that needs to be done is the expressing of $\nabla^2 \phi$ in either cylindrical coordinates (r, θ, z) or spherical coordinates (r, θ, ψ). For example, Figure 7.3 shows the spherical system where

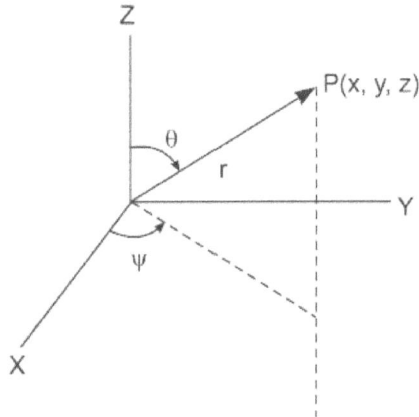

FIGURE 7.3. Spherical coordinates.

$$x = r \sin\theta \cos\psi \tag{7.54a}$$

$$y = r \sin\theta \sin\psi \tag{7.54b}$$

$$z = r \cos\theta \tag{7.54c}$$

and

$$r = \sqrt{x^2 + y^2 + z^2} \tag{7.55a}$$

$$\theta = \tan^{-1}\left(\frac{\sqrt{x^2 + y^2}}{z}\right) \tag{7.55b}$$

$$\psi = \tan^{-1}\left(\frac{y}{x}\right) \tag{7.55c}$$

If ϕ is a function of r and t only, the system is said to have spherical symmetry. We shall then have

$$\nabla^2 \phi = \frac{\partial^2 \phi}{\partial t^2} + \frac{2}{r}\frac{\partial \phi}{\partial t} + \frac{1}{r^2 \sin\theta}\frac{\partial}{\partial \theta}\left(\sin\theta\frac{\partial \phi}{\partial \theta}\right) + \frac{1}{r^2 \sin^2\theta}\frac{\partial^2 \phi}{\partial \psi^2} \tag{7.56}$$

and

$$\nabla^2 \phi = \frac{\partial^2 \phi}{\partial r^2} + \frac{2}{r}\frac{\partial \phi}{\partial r} = \frac{1}{r}\frac{\partial^2}{\partial r^2}(r\phi) \tag{7.57}$$

and Equation (7.57) reduces to

$$\frac{1}{r}\frac{\partial^2}{\partial t^2}(r\phi) = \frac{C^2}{r}\frac{\partial^2}{\partial r^2}(r\phi) \tag{7.58}$$

where the factor "r" on the left-hand side of this equation is not a function of time so that

$$\frac{1}{r}\frac{\partial^2}{\partial t^2}(r\phi) = \frac{\partial^2 \phi}{\partial t^2} \tag{7.59}$$

The differential equation of spherical waves is therefore

$$\frac{\partial^2}{\partial t^2}(r\phi) = C^2 \frac{\partial^2}{\partial r^2}(r\phi) \tag{7.60}$$

The solution of Equation (7.60) can be written at once if we consider "$r\phi$" as a single variable so that

$$r\phi = f_1(ct - r) + f_2(ct + r) \text{ or}$$

$$\phi = \frac{1}{r}\left(f_1(ct - r) + f_2(ct + r)\right) \tag{7.61}$$

where the first term of the right-hand side represents a wave diverging from a central point and the other another wave converging on this point. A harmonic solution of the wave equation can be written as

$$\phi = \frac{1}{r}\left(Ae^{-i(\omega t - kr)} + Be^{-i(\omega t + kr)}\right) \tag{7.62}$$

As before we can derive expressions for energy density, specific acoustic impedance, and so on for spherical waves. It may be mentioned that spherical waves tend to behave like plane waves at considerable distances from the source.

In the derivation and discussions of the wave equations and (7.61) we had assumed that there were no dissipative forces present in the medium. In the presence of such dissipative forces, we have to take into account the effect of damping by introducing a term $\left(r \dfrac{\partial \phi}{\partial t} \right)$ in the wave equation. Thus, we write

$$\nabla^2 \phi = \frac{1}{C^2} \left(\frac{\partial^2 \phi}{\partial t^2} + r \frac{\partial \phi}{\partial t} \right) \tag{7.63}$$

An equation similar in form to Equation (7.63) is of great importance in electromagnetic theory. This equation can be solved by proper substitutions and gives an amplitude factor decaying exponentially with time. Thus, the solution for a one-dimensional equation is of the form

$$\phi = e^{-\left(\frac{rt}{2} \right)} f_1 \left(x - ct \right) \tag{7.64}$$

We shall be dealing mostly with harmonic waves, and therefore it is appropriate at this stage to understand a notation used for representing such waves. We have seen in Equation (7.30) that a plane progressive wave may be represented by

$$\phi = a \cos \left(\omega t - kx \right) \tag{7.65}$$

where the velocity potential ϕ has an amplitude "a." If we also introduce a phase δ we can write

$$\phi = a \cos \left(\left(\omega t - kx \right) + \delta \right) \tag{7.66}$$

The trigonometric function on the right-hand side is seen to be the real part of the *complex quantity*

$$\phi = a e^{\pm i \left(\left(\omega t - kx \right) + \delta \right)} \tag{7.67}$$

It is more convenient to use the negative sign in the previous expression. We can also write Equation (7.67) as

$$\phi = a e^{-i\delta} e^{-i \left(\omega t - kx \right)} \tag{7.68a}$$

or

$$\phi = Ae^{-i(\omega t - kx)} \qquad (7.68b)$$

where the apparent amplitude "A" is a complex quantity defined by

$$A = ae^{-i\delta} \qquad (7.69)$$

So that the true amplitude is $|A| = a$, and the negative argument of A i.e., δ gives the phase. The intensity which is proportional to a^2 can be obtained by finding the product of A and its complex conjugate A^*. Thus

$$AA^* = \left(ae^{-i\delta}\right)\left(ae^{+i\delta}\right) = a^2 \qquad (7.70)$$

The scalar quantity ϕ expressed as in Equation (7.67) is itself a solution of the one dimensional wave equation (this we can easily verify by direct substitution). Hence, in any equation in which ϕ appears in the first degree, we can use Equation (7.68), as it is without reference to its real or imaginary parts, and later on extract the part of the final result of interest to us.

The great advantage of using Equation (7.68) is that if we want to compose a large number of coherent disturbances (disturbances in the same phase or having a constant phase difference), we have simply to add up the complex amplitude, because the other term $e^{-i(\omega t - kx)}$ will be common to all of them. But for composing incoherent disturbances, we shall have to add individual intensities.

7.7 REFLECTION AND REFRACTION OF WAVES

Normal Incidence. When a plane progressive wave propagating in a fluid medium strikes its boundary with a second medium, a reflected wave is produced in the first medium and a transmitted or refracted wave is produced in the second. The ratio of the intensities of the reflected and transmitted waves is determined by the properties (say acoustic impedance in the case of pressure waves) of the two media and the angle of incidence at the boundary. Figure 7.4 illustrates reflection at normal incidence. Suppose BB' represents the boundary between two media *I* and *II* at which an acoustic plane wave is incident normally.

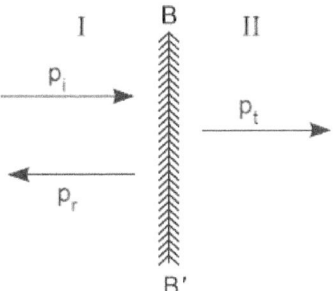

FIGURE 7.4. Reflection at normal incidence.

This wave may be represented in terms of the *acoustic pressure* p_i (rather than velocity potential) by an equation of the form

$$p_i = A_i e^{-i(\omega t - k_1 x)} \tag{7.71a}$$

where k_1 is the reduced wave number and A_i the pressure amplitude in the first medium. The reflected and transmitted waves will then be given by

$$p_r = A_r e^{-i(\omega t + k_1 x)} \tag{7.71b}$$

and

$$p_t = A_t e^{-i(\omega t - k_2 x)} \tag{7.71c}$$

where k_2 is the reduced wave number in the second medium, and the subscripts r and t stand for the reflected and transmitted waves. It may be noted that the frequency of the transmitted wave is the same as that of incident wave; it is only the wavelength $\lambda = \dfrac{2\pi}{k}$ which undergoes a change with the change in medium.

Let us suppose that the boundary coincides with the coordinate $x = 0$. The conditions at the boundary can then be written as

$$p_i = A_i e^{-i\omega t} \tag{7.72a}$$

$$p_r = A_r e^{-i\omega t} \tag{7.72b}$$

$$p_t = A_t e^{-i\omega t} \tag{7.72c}$$

Since there cannot exist any discontinuity of pressure in a fluid medium, the pressure in the first medium must be equal to that in the second medium at the surface of separation between the two. We thus have

$$p_i + p_r = p_t \qquad (7.73a)$$

or

$$A_i + A_r = A_t \qquad (7.73b)$$

Also, if the two media are to remain in contact with each other, the resultant particle velocity in the first medium should be equal to that in the second. Thus

$$u_i + u_r = u_t \qquad (7.74a)$$

or

$$\frac{1}{\rho_1 c_1}(p_i - p_r) = \frac{1}{\rho_2 c_2} p_t \qquad (7.74b)$$

where $\rho_1 c_1$ and $\rho_2 c_2$ are respectively the specific acoustic impedances in media I and II. Equation (7.74b) can also be written as

$$r_{12} = (A_i - A_r) = A_t \qquad (7.75)$$

where $r_{12} = \dfrac{\rho_1 c_1}{\rho_2 c_2}$ is the relative impedance of the second medium with respect to the first. Combining Equations (7.73b) and (7.75), we get

$$\frac{A_r}{A_i} = \frac{r_{12} - 1}{r_{12} + 1} \qquad (7.76a)$$

and

$$\frac{A_t}{A_i} = \frac{2r_{12}}{r_{12} + 1} \qquad (7.76b)$$

If $r_{12} > 1$, the reflected wave at the boundary is in the same phase as the incident wave, but if $r_{12} < 1$, the two waves will be in opposite phase. Expressed differently, it means that if the characteristic impedance of the second medium is greater than that of the first, reflection will take place without any change of phase at the boundary. This condition is encountered when a wave is reflected at the boundary while going from a rarer to a denser medium (e.g., air and water). On the contrary, if reflection takes place when a wave traveling from a denser to a rarer medium encounters the boundary, a 180° change of phase is introduced.

There are a few special cases of the phenomenon. If $r_{12} \rightarrow 0$, the amplitudes of the incident and reflected waves tend to be the same, and the two waves will be 180° out of phase. If $r_{12} \to \infty$, as for example in the case of reflection at a very rigid boundary, the two waves will again have the same amplitude and will be in the same phase as well. In both these cases standing waves will be produced, but the state of disturbance at the boundary will not be the same. If $r_{12} = 1$, that is, the two media have the same acoustic impedance, there is no reflection at all. Under this condition all the incident energy will be totally transmitted into the second medium. This important result is often made use of, for example, in developing special rubbers for protecting the surface of transducers against the corrosive action of water in the equipment for underwater sounding. The transmitted wave, however, remains always in the same phase as the incident wave, and its amplitude may range from $2A$ to zero depending upon the value of r_{12}.

7.8   OBLIQUE INCIDENCE

Let us now take the general case. Suppose AO is the direction of the propagation of plane waves which lies in the XY plane as shown in Figure 7.5.

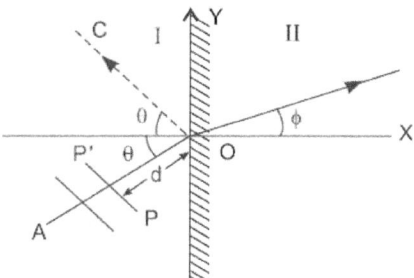

FIGURE 7.5. Reflection at oblique incidence.

A plane such as PP', perpendicular to AO, will then give the locus of all points in the same phase of vibration; this is called a *wavefront*. The equation of this plane is

$$d = x\cos\theta + y\sin\theta \qquad (7.77)$$

where d is the distance of the plane from the Z-axis measured in the direction of the propagation of the waves. The excess pressure in the *wavefront* can then be written as

$$p_i = A_i e^{-i(\omega t - k_i d)} \qquad (7.78a)$$

or

$$p_i = A_i e^{-i\left(\omega t - k_1 x \cos\theta - k_1 y \sin\theta\right)} \tag{7.78b}$$

where k_1 is the reduced wave number in the first medium. This plane *wavefront* on striking the boundary between the two media will be reflected and transmitted.

Let the excess pressure p_r and p_t in the reflected and transmitted waves be represented by

$$p_r = A_r e^{-i\left(\omega t - k_1 d_r\right)} \tag{7.78c}$$

and

$$p_t = A_t e^{-i\left(\omega t - k_2 d_t\right)} \tag{7.78d}$$

where the reflected and the transmitted plane *wavefronts* have the equations

$$d_r = l_1 x + m_1 y + n_1 z \tag{7.79a}$$

and

$$d_t = l_2 x + m_2 y + n_2 z \tag{7.79b}$$

respectively. We then have

$$p_r = A_r e^{-i\left(\omega t - k_1 \left(l_1 x + m_1 y + n_1 z\right)\right)} \tag{7.80a}$$

and

$$p_t = A_t e^{-i\left(\omega t - k_2 \left(l_2 x + m_2 y + n_2 z\right)\right)} \tag{7.80b}$$

for the reflected and transmitted components.

Applying the condition for continuity of pressure at the boundary ($x = 0$), we have

$$A_i e^{-i\left(\omega t - k_1 y \sin\theta\right)} + A_r e^{-i\left(\omega t - k_1 \left(m_1 y + n_1 z\right)\right)} = A_t e^{-i\left(\omega t - k_2 \left(m_2 y + n_2 z\right)\right)} \tag{7.81}$$

This identity can only be satisfied if the indices of all three terms are identical, that is:

$$\omega t - k_1 y \sin\theta = \omega t - k_1 \left(m_1 y + n_1 z\right) = \omega t - k_2 \left(m_2 y + n_2 z\right) \tag{7.82}$$

or

$$k_1 \sin\theta = k_1 m_1 = k_2 m_2 \tag{7.83}$$

or

$$0 = k_1 n_1 = k_2 n_2 \tag{7.84}$$

Equation (7.84) shows that $n_1 = n_2 = 0$, so the reflected and transmitted rays are found to be in the plane of incidence XY. The first part of Equation (7.83) shows that

$$m_1 = \sin\theta \tag{7.85}$$

Hence, the angles of incidence and reflection are seen to be equal. The second of Equation (7.83) reveals that

$$k_1 \sin\theta = k_2 \sin\phi \tag{7.86}$$

which is the well-known *Snell's law of refraction*.

Under these conditions, Equation (7.81) reduces to

$$A_i + A_r = A_t \tag{7.87}$$

As before we can write the condition for continuity of the X-component of velocity at the boundary

$$U_i \cos\theta + U_r \cos(\pi - \theta) = U_t \cos\phi \tag{7.88}$$

where the particle velocities can be written in terms of the acoustic impedances of the media. Equations (7.87) and (7.88) may then be combined to give the reflected and transmitted fractions of the incident pressure amplitude.

7.9 SUPERPOSITION OF WAVES

It is the property of linear differential equations that the algebraic sum of all its solutions is also one of its solutions. We shall prove this interesting result for harmonic waves. Such waves satisfy the differential equation

$$\frac{\partial^2 \phi}{\partial t^2} = c^2 \nabla^2 \phi \tag{7.89}$$

which is linear, as it does not contain any powers higher than unity of the velocity potential ϕ or any of its derivatives, namely $\dfrac{\partial \phi}{\partial t}, \dfrac{\partial^2 \phi}{\partial t^2}$, or, $\dfrac{\partial \phi}{\partial x}, \dfrac{\partial^2 \phi}{\partial x^2}$, and so on. In addition, it is a homogeneous equation, as it does not contain any term independent of ϕ. If ϕ_1, ϕ_2, and so forth are the solutions of this equation, we can write

$$\phi_1 = A_1 e^{-i(\omega_1 t - k_1 x)} \tag{7.90a}$$

$$\phi_2 = A_2 e^{-i(\omega_2 t - k_2 x)} \tag{7.90b}$$

If the individual harmonic disturbances ϕ_1, ϕ_2, and so on travel through the medium with the same velocity

$$\frac{\omega_1}{k_1} = \frac{\omega_2}{k_2} = \cdots = \frac{\omega_n}{k_n} = c \tag{7.91}$$

Equation (7.90) can then be written as

$$\phi_1 = A_1 e^{-ik_1(ct-x)} \tag{7.92a}$$

$$\phi_2 = A_2 e^{-ik_2(ct-x)} \tag{7.92b}$$

and so on. Therefore, the algebraic sum of all the individual solutions may be written as

$$\phi_1 + \phi_2 + \cdots = A_1 e^{-ik_1(ct-x)} + A_2 e^{-ik_2(ct-x)} + \cdots \tag{7.93}$$

If the left-hand side of this equation is represented by ψ and the right-hand side by a function $F(ct-x)$, we can write

$$\psi = F(ct - x) \tag{7.94}$$

We now have to show that given by Equation (7.94) is a solution of the classical wave Equation (7.89). Successive partial differentiations of Equation (7.94) lead to the following results.

$$\frac{\partial \psi}{\partial t} = \frac{dF}{d(ct-x)} \times \frac{\partial(ct-x)}{\partial t} = c\frac{dF}{d(ct-x)} \tag{7.95}$$

$$\frac{\partial^2 \psi}{\partial t^2} = c\frac{d^2 F}{d(ct-x)^2} \times \frac{\partial(ct-x)}{\partial t} = c^2 \frac{d^2 F}{d(ct-x)^2} \tag{7.96}$$

Similarly,

$$\frac{\partial \psi}{\partial x} = \frac{dF}{d(ct-x)} \times \frac{\partial(ct-x)}{\partial x} = -\frac{dF}{d(ct-x)} \tag{7.97}$$

and

$$\frac{\partial^2 \psi}{\partial x^2} = -\frac{d^2 F}{d(ct-x)^2} \times \frac{\partial(ct-x)}{\partial x} = \frac{d^2 F}{d(ct-x)^2} \tag{7.98}$$

We thus find that

$$\frac{\partial^2 \psi}{\partial t^2} = c^2 \frac{\partial^2 \psi}{\partial x^2} \qquad (7.99)$$

so that ψ satisfies the wave Equation (7.89). Hence, we arrive at the interesting result that if two or more disturbances propagate through the same medium simultaneously with the same velocity, the algebraic sum of the individual disturbances shall also travel through the medium with the same velocity, and hence may be taken as the resultant disturbance. This is the general enunciation of the *principle of superposition* first propounded by Young. In other words, when a non-dispersive medium is disturbed simultaneously by any number of waves, the instantaneous resultant displacement of the medium at every point at every instant is the algebraic sum of the displacements taken independently.

7.10 SUPERPOSITION OF WAVES OF THE SAME FREQUENCY

(i) Traveling along the same direction. Consider two plane progressive waves of the same frequency and wavelength traveling approximately along the same direction in a medium as shown in Figure 7.6. Let their plane wave fronts be located at any instance at A and B.

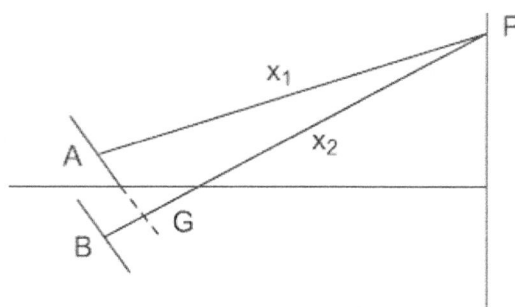

FIGURE 7.6. Superposition of waves traveling along the same direction.

For reaching P, they have to travel distances x_1 and x_2, and hence the disturbance at P due to the individual waves may be written as

$$\xi_1 = A\cos(\omega t - kx_1) \qquad (7.100a)$$

and

$$\xi_2 = A\cos(\omega t - kx_2)$$ (7.100b)

Where ξ represents the instantaneous displacement at any point of the medium whose maximum value A has been taken to be the same for both waves for the sake of simplicity. According to the principle of superposition, the resultant displacement at P is given by

$$\xi = \xi_1 + \xi_2 = R\left(Ae^{-i(\omega t - kx_1)} + Ae^{-i(\omega t - kx_2)}\right) = R\left(Ae^{-i\omega t}\left(e^{-ikx_1} + e^{-ikx_2}\right)\right)$$ (7.101)

If $x = \dfrac{1}{2}(x_1 + x_2)$, Equation (7.101) reduces to

$$\xi = R\left[Ae^{-i(\omega t - kx)}\left(e^{+i\frac{1}{2}k(x_2 - x_1)} + e^{-i\frac{1}{2}k(x_2 - x_1)}\right)\right]$$ (7.102)

or

$$\xi = R\left[2A\cos\frac{1}{2}k(x_2 - x_1)e^{-i(\omega t - kx)}\right]$$ (7.103)

The resultant displacement at P is therefore obtained by extracting the real part from Equation (7.103), and may be written as

$$\xi = 2A\cos\frac{1}{2}k(x_2 - x_1)\cos(\omega t - kx)$$ (7.104)

Equation (7.104) represents a plane harmonic wave of the same frequency and wavelength as the individual waves. Its amplitude, however, varies periodically due to the cosine term. The resultant amplitude at P, given by

$$Resultant\ Amplitude = 2A\cos\frac{1}{2}k(x_2 - x_1)$$ (7.105)

will therefore depend upon the two path lengths which the wave trains cover to reach P.

If $\dfrac{k}{2(x_2 - x_1)} = \left(\dfrac{\pi}{\lambda}\right)(x_2 - x_1) = n\pi$, or $x_2 - x_1 = n\lambda$, where $n = 0, 1, 2, \ldots$

the amplitude of disturbance will be a maximum. But if

$$\dfrac{k}{2(x_2 - x_1)} = \left(\dfrac{\pi}{\lambda}\right)(x_2 - x_1) = (2n + 1)\left(\dfrac{\pi}{2}\right), \text{ or } x_2 - x_1 = \left(n + \dfrac{1}{2}\right)\lambda,$$

the amplitude of disturbance will be zero at the point under reference.

Thus, we find that under certain conditions, two disturbances traveling through a medium can completely destroy each other. This is a very important property of all wave motions and is frequently encountered in acoustics and optics.

(ii) Stationary Waves. Another important example of the superposition principle is that of two waves traveling in the same medium simultaneously but in opposite directions. This may be the case when a direct wave is reflected at either a rigid surface (closed end) or a free surface (open end). In terms of the particle displacement, these waves may be represented as

$$\xi_1 = R\left(ae^{-i(\omega t - kx)}\right) \tag{7.106a}$$

and

$$\xi_2 = \pm R\left(ae^{-i(\omega t + kx)}\right) \tag{7.106b}$$

where the positive sign in Equation (7.106b) is to be taken in the case of reflection at an open end and negative in the case of reflection at a closed end. It may be recalled that while acoustic pressure in the direct and reflected waves remains in the same phase on reflection at a rigid surface, the particle displacement becomes 180° out of phase. We shall take the case of reflection at a rigid surface. The displacement at any point in front of the surface under the simultaneous action of the two waves will be given by

$$\xi = \xi_1 + \xi_2$$

$$\xi = R\left\{\left(ae^{-i\omega t}\right)\left(e^{ikx} - e^{-ikx}\right)\right\} = R\left\{\left(2ia\sin kx\right)e^{-i\omega t}\right\}$$

$$\xi = +2a\sin kx \sin \omega t \tag{7.107}$$

The particle velocity u and condensation s will then be given by

$$u = \xi' = 2a\omega \sin kx \cos \omega t \tag{7.108}$$

and

$$s = -\frac{d\xi}{dx} = -2ak\cos kx \sin \omega t \tag{7.109}$$

It is thus seen from Equations (7.107, 7.108, 7.109) that the displacement, particle velocity, and condensation at various points of the medium change harmonically with time. Also, at any instant, these physical properties have different values at different points and repeat themselves at regular intervals.

Whatever the value of t, the particle velocity "u" is always zero for points at which

$$\sin kx = 0 \text{ or } \cos kx = 1$$

or

$$kx = \pm n\pi, \text{ where } n = 0, 1, 2, 3, \ldots$$

That is, $x = 0, \pm\dfrac{\lambda}{2}, \pm\lambda, \pm\dfrac{3\lambda}{2}, \ldots$

Figure 7.7 illustrates the phase relationship for plane waves traveling in the positive and negative X-directions respectively.

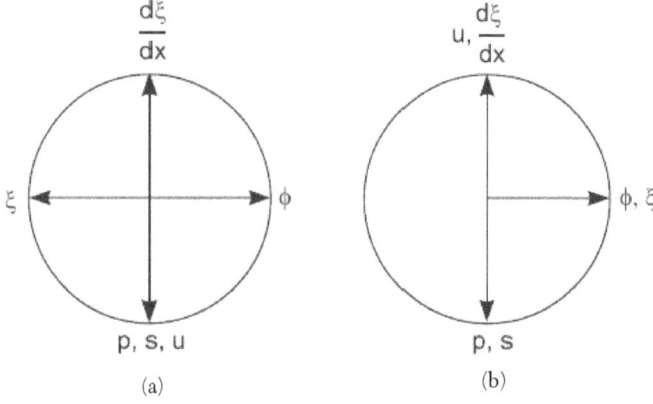

FIGURE 7.7. Phase relationship for plane waves traveling in (a) positive and (b) negative X-directions.

If the distance x is to be measured from the rigid surface, the particle velocity u and also the displacement ξ will be zero at the surface itself and at points $\left(-\dfrac{n\lambda}{2}\right)$ in front of the surface. These values and positions are independent of time; such points are called *nodes*, and the planes passing through them in a direction perpendicular to the direction of propagation of the incident wave are called *nodal planes*.

7.11 WAVE PACKETS

Let consider an oscillator switched on at any instant. It takes some time, however small, in building up to constant amplitude. Suppose it is allowed to oscillate for some time and is then switched off so that the oscillations die out. This process is illustrated in Figure 7.8.

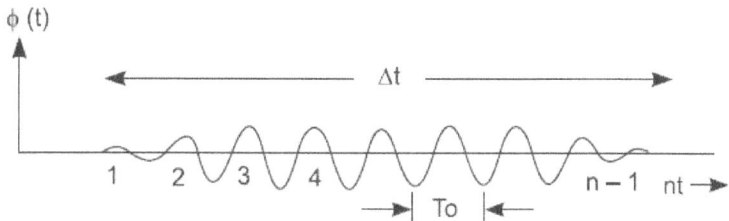

FIGURE 7.8. The process of an oscillator switched on at any instant then switched off.

It is seen that the steady angular frequency "$\omega = \omega_o$" predominates. This can be determined by counting the number of cycles "n" during the period "Δt" for which the oscillator was on.

Thus,

$$\frac{\omega_o}{2\pi} = v_o = \frac{n}{\Delta t} \tag{7.110}$$

Nevertheless, there is an inherent uncertainty in the determination of "n." Looking at Figure 7.8, it is seen that the uncertainty in the determination of "n" amounts to $\left(\pm\dfrac{1}{2}\right)$ cycles at each end of the pulse. Thus, an uncertainty bandwidth "Δv" creeps in, and this is given by

$$\Delta v = \frac{\Delta n}{\Delta t} \tag{7.111}$$

and because "$\Delta n v$" is of the order of one cycle, we have

$$\Delta v \sim \frac{1}{\Delta t} \tag{7.112}$$

The traveling waves propagating in the medium in consequence of the pulse shown in Figure 7.8 are said to form a wave packet. Such wave packets travel with the group velocity "U." Due to the presence of a band of frequencies "$\Delta\omega$" emitted by the source, the wave packet will contain a band of wave numbers "Δk" if the medium is dispersing. This band is centered about "k_o," the wave number corresponding to the dominant frequency "ω_o," and has a width

$$\Delta k = \frac{\Delta\omega}{V_g} \tag{7.113}$$

A wave packet of length "Δx" moves across a stationary observer in time "Δt" such that

$$\Delta x = V_g \Delta t \tag{7.114}$$

It is thus seen that

$$(\Delta k)(\Delta x) = (\Delta \omega)(\Delta t) \tag{7.115}$$

or

$$\Delta\left(\frac{k}{2\pi}\right)(\Delta x) = \Delta\left(\frac{\omega}{2\pi}\right)(\Delta t) \tag{7.116}$$

or

$$\Delta\left(\frac{1}{\lambda}\right)(\Delta x) = \Delta(v)(\Delta t) \sim 1 \tag{7.117}$$

We once again get the uncertainty relation; this is seen with a wave packet in space.

7.12 ELECTROMAGNETIC WAVES

Prior to the birth of electromagnetic theory, light waves were thought to be associated with transport of mechanical energy through an all-pervading medium called *luminiferous ether* endowed with (properties of) inertia, density, and elasticity. *Maxwell's theory* brought out for the first time the probable nature of light. But even he tried at first to explain his theory in terms of mechanical models. Subsequently, it was discovered that in fact no medium is necessary for the propagation of the type of waves postulated by Maxwell. Indeed, his electromagnetic waves can travel through empty space or a vacuum, and the velocity of their propagation is found to be identical with the velocity of light in a vacuum.

Maxwell formulated his theory for a homogeneous, isotropic medium on the basis of four physical laws known in his time. These are:

1. *Gauss's theorem* of total normal induction.

2. The non-existence of isolated magnetic poles.

3. *Ampere's rule* about the work done in taking a unit pole around a closed circuit carrying current.

4. *Lenz's law* of electromagnetic induction.

Each of these laws can be written down in the form of mathematical equations. On combination with constitutive relations giving certain physical properties of a dielectric medium, the set of equations leads to the following results:

1. In an electromagnetic field, the strengths of the electric and magnetic fields are given by

$$\nabla^2 \vec{E} = \left(\frac{\varepsilon\mu}{c^2}\right)\left(\frac{\partial^2 \vec{E}}{\partial t^2}\right) \tag{7.118}$$

$$\nabla^2 \vec{H} = \left(\frac{\varepsilon\mu}{c^2}\right)\left(\frac{\partial^2 \vec{H}}{\partial t^2}\right) \tag{7.119}$$

where E and H are respectively measured in electrostatic and electromagnetic units, and "c" is the ratio between the two sets of units.

An electromagnetic wave propagates in a homogeneous, isotropic, dielectric medium with a velocity $\left(\dfrac{c}{\varepsilon\mu}\right)$, where ε and μ are respectively the dielectric constant and permeability of the medium. For free space or a vacuum, $\varepsilon = \mu = 1$; hence, the velocity of propagation of electromagnetic waves in free space should just equal the ratio of, say, the "e. m." and "e. s." units of charge. On determination, this ratio in fact comes out to be 2.998×10^{10} $cms\,/\,sec$, which is also the velocity of light in free space. It was this fact that led Maxwell to believe that light was a form of electromagnetic radiation. This belief has since been confirmed in many ways, and it is now known that X-rays, γ-rays, ultraviolet and infra-red rays, and radio waves are all electromagnetic in character; the entire electromagnetic spectrum is propagated with the same velocity, namely 3×10^{10} $cms\,/\,sec$ through free space; only their wavelengths are different. For other dielectric media, ε is not equal to unity; also, μ depends upon the frequency of waves. For the visible region, however, μ may be taken to be unity. Hence, the velocity of light in these media is $\left(\dfrac{c}{\varepsilon}\right)$ and, therefore, the refractive index of such media should be ε. Many substances exhibit this relationship, but many others don't. The breakdown of this relation occurs because the electromagnetic theory does not take into account the detailed atomic structure of the dielectric.

2. In an electromagnetic field, the rate of flow of energy per unit area is represented by a vector, called a *Poynting vector*, perpendicular to the area under reference. The magnitude and direction of this vector is given by

$$\vec{N} = \frac{c}{4\pi}\left(\vec{E} \times \vec{H}\right) \tag{7.120}$$

Equation (7.120) shows that if there is a flow of electromagnetic energy, the X-direction as in Figure 7.9, it is always accompanied by an electric as well as a magnetic field perpendicular to the direction of energy flow as well as perpendicular to each other.

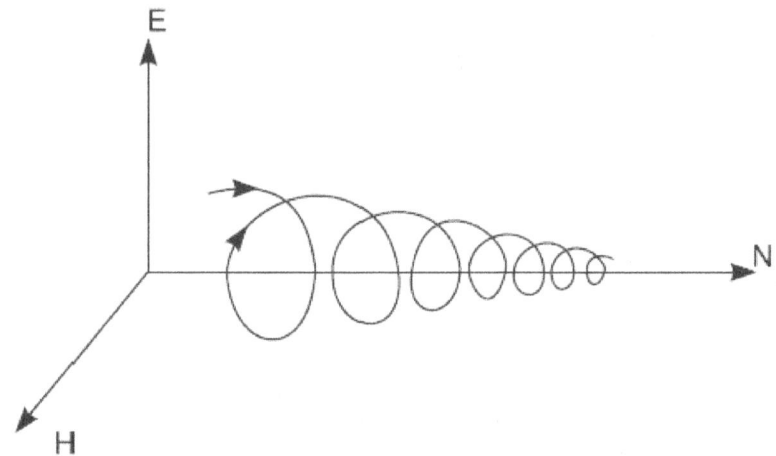

FIGURE 7.9. The X-direction for magnitude and direction of a *Poynting vector*.

In fact the three mutually perpendicular vectors are so oriented that if a right-handed screw is turned from the direction E to H, it will advance in the direction of the *Poynting vector*. We shall see from the following pages that E and H are quantities varying rapidly in magnitude as well direction, but the angles between the three vectors are each equal to ninety degrees at all times.

3. The electric vector of electromagnetic radiation is responsible for all observed optic phenomena. Since this vector is always directed perpendicular to the direction of the propagation of energy, it is inferred that light waves are transverse in nature. The differential equation of interest in *Optics*, equation (7.118), may also be written in another form by substituting

$$\vec{E} = -\frac{1}{c}\frac{\partial \vec{A}}{\partial t} \tag{7.121}$$

Thus, the quantity analogous to the velocity potential ϕ is the electric intensity $\vec{E}$, and that analogous to particle displacement ξ is the vector potential $\vec{A}$ defined by Equation (7.121). We may therefore use the equations developed in that chapter with this understanding, keeping in mind the transverse nature of electromagnetic waves. In addition, if the electromagnetic wave encounters a boundary between two dielectric media at which it may undergo reflection or refraction, due attention also has to be paid to the conditions at the interface called boundary conditions.

4. The electromagnetic field stores energy, and the density of this energy per unit volume is given by

$$\frac{1}{8\pi}\left(\varepsilon E^2 + \mu H^2\right) \tag{7.122}$$

Thus, the *classical electromagnetic theory* of light yields a definite expression for the energy density of such waves.

7.13 IMPLICATIONS OF MAXWELL'S THEORY

If, as predicted by the classical electromagnetic theory (*Maxwell's*), light is indeed a form of electromagnetic radiation, the following consequences would seem to follow from the assumptions implied in the derivation of the theory.

1. The propagation of light through dielectric media should be affected on the application of electric and magnetic fields across them. Such effects (e.g., Kerr effect, Faraday effect) have been observed for a large number of media.

2. The electromagnetic waves carry a continuous energy flux of density given by Equation (7.120). The classical theory therefore does not permit the propagation of any packets or lumps of energy through a dielectric medium. On the contrary, the modern quantum theory postulates the existence of "quanta" or packets of energy. The concept of energy quanta has also been borne out by experimental observations (e.g., Raman effect, photoelectric effect, and compton effect).

This apparent difficulty has been removed by reinterpreting the expression for energy flux in the light of *quantum theory*. At the same time, the results of the two theories may be reconciled if it is recognized that the classical theory leads to results depicting the average effects associated with wave propagation through dielectric media. As soon as we start examining the details of the phenomena, for example, the emission and absorption of electromagnetic radiations, the classical theory is found wanting in many ways.

7.14 PRACTICAL CONSIDERATIONS REGARDING LIGHT WAVES

Let us consider the phenomenon of emission of light as postulated by the classical theory on the understanding that it gives only the overall picture and is found wanting in finer details. Consider an electron bound to a heavy nucleus emitting electromagnetic radiations by way of oscillations induced by forces external to the atom. This is of course a very coarse picture of the probable state of affairs, but since at this stage we are not interested in the actual processes of the emission of light, we may as well omit the details. These oscillations do not build up instantaneously but take a finite interval of time to attain constant amplitude. This interval may nevertheless be very small as compared to the time for which the electron keeps oscillating and emitting electromagnetic radiations. Since an oscillating electron continuously loses energy by radiation, the amplitude of oscillations goes on decreasing till a stage arises when it becomes hardly detectable; subsequently, the oscillations die out completely with the amplitude of oscillation. This variation of amplitude with time can be depicted by a curve of the form shown in Figure 7.10.

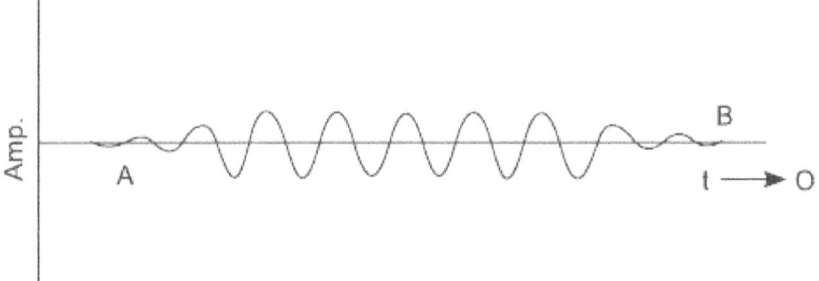

FIGURE 7.10. The curve for variation of amplitude with time.

If we wish to determine the frequency of these oscillations, we are immediately faced with a difficulty. If we knew the precise instants at which the oscillations had started and died out, the determination of frequency would have been an easy matter. We would then have been in a position to locate points *A* and *B* accurately and then to count the number of waves during the time interval *AB* and obtain the desired result. But due to the inherent uncertainty in locating the onset and dying out of the oscillations, their frequency cannot be determined accurately. Conversely, if the time interval somehow could be determined precisely, the number of waves during this period could not be counted accurately from the graph. No matter what we do we can never do better than counting the number of waves to within an accuracy of $(\pm \frac{1}{2})$ waves at each end of the pulse. Thus, if ΔT is the time for which the electron had emitted radiations, the uncertainty in determining the frequency v is given by

$$\Delta v \sim \left(\frac{1}{\Delta t} \right) \tag{7.123}$$

In view of this uncertainty, we should be naturally curious about knowing the frequency of oscillations. Of course, there is no exact value for this frequency. The best we can do is to count the number of waves "n" without bothering about the uncertainties at the two extremes and define a dominant frequency, "v_o," by the equation

$$v_o = \left(\frac{n}{\Delta t} \right) \tag{7.124}$$

Thus, we find that the radiation emitted out is not strictly monochromatic but contains a band of width Δv centered about the dominant frequency v_o. This bandwidth may be estimated by determining the time interval Δt for which the radiating electron remains in the incited state. An order of magnitude calculation is usually made by expressing Δt in terms of the mean life τ of the excited electron. Thus, if we suppose that the amplitude of oscillations falls to $\dfrac{1}{e\text{th}}$ of its steady value in time τ, it may be reasonable to assume that these oscillations die out for all practical purposes in a time which is a small multiple of τ. In other words, the orders of magnitude of Δt and τ may be taken to be the same. Equation (7.123) then yields a measure of the bandwidth. Thus, we find that

$$\Delta v \sim \left(\frac{1}{\tau} \right) \tag{7.125}$$

The actual state of affairs, however, does not even correspond to this highly simplified picture. This is on account of several complicating factors which may be mentioned to indicate the complexity of the phenomenon of emission. Firs, it never happens that atoms take their turn in emitting radiations. In fact, there may be several atoms in the excited state simultaneously, and hence many electrons undergoing oscillations leading to the generation of electromagnetic radiations. Second, none of these atoms are at rest in gaseous discharge tubes—the most commonly used sources of electromagnetic radiations. They in fact travel with velocities which may be as high as $10^5 cm/sec$. On account of this motion of the radiating source, the bandwidth may be broadened by at least two orders of magnitude. The Doppler broadening is usually accompanied by another type called "*Collision broadening.*" This is brought about when the radiating atom suffers a collision with others in the discharge tube. The combined effect of all these factors and a few more amounts to considerable frequency broadening of the emitted radiations. In real sources, therefore, the bandwidth is no longer of the order of $\left(\dfrac{1}{\tau}\right)$ but has a much greater value. If Δv_o is the actual bandwidth of a spectral line with a dominant frequency v_o, Equation (7.125) has to be replaced by the expression

$$\Delta v_o \sim \left(\frac{1}{t_{coh}}\right) \tag{7.126}$$

where the mean life "T" has been replaced by a much smaller time interval "t_{coh}," called *coherence time*. The importance of this time interval becomes apparent from the alternative expression

$$2\pi\Delta v_o \sim \frac{2\pi}{t_{coh}} \tag{7.127}$$

or

$$\left(\Delta\omega_o\right)\left(t_{coh}\right) \sim 2\pi \tag{7.128}$$

Hence, coherence time may be defined as the time interval needed for the extreme frequencies of a band to get out of phase by 2π.

It may be mentioned that in the foregoing discussion regarding bandwidth and coherence time, the emission has been assumed to be spontaneous rather than *stimulated*.

7.15 COHERENCE TIME AND INTERFERENCE OF LIGHT

We thus find that the so-called line spectrum spontaneously emitted from practical light sources does not consist of a series of strictly monochromatic lines. On the contrary, each line has an unavoidable bandwidth which plays a significant role in determining the properties of the emitted radiation. For a discussion of these, let us consider one such line of dominant frequency v_o and bandwidth Δv_o. Further, let us suppose that $\left(\dfrac{\Delta v_o}{v_o}\right) \gg 1$.

Such line sources are called quasi-monochromatic. Therefore, during the time interval $\left(\Delta v_o\right)^{-1}$, that is, the coherence time, there will be $\left(\dfrac{\Delta v_o}{v_o}\right)$ oscillations, a number which is quite large in view of the previous inequality. Let us now consider two independent point sources illuminated by light of the dominant frequency v_o. These independent sources may either be two pinholes in a screen surrounding a laboratory source or even two locations within this source at which the atoms are emitting radiations independent of each other. This is usually true if they are separated by a distance d larger than the wavelength under discussion. If a screen is placed at a distance much larger than d, the observed intensity distribution will be determined by the relative phases of the wave trains arriving at various points. For any point P as in Figure (7.11), the path difference between the waves arriving from A and B can be shown to be "$d\sin\theta$," and hence the phase difference will be

$$\left(\frac{2\pi}{\lambda}\right)d\sin\theta \tag{7.129}$$

In view of the finite bandwidth of the spectral frequency, this phase difference between the waves arriving at P will change with time. If at any instant the phase difference

$$\left(\frac{2\pi}{\lambda}\right)d\sin\theta = 2n\pi, \quad n = 0,1,2,\ldots \tag{7.130}$$

then within the time interval "t_{coh}" this value will be essentially constant. After a greater time, the value of phase difference, however, undergoes a change (the geometry of the setup remaining unaltered). Thus, if initially the intensity distribution on the screen was as shown at Figure (7.11a), after a time larger than "t_{coh}" it may change to that shown in Figure (7.11b), for which

$$\left(\frac{2\pi}{\lambda}\right)d\sin\theta = (2n+1)\pi, \quad n = 0,1,2,\ldots \tag{7.131}$$

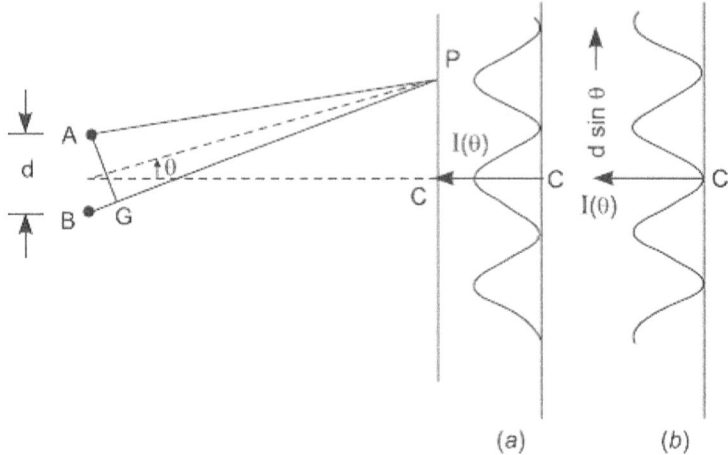

FIGURE 7.11. The path difference between the waves arriving from *A* and *B* and the phase difference of (a) the intensity distribution on the screen (b) after a time larger than "t_{coh}."

Under this situation the initial maxima intensity gives place to minima and vice versa. We therefore find that if our period of observation is larger than the coherence time of the quasi-monochromatic light, the intensity distribution on the screen will on the average be uniform. But if this period is shorter than "t_{coh}," the interference pattern exhibiting maxima and minima will be observed. Thus, the probability that the sources *A* and *B* will be coherent (i.e., maintain a constant phase difference at an external point) or incoherent will be entirely determined by the time taken by the receptor for recording the distribution of intensity on the screen. Since for usual laboratory sources "t_{coh}" is of the order of 10^{-8} or 10^{-9} seconds, the human eye with a much slower response always observes incoherent (i.e., uniform) illumination on the screen, even if the two independent sources are quasi-monochromatic. Thus, there are in fact no coherent or incoherent sources intrinsically; the distinction becomes necessary only due to the limitations of the tools of observation.

7.16 COHERENCE TIME AND POLARIZATION

The frequency bandwidth and coherence time also play a significant role in determining the state of polarization of electromagnetic radiations. Suppose an electron at a given location in the source of light is thrown into vibrations. Confining our attention to the radiation propagating along the direction $\vec{X}$, we have to consider only the $\vec{Y}$ and $\vec{Z}$ components of the electrons' motion for determining the nature of the radiation. These components of motion retain a constant relative phase only for the radiating lifetime of the excited electron. After some time, this same electron may suffer a second collision, again be excited, and start emitting radiations. The amplitudes and phase constants of the $\vec{Y}$ and $\vec{Z}$ motions of the electron will, as before, depend upon the circumstances of the collision. Invariably, as in the case of gaseous discharge tubes, these variables of electron motion are in no way correlated between successive collisions.

In actual practice, however, we should consider not one but many excited electrons for getting an insight into the nature of the emitted radiation the $\vec{Y}$ and $\vec{Z}$ components describing the radiation proceeding along. The direction $\vec{X}$ will now be the resultants of the corresponding motions of all electrons in the excited state. In practice, the components e_y and e_z associated with the vibrations of every excited electron may be combined to give the resultants E_y and E_z, whose amplitudes and phase constants depend upon those of the contributing atoms; of course, their frequencies will be the same. During a time interval shorter than the coherence time, "t_{coh}," the amplitude or phase of either E_y or E_z does not change appreciably in spite of the very large number of oscillations taking place during the period. Technically, this state of affairs means that the polarization state of the emitted electromagnetic wave remains unchanged during a time interval shorter than the coherence time.

If we examine the situation after some time, we shall find a new set of atoms in the excited state. The resultants E_y and E_z associated with the excited electron vibrations in the new set of atoms will, on the average, have the same amplitudes as before, but the phase constants will in no way be related to their corresponding values earlier on. Thus, we find that although the polarization state of the emitted radiation (described in terms of the relative phase constants of E_y and E_z) remains unchanged during a time interval shorter than the coherence time, it drifts in an entirely unpredictable manner over periods of observation larger than it. Therefore, coherence time is seen to be a fundamental property of light waves.

7.17 EXERCISES

1. Write the alternative forms of a wave equation.

2. Write the formula for energy contained in a length C of the medium that passes through a unit area in one second and its consequential sound intensity.

3. Derive an expression for the intensity of a spherical wave at a distance 2 m from its center in terms of the optical power of 150 W.

4. An electromagnetic wave in free space has $\vec{E} = f\left(\dfrac{t+x}{c_o}\right)\vec{a}_x$, where $\vec{a}_x$ is a unit vector in the x direction, $f(t) = e^{\left(\frac{t}{\tau}\right)^2} e^{(2\pi j v_o t)}$, where τ is a constant. What is the physical nature of the wave, and find an equation for the magnetic field vector.

5. Determine the velocity of propagation C in the form of bulk modulus and density.

6. Define a Poynting vector.

7. What is a wavefront?

8. What is collision broadening?

CHAPTER 8

HUYGENS'S PRINCIPLE

Chapter Outline

8.0 INTRODUCTION

One of the greatest difficulties encountered in the early development of the wave theory of light lay in its inability to explain the rectilinear propagation of light—a fact which obviously follows on the basis of the corpuscular conception of light. To overcome this difficulty in the progress of wave theory, *Huygens* in his work on light enunciated a principle for explaining wave propagation, which was later used by *Fresnel* in conjunction with the principles of interference to provide a satisfactory explanation of the rectilinear propagation of light, but at the same time he also showed that this propagation is only approximately rectilinear.

8.1 HUYGENS'S PRINCIPLE

One of the most important contributions of Huygens to the theory of light and to wave motion in general lies in his enunciation of a principle, known by his name, which governs the wave propagation in any medium. He declares in his work on light as follows:

> In considering the propagation of waves, we must remember that each particle of the medium through which the wave spreads does not communicate its motion only to that neighbor which lies in a straight line drawn from the luminous point but shares it also with all particles which touch it and resist its motion. Each particle is thus to be considered as the centre of a wave.

This statement, known as Huygens's principle, affords a geometrical construction for determining, from the given shape and position of the wavefront at any instant, the subsequent position and form of the new wavefront at some later instant. According to Huygens, every point of a wavefront may be regarded as the origin of a new disturbance, which in its turn emits small *secondary wavelets*, propagated in all directions from their respective origins with a speed equal to the speed of propagation of the wave in that medium. The new wavefront at a later stage is obtained by drawing a surface tangential to then-existing positions of these elementary secondary wavelets or, as it is often described, the *new wavefront is simply the envelope of these wavefronts*. He arrived at this principle of wave propagation from the following considerations:

It will be recalled that Huygens's conception of a beam of light was simply a succession of a great number of longitudinal pulses propagated at high velocity as condensations and rarefactions. Now, suppose that a luminous point emits a light pulse at a given instant. In a homogeneous isotropic medium, this pulse will spread in all directions with a constant velocity, so that after an interval of time t it will have reached a spherical surface $ABC...N$ of radius V_t, where V represents the velocity of the propagation of light in the medium under consideration. Thus, at this instant, the form of the wavefront is spherical. A wavefront, it will be recalled, is a surface in which all points are always in an equal phase of vibration. After a lapse of time t' from this moment, the light pulse will have reached the spherical surface $A'B'C'...N'$ of radius $V(t + t')$. Huygens now argued that by virtue of ether on the wavefront $ABC...N$ being disturbed by the light pulse originating from O, one finds a disturbance at every point of this wavefront exactly similar to that which originated t seconds earlier at O. Accordingly, every point of this wavefront can be regarded as the origin of ether disturbance, exactly similar to that which originated at O. Thus, around every point, after a time t', a new elementary wavefront commonly called a *secondary spherical wavelet* will have formed with a radius V_t'. The resultant effect of these secondary wavelets must be that the wavefront *ABC...N* has moved to a position *A'B'C'...N'* in time t'. Now, it is easy to see that the new wavefront is simply the envelope of the individual secondary wavelets. This explanation clearly establishes the *Huygens principle* of wave propagation. We therefore conclude that in the space in front of the wavefront *ABC...N*, everything takes place in an exactly similar manner as though the original light source were absent but only a sheet of secondary sources were present in the surface *ABC...N*. Kirchhoff has shown that this new interpretation of *Huygens's principle* helps us to discuss quantitatively diffraction and interference phenomena.

Figure 8.1 illustrates Huygens's principle of the propagation of wavefronts, while Figure 8.2 illustrates the application of the Huygens principle for the propagation of the wavefront from the luminous point at infinity. In this case, the wavefronts are plane. The lines *AA'*, *BB'*, and so forth joining the corresponding points of two wavefronts can be taken as the path traversed by the light or ray of light. In a homogeneous medium, rays are normal to the wavefront. If the medium is not homogeneous, the velocity of propagation is different in different directions. Consequently, the appropriate velocity must be used at every point of the wavefront to find the new position and shape of the wavefront. The velocity of light, *V*, has been assumed to be the same at all points and in all directions in Figures 8.1 and 8.2.

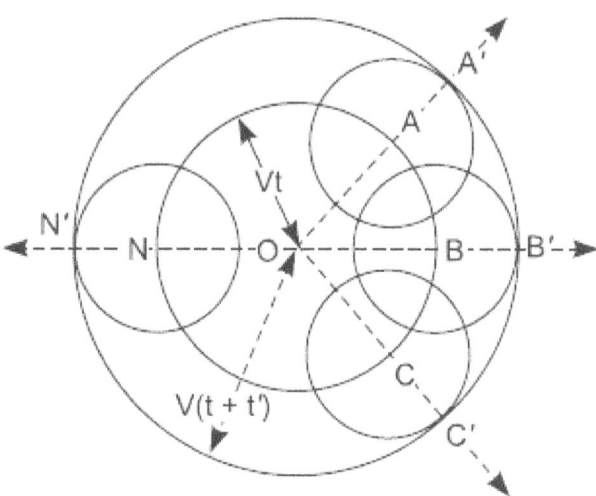

FIGURE 8.1. Huygens's principle of the propagation of wavefronts.

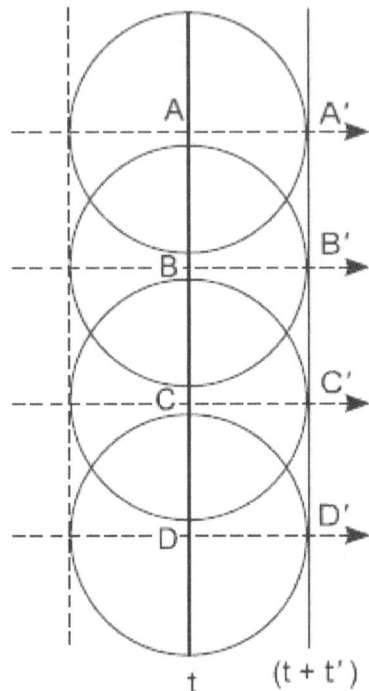

FIGURE 8.2. The application of Huygens's principle for the propagation of the wavefront from the luminous point at infinity.

In its simple form, the principle discussed previously is, however, not fully satisfactory. The reasons are not far to seek. In the first place, the secondary wavelets, if they spread out in all directions, should also combine to form a backward wave moving towards the source O. This, of course, is never observed experimentally. Second, the secondary wavelet which originates from A also reaches the points B', C', D',...., and accordingly light would be present at these points, which had traveled via A, contrary to the rectilinear propagation of light. Last, according to this principle, only one point of the secondary wavelet is effective, namely the point at which they touch their envelope in the forward direction of propagation. There is no direct physical or mathematical explanation in Huygens's original presentation of the theory for this arbitrary decision to ignore all the unwanted parts of the secondary wavelets.

We will give a satisfactory explanation of the absence of the back wave in the chapter on *Diffraction of Light—Fresnel Class* by following Fresnel's modification of Huygens's idea of secondary wavelets based on the principles of interference and superposition of wave motions applied to secondary wavelets.

8.2 REFLECTION OF A PLANE WAVE AT A PLANE REFLECTING SURFACE

Consider a plane wavefront ABC advancing in the direction depicted by arrows. Its position is marked in Figure 8.3 at an instant, say $t = 0$, at which its lower edge just meets a plane reflecting surface MM along a line through A perpendicular to the plane of the diagram. The planes of the wavefront and the reflecting surface are also supposed to be at right angles to the plane of the figure. At this instant, the point A and all points of the surface on the line through A normal to Figure 8.3, according to Huygens, will just become origins of secondary wavelets. But as the wavefront ABC advances further, the points of the reflecting surface between A and E, successively struck by corresponding points on the advancing wavefront, become the origins of secondary wavelets which spread out in the upper medium with a speed V. Let us find the shape of the reflected wavefront after a time interval t—the time taken by the end C of the incident wavefront to advance a distance CE, where $CE = V_t$.

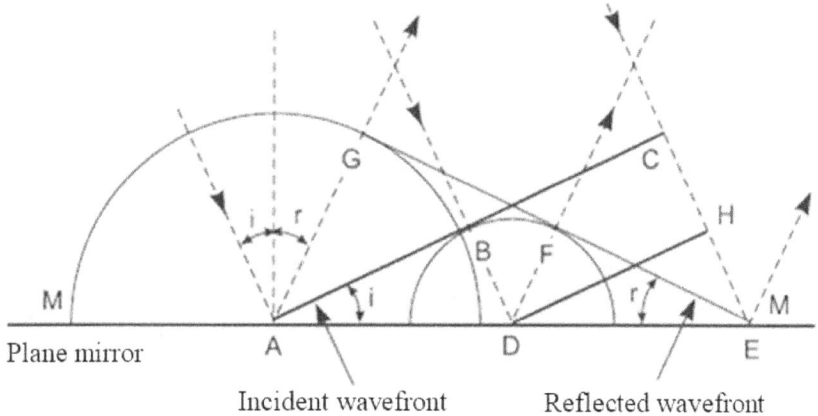

FIGURE 8.3. The laws of reflection by means of Huygens's undulation theory.

At this moment, when the point E of the reflecting surface just becomes the origin of the secondary wavelet, the radius acquired by the secondary wavelet which originated t seconds earlier from A would be

$$AG = Vt = CE \tag{8.1}$$

and the radius of the secondary wavelet which originated from the point D would be

$$DF = Vt' = HE \tag{8.2}$$

where t' is the time interval in which the end H of the incident wavefront from the position DH would just reach E. With points A,D as centers, the hemispheres of radii $AG = CE$ and $DF = HE$ are drawn in the upper medium, which consequently represent the spread of the secondary wavelets which originated from A and D respectively, by the time the end C of the incident wave has reached E. A plane drawn though E, normally to the plane of the diagram so as to be tangential to the secondary wavelet from A at G, can be easily shown to touch all the secondary wavelets which originated from the points of the surface between A and E. The proof is as follows:

It is easy to see that triangles ACE and DHE are similar. Hence, we have

$$\frac{AE}{DE} = \frac{CE}{HE} \tag{8.3}$$

Also, by the help of Equations (8.1) and (8.2), the previous relation can be easily written as

$$AE/DE = AG/DF \tag{8.4}$$

or

$$\frac{AE}{DE} = \frac{AG}{DF} = \sin r \tag{8.5}$$

Hence, in triangle DEF the angle DFE must be 90°. But the radius of the secondary wavelet from D, at the moment when E just becomes the origin of the secondary wavelet, is DF. Thus, the wavelet from D will touch the line GE at F. The plane GE normal to the plane of the diagram is, therefore, the envelope of all secondary wavelets generated from A to E by the time the incident wave has reached E. The plane GE, therefore, represents the reflected wavefront.

The angle i between the incident wavefront and the reflecting surface is called the angle of incidence; the angle between the reflected wavefront and the reflecting surface r is called the angle of reflection. The relation between them can be easily obtained as follows:

$$\sin i = \frac{CE}{AE} \text{ and } \sin r = \frac{AG}{AE} \tag{8.6}$$

But as proven previously, $AG = CE$, and hence we have

$$\sin i = \sin r \text{ or } i = r \tag{8.7}$$

Equation (8.7) is the law of reflection. Thus, a plane wave is reflected from a plane surface with an angle of reflection equal to the angle of incidence.

8.3 REFRACTION OF A PLANE WAVE THROUGH A PLANE REFRACTING SURFACE

Consider a plane wavefront *ABC* advancing in the direction depicted by arrows. Its position is marked in Figure 8.4 at the instant, say $t = 0$, at which its lower edge just meets a plane surface *MM* separating two transparent media (e.g., air and glass), along a line through A perpendicular to the diagram. The speeds of propagation in the two media are V_a and V_g respectively, and the refractive indices are $\mu_a = \dfrac{c}{V_a}$ and $\mu_g = \dfrac{c}{V_g}$ respectively.

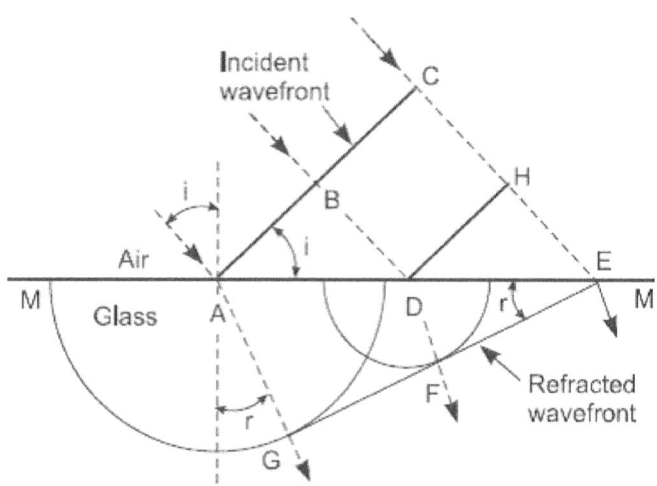

FIGURE 8.4. Explanation of Snell's Laws by means of Huygens's wave theory.

The plane of the wavefront and that of the refracting surface are also supposed to be perpendicular to the plane of the diagram. At this instant, according to *Huygens's principle*, two secondary wavelets just originate from the point A of the interface; one, above the interface spreads into the original medium with speed V_a, and the second, below the interface, spreads into the second medium with speed V_g, as the incident wavefront advances further. The direction of the reflected wavefront can be obtained by the law of reflection. We now proceed to find the direction of propagation of the refracted wavefront, it being assumed that $V_a > V_g$ and $\mu_g > \mu_a$. In what follows we shall simply confine ourselves to wavelets in the lower medium.

Now, as the incident wavefront *ABC* proceeds further, the points $A..D..E$ of the interface are successively disturbed by the advancing wavefront. As a consequence, they successively become the origins of secondary wavelets.

Obviously, the point A is the first to act as such and E the last. Let us find the shape of the wavefront after a time interval t, the time taken by the end C of the incident wavefront to advance a distance CE where

$$CE = V_a\, t \tag{8.8}$$

At the end of this time interval, when the point E of the refracting surface is just struck by the incident wavefront and therefore just becomes the origin of secondary wavelets, the radius acquired by the wavelet, which originated t seconds earlier from A in the lower medium, would be

$$AG = V_a\, t \tag{8.9}$$

By the use of Equation (8.9), this reduces to

$$AG = \left(\frac{V_g}{V_a}\right) CE \tag{8.10}$$

Similarly, the radius of the secondary wavelet which originated from D would be

$$DF = \left(\frac{V_g}{V_a}\right) HE \tag{8.11}$$

Since the secondary wavelets from points intermediate between A and E originate at successively somewhat later times than that which originated at A, their radii, at the instant when E just becomes the origin of wavelets, are successively smaller than that of the wavelet which originated at A.

With points A and D as centers, hemispheres of radius AG given by Equation (8.10) and radius DF given by Equation (8.11) are drawn in the lower medium, which consequently represent the spread of the secondary wavelets which originated from A and D respectively, by the time the end C of the incident wave has reached E. A plane drawn through E normally to the plane of the diagram so as to be tangential to the secondary wavelet from A at G can be easily shown to touch all secondary wavelets, which originated from the points of the refracting surface between A and E. The proof may be given as follows:

It is easy to see that triangles ACE and DHE are similar, and hence we get

$$\frac{AE}{DE} = \frac{CE}{HE} \tag{8.12}$$

By the help of Equations (8.10) and (8.11), the previous relation can be easily expressed as

$$\frac{AE}{DE} = \frac{AG}{DF} \tag{8.13}$$

or

$$\frac{DF}{DE} = \frac{AG}{AE} \tag{8.14}$$

From this relation it is obvious that $\angle DFE = 90°$. But DF is the radius of the secondary wavelet from D; thus, the line EG just touches the wavelet form D at F. We therefore conclude that the plane through EG, normal to the plane of the diagram, is the envelope of all secondary wavelets generated from A to E by the time the incident wave has reached E. The plane EG therefore represents the refracted wavefront in the lower medium.

Snell's law of refraction can now be easily derived as follows:

$$\frac{\sin i}{\sin r} = \frac{\left(\dfrac{CE}{AE}\right)}{\left(\dfrac{AG}{AE}\right)} = \frac{CE}{AG} \tag{8.15}$$

By the help of Equation (8.10), the previous equation becomes

$$\frac{\sin i}{\sin r} = \frac{V_a}{V_g} = \frac{\left(\dfrac{c}{V_g}\right)}{\left(\dfrac{c}{V_a}\right)} = \frac{\mu_g}{\mu_a} = \mu_g^a \tag{8.16}$$

Equation (8.16) is called *Snell's law*. With the help of wave theory, we have proven that the refractive index of glass with respect to air must be equal to $\left(\dfrac{V_{air}}{V_{glass}}\right)$. The result differs fundamentally from that obtained with *Newton's corpuscular theory of light*, which gives $\mu_g^a = \left(\dfrac{V_{glass}}{V_{air}}\right)$. As a consequence, according to *wave theory*, the speed of light propagation in a denser medium must be less as compared to its value in the rarer medium. The corpuscular theory, however, claims just the reverse effect. In 1850 Foucault experimentally showed that the speed of light in water is indeed smaller than its speed in air. This resulted in a resounding victory for *Huygens's wave theory*.

8.4 TOTAL REFLECTION

Huygens's principle can be very easily applied to give a physical explanation of the phenomenon of total reflection. We may first briefly describe this phenomenon on the basis of the ray theory. Figure 8.5 shows a number of rays diverging from a monochromatic point source *P* in glass and striking the interface separating it from air, the refractive index of glass with respect to air being μ_g^a.

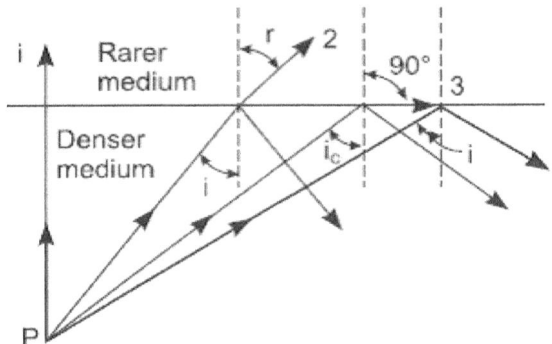

FIGURE 8.5. The number of rays diverging from a monochromatic point source *P* in glass and striking the interface separating it from air.

From *Snell's law*, since light is refracted from glass in air, is follows that

$$\frac{\sin i}{\sin r} = \mu_a^g = \frac{1}{\mu_g^a} \tag{8.17}$$

or

$$\sin r = \mu_g^a \sin i \tag{8.18}$$

Since $\mu_g^a > 1$, the angle of refraction *r* in the case under discussion is always greater than the angle of incidence *i*. Therefore, for some value of *i* less than 90°, the angle of refraction *r* becomes equal to 90°. In the latter case, as illustrated, the refracted ray 3 just goes grazing the interface and separating the two media. The particular angle of incidence corresponding to *r* = 90° is called the *critical angle* for the pair of media under consideration, and it is usually denoted by i_C. Equation (8.17) or (8.18), therefore, gives

$$\sin i_C = \frac{1}{\mu_g^a} \tag{8.19}$$

If the angle of incidence is greater than i_C, then it follows from Equation (8.17) or (8.18) that the sine of the angle of refraction becomes greater than 1, which is obviously impossible. This is interpreted as meaning that for an angle of incidence greater than the critical angle i_C, the refracted ray does not exist; only the incident ray is reflected back in the same medium. This is known as the *phenomenon of total reflection*.

We now give the explanation of the previous phenomena according to wave theory. In Figure 8.6, A_1C_1 represents a plane wavefront in a denser medium (glass), just incident at A_1 on the interface separating it from the rarer medium (air). According to *Huygens's principle*, the points $A_1...C_2$ become the origins of two secondary wavelets which spread out respectively in the upper and the lower medium as the incident wavefront sweeps the interface. The secondary wavelet which originated from A_1 would acquire a radius $A_1A'_2$ in the upper medium, given by the relation

$$A_1A'_2 = \left(\frac{V_a}{V_g}\right)C_1C_2 = \mu_g^a C_1 C_2 \tag{8.20}$$

at the instant when C_2 just becomes the origin of secondary wavelets.

But

$$C_1C_2 = A_1C_2 \sin i \tag{8.21}$$

and hence, Equation (8.20) may be written as

$$A_1A'_2 = \mu_g^a A_1 C_2 \sin i \tag{8.22}$$

Now, if we vary the angle of incidence i, keeping A_1C_2 constant, three possible cases arise:

1. The angle of incidence i is such that we have an inequality,

$$\mu_g^a \sin i < 1 \tag{8.23}$$

Hence, Equation (8.23) yields an inequality

$$A_1A'_2 < A_1C_2 \tag{8.24}$$

In this case, the radius of the secondary wavelet which originated at A_1 in the upper medium is less than A_1C_2 at the instant when C_2 just becomes the origin of secondary wavelets as shown in Figure 8.6. The refracted wavefront and the ray $A_1A'_2$, normal to $A_2C'_2$, is the refracted ray.

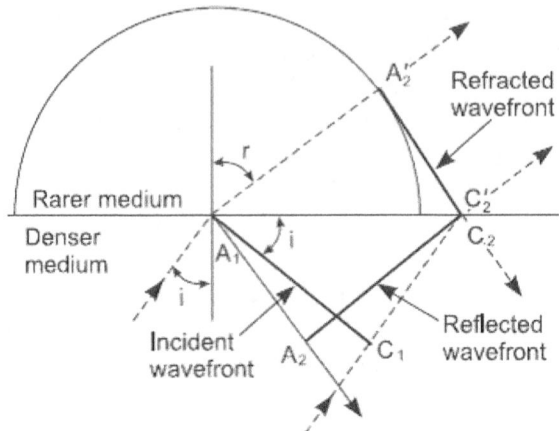

FIGURE 8.6. Reflected and refracted wavefronts.

2. If the angle of incidence increases to i_C so that we have

$$\mu_g^a \sin i_C = 1, \text{ then } A_1 A_2' < A_1 C_2$$

In this case, the secondary wavelet which originates from A_1 will pass through C_2 at the instant when the latter just becomes the origin of secondary wavelets, and this obviously holds good for all wavelets generated between A_1 and C_2. In other words, all the secondary wavelets which originate at various points touch each other at C_2 and therefore reinforce each other at C_2. As a consequence, the resultant disturbance is propagated just grazing the interface of the two media. Thus, the refracted ray goes just grazing the interface, and the angle i_C is the critical angle for the two media.

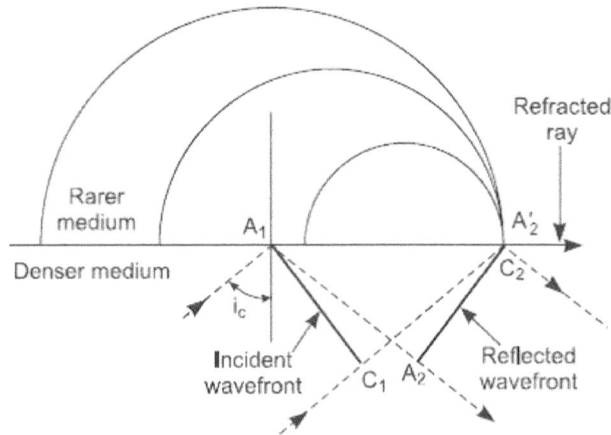

FIGURE 8.7. Refracted ray grazing the refracting surface.

3. If the angle of incidence, say i', is greater than i_C, so that we have an inequality

$$\mu_g^a \sin i' > 1$$

then it follows from Equation (8.22) that

$$A_1 A_2' < A_1 C_2$$

This simply means that the point C_2 will lie within the secondary wavelet which originated from A_1, and this property holds for all other wavelets which originated from points intermediate between A_1 and C_2 at the instant when the end C_1 of the incident wavefront just strikes C_2. Obviously, it is impossible to draw a plane through C_2 tangential to these wavelets. Thus, the refracted wavefront does not exist in this case; only the reflected wavefront exists. In other words, light is totally reflected within the same medium.

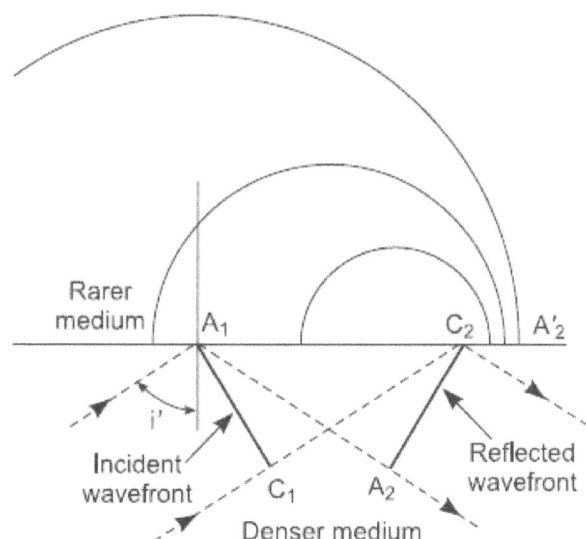

FIGURE 8.8. Total internal reflection.

8.5 REFRACTION THROUGH A SPHERICAL SURFACE

Suppose in Figure 8.9 that the circular arc AOA' represents the trace, in the plane of paper, of the spherical refracting surface convex toward the rarer medium.

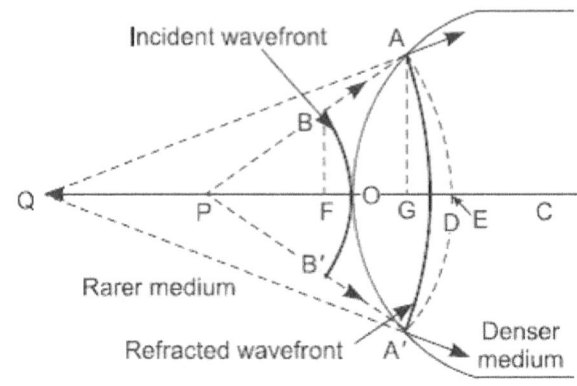

FIGURE 8.9. Formation of a virtual image by a convex refracting surface.

Let *BOB′* represent the trace of spherical wavefront diverging from the monochromatic luminous point source *P*. Now, if *P* is so close to the refracting surface that at the instant when *A* and *A′* become the origins of wavelets, the radius *OD* of the wavelet which originated at *O* (when *BOB′* was the position of the incident wavefront) becomes greater than *OG*, then the refracted wavefront *ADA′* will be a spherical surface apparently diverging from the axial point *Q*. Thus, the image *Q* of the real axial point *P* is *virtual*.

The optical path from *O* to *D* must be equal to that from *B* to *A*. Thus, the refracted wavefront *ADA′* satisfies the condition

$$\mu_2 OD = \mu_1 BA \qquad (8.25)$$

where μ_1 is the refractive index of the rarer medium and μ_2 that of the denser medium. In the paraxial region, *BA = FG* approximately, hence

$$\mu_2 OD = \mu_1 FG \qquad (8.26)$$

or

$$\mu_2 \left(OD + GD \right) = \mu_1 \left(FO + OG \right) \qquad (8.27)$$

or

$$\mu_1 FO - \mu_2 GD = \left(\mu_2 - \mu_1 \right) OG \qquad (8.28)$$

Since *P* and *Q* are to the left of *O* while *C* is to the right of *O*, according to our sign convention, we have *OP* = −*u*, *OQ* = −*v*, and *OC* = *R*. Therefore, to a close approximation we write

$$FO = \frac{y^2}{(-2u)} \tag{8.29}$$

$$GD = \frac{y^2}{(-2v)} \tag{8.30}$$

$$OG = \frac{y^2}{2R} \tag{8.31}$$

where we have written $AG = y = BF$ approximately. Substituting the values of FO, GD, and OG in Equation (8.28)

$$\frac{\mu_2}{v} - \frac{\mu_1}{u} = \frac{\mu_2 - \mu_1}{R} \tag{8.32}$$

Dividing Equation (8.32) by μ_1 and writing μ_2^1 for $\frac{\mu_2}{\mu_1}$, the refractive index of the second medium (denser) with respect to the first (rarer) medium, we get

$$\frac{\mu_2^1}{v} - \frac{\mu_1}{u} = \frac{\mu_2^1 - 1}{R} \tag{8.33}$$

8.6 REFRACTION THROUGH A THIN LENS

1. Spherical Waves Incident on a Thin Convex Lens

Let C_1BC_2 and C_1DC_2 in Figure (8.10) represent two spherical surfaces of a thin convex lens, the radii of curvature of the surfaces being R_1 and R_2 respectively.

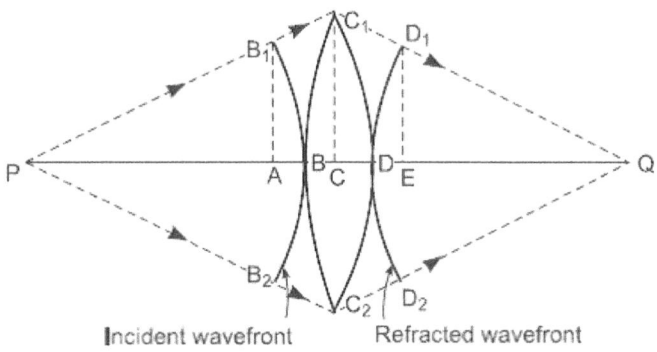

FIGURE 8.10. Refraction through a convex lens.

We suppose the lens material to be denser than the surrounding medium. Let B_1BB_2 represent a spherical wavefront, shown diverging from a luminous axial point P emitting light of one wavelength only, and to have reached a position so as to just touch the pole B of the surface C_1BC_2 of the lens. According to Huygens's principle, at this instant, the pole B just becomes the origin of a secondary wavelet, and this wavelet traverses an axial thickness BD within the lens medium in the time interval t. On the other hand, by Huygens's principle of propagation of the wavefront, the ends B_1 and B_2 in this interval t practically travel in a rarer medium (air) via C_1 and C_2 respectively to respective positions D_1 and D_2. Therefore Optical path $B_1C_1D_1$ = Optical path BD
i.e.,

$$B_1C_1 + C_1D_1 = \mu BD \tag{8.34}$$

where μ is the refractive index of the lens material with respect to the surrounding medium.

In reality, in order to trace the formation of the image, we should regard every point of the back surface of the lens as new origins of secondary wavelets which are generated when the points are successively struck by the advancing wavefronts from the left. Figure 8.11 represents a particular case where wavefronts in the lens medium are plane.

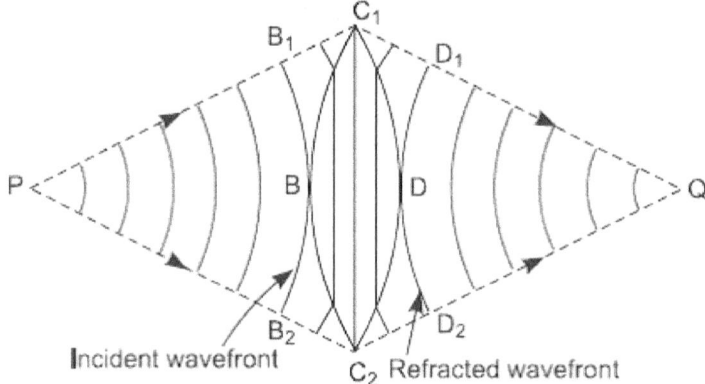

FIGURE 8.11. Refraction of waves through a convex lens.

This has been assumed for simplicity of the figure; nevertheless, the conclusions which will be drawn are quite general and apply to account for the formation of the real image by the lens. If we now confine our attention to the advancing wavefront when in the position C_1CC_2 within the lens as in

Figure 8.10, then by the time interval in which this wavefront reaches D, the secondary wavelets from C_1 and C_2 would acquire the radii, to a close approximation, given by

$$C_1 D_1 = C_2 D_2 = \mu CD \qquad (8.35)$$

The lens is denser than the surrounding medium, that is, $\mu > 1$, and therefore it follows from the previous relation that $C_1 D_1$ as well as $C_2 D_2$ both are greater than CD. Therefore, in the paraxial region, the points D_1 and D_2 of the secondary wavelets from C_1 and C_2 lie to the right of D, and this applies to every secondary wavelet originated at the back surface except the point D, which has just become the origin of the secondary wavelets. The envelope of these secondary wavelets in the paraxial region is a spherical surface, converging toward the axial point Q. Thus, a real image of P is formed at Q, due to refraction of monochromatic waves through the lens. In short, the action of the lens is to retard the central portion of the incident wavefront relative to its peripheral parts, thereby producing a change in its curvature.

The refracted spherical wavefront $D_1 D D_2$, as explained earlier, satisfies the condition

$$B_1 C_1 + C_1 D_1 = \mu BD \qquad (8.36)$$

For a sufficiently small aperture of the lens, we may put

$$B_1 C_1 = AC \ \text{ and } \ C_1 D_1 = CE \qquad (8.37)$$

With this approximation, Equation (8.36) becomes

$$AC + CE = \mu BD \qquad (8.38)$$

or

$$AB + BC + CD + DE = \mu (BC + CD) \qquad (8.39)$$

or

$$AB + DE = (\mu - 1)(BC + CD) \qquad (8.40)$$

The object point P is to the left of B, while the image point Q is to the right of D. Therefore, we write

$$BP = -u \ \text{ and } \ DQ = +v \qquad (8.41)$$

Further, the radius of curvature R_1 of the surface C_1BC_2 is positive, while R_2, the radius of curvature of C_1DC_2, is negative. Then we write to a first approximation

let $AB1 = CC1 = ED1 = y$, then

$$AB = \frac{y^2}{(-2u)}; \ DE = \frac{y^2}{2v}; \ BC = \frac{y^2}{(+2R_1)}, \text{ and } DC = \frac{y^2}{(-2R_2)}$$

Substitution of these values in Equation (8.40) gives

$$\frac{1}{v} - \frac{1}{u} = (\mu - 1)\left(\frac{1}{R_1} - \frac{1}{R_2}\right) \tag{8.42}$$

which is called the *thin lens equation*. The usual sign convention applies to this equation.

2. Spherical Waves Incident on a Thin Concave Lens

Let B_1BB_2 represent a spherical wavefront, shown diverging from a monochromatic luminous point P, and having reached a position just incident on a concave lens as illustrated in Figure 8.12.

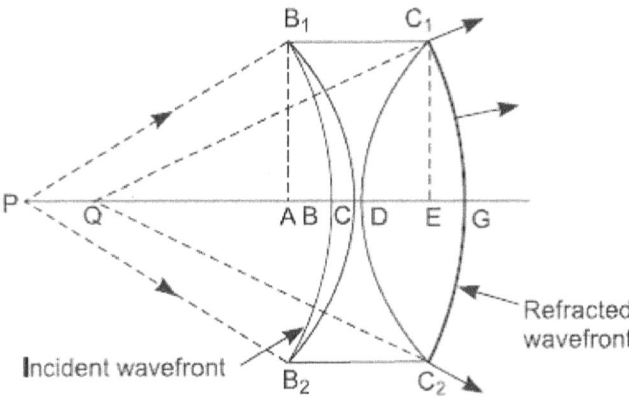

FIGURE 8.12. Refraction through a concave lens.

Since the lens is thin in the middle as compared to its peripheral parts, it therefore retards the peripheral parts of the incident wavefront relative to its central portion provided the lens material is denser than the surrounding medium, which we assume to be the case under consideration. As a consequence, the emergent spherical wavefront C_1GC_2 is a divergent one with a curvature greater than that of the incident wavefront B_1BB_2. The emergent wavefront appears to diverge from an axial point Q, situated on the same side of the lens as luminous object point P and at a distance from the lens less than that of P. Thus, a virtual image Q of P is formed by a concave lens.

The optical path from B_1 to C_1 must be equal to that between B and G, i.e.,

$$BC + \mu CD + DG = \mu B_1 C_1 \tag{8.43}$$

or

$$(AC - AB) + \mu CD + DE + EF \ \mu(AC + CD + DE) \tag{8.44}$$

or

$$EG - AB = (\mu - 1)(AC + DE) \tag{8.45}$$

Let $AB_1 = EC_1 = y$. We therefore write to a close approximation,

$$EG = \frac{y^2}{(-2v)}; \ AB = \frac{y^2}{(-2u)}; \ AC = \frac{y^2}{(-2R_1)}, \text{ and } DE = \frac{y^2}{2R_2}$$

Substituting these values in Equation (8.45), we get the familiar relation,

$$\frac{1}{v} - \frac{1}{u} = (\mu - 1)\left(\frac{1}{R_1} - \frac{1}{R_2}\right)$$

8.7 FOCAL LENGTH OF THE COMBINATION OF TWO THIN LENSES IN CONTACT

When two thin lenses of focal lengths f_1' and f_2' are placed in close contact, the focal length f' of this combination, as obtained by the ray theory, is given by the relation

$$\frac{1}{f'} = \frac{1}{f_1'} + \frac{1}{f_2'} \tag{8.46}$$

We shall now derive this formula according to wave theory.

Consider a plane wavefront B_1AB_2 of monochromatic light, just incident on a combination formed of two convergent lenses L_1 and L_2, placed in contact as in Figure 8.13.

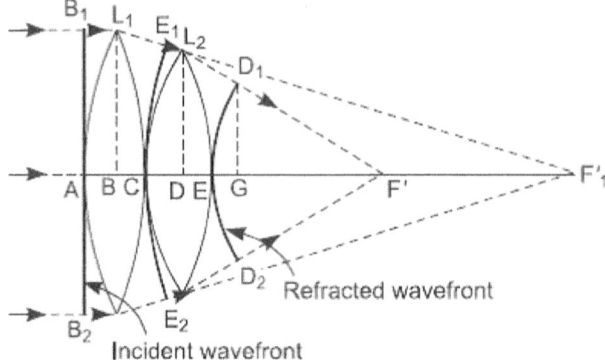

FIGURE 8.13. Refraction through two lenses in contact.

The first lens retards the central part of the incident wavefront relative to its peripheral portions, thereby changing it to a spherical wavefront E_1CE_2, converging toward the second principal focal point F_1' of the first lens. The second lens further retards the central portion of E_1CE_2 with respect to the peripheral parts. As a consequence, the emergent wavefront D_1ED_2 from the combination is spherical (in the paraxial region), this time, however, converging toward the axial point F', which is called the second principal focal point of the combination of lenses in contact. It is now easy to see that light disturbance from B_1 must traverse the path $(B_1L_1 + L_1L_2 + L_2D_1)$ in air, in the time interval required for the light disturbance from A to reach E, traveling through the lens material.

Therefore, if μ is the refractive index of the first lens and μ' that of the second lens, both with respect to the surrounding medium, we can easily get

$$B_1L_1 + L_1L_2 + L_2D_1 = \mu AC + \mu'CE \tag{8.47}$$

For a sufficiently small aperture of the combination, we may put to a close approximation,

$$L_1L_2 = BD \text{ and } L_2D_1 = DG \tag{8.48}$$

With this approximation, Equation (8.47) becomes

$$AB + BD + DG = \mu AC + \mu'CE \tag{8.49}$$

or

$$AB + BC + CD + DE + EG = \mu(AB + BC) + \mu'(CD + DE) \tag{8.50}$$

or

$$EG = (\mu - 1)(AB + BC) + (\mu' - 1)(CD + DE) \tag{8.51}$$

Let the radii of curvature of the surfaces of the first lens be denoted by R_1 and R_2 with that of the second lens by r_1 and r_2. Also, suppose $EF' = f'$. Further, we write to a close approximation,

Let $BL_1 = DL_2 = GD_1 = y$.

Now, considering the spherical wavefront D_1ED_2 of radius of curvature f' and centered at F', we have

$$(GD_1)^2 = EG(2f' - EG)$$ (8.52)

or

$$EG = \frac{y^2}{2f'} \text{ approximately}$$

Similarly, to a close approximation, we have

$$AB = \frac{y^2}{2R_1}; \quad BC = \frac{-y^2}{2R_2}; \quad CD = \frac{y^2}{2r_1}; \quad DE = \frac{-y^2}{2r_2}$$

Substituting these values in Equation (8.51), we get

$$\frac{1}{f'} = (\mu - 1)\left(\frac{1}{R_1} - \frac{1}{R_2}\right) + (\mu' - 1)\left(\frac{1}{r_1} - \frac{1}{r_2}\right)$$ (8.53)

or

$$\frac{1}{f'} = \frac{1}{f_1'} + \frac{1}{f_2'}$$

8.8 EXERCISES

1. Explain clearly Huygens's principle of wave propagation. Deduce the laws of reflection with the help of Huygens's wave theory.

2. (a) State *Huygens's principle*. Explain on its basis the phenomenon of refraction and obtain the law of refraction. Give the physical significance of refractive index.

   (b) Explain total internal reflection on the basis of wave theory and obtain the value of the critical angle.

3. Apply Huygens's theory of light to obtain an expression for refraction of a spherical wave at a spherical surface.

4. Account for the formation of images by refraction through lenses on wave theory and prove the relation

$$\frac{1}{v} - \frac{1}{u} = (\mu - 1)\left(\frac{1}{R_1} - \frac{1}{R_2}\right)$$

where the symbols have their usual meanings.

INTERFERENCE OF LIGHT

9.0 INTRODUCTION

It is a matter of common experience that two trains of ripples on the surface of water cross each other and proceed onward in their directions undisturbed. In the same way, two beams of light cross each other and proceed without being influenced by each other in any way. For example, different observers can view at the same instant different objects through the same narrow aperture with perfect clarity, the beams of light having crossed at the aperture in reaching the observers. However, in the region of crossing where both the beams are acting simultaneously, a modification in their intensity is expected, which should be either less or greater than that which would be given by one beam alone. This modification of intensity due to superposition of two or more beams of light is spoken of as *interference of light*. This phenomenon demands for its explanation that light must be a wave motion.

9.1   PRINCIPLE OF SUPERPOSITION

The fundamental basis of the explanation of the phenomenon of *interference of light* is the principle of superposition of wave motions first enunciated by Thomas Young in 1801. This principle states that when a medium is disturbed simultaneously by any number of waves, the instantaneous resultant displacement of the medium at every point at every instant is the algebraic sum of the displacements of the medium due to individual waves, in the absence of the others. In Young's own words: *"When two undulations from different origins coincide, either perfectly or very nearly, in direction, their joint effect is a combination of the motions belonging to each."* Suppose due to a single wave train the displacement of the medium at a certain point at any instant is y_1 in a given direction and that due to another wave train, in the absence of the first, the displacement is y_2. According to the previous principle, the instantaneous resultant displacement R of the medium at that point due to two waves acting together is expressed by

$$R = y_1 + y_2 \qquad (9.1)$$

when the two separate displacements y_1 and y_2, as depicted in Figure 9.1 (a), are in the same direction. However, when the two individual displacements y_1 and y_2 are in opposite directions, the instantaneous resultant displacement due to two waves acting together is now given by

$$R = y_1 - y_2 \qquad (9.2)$$

In Figure 9.1 (b), y_1 is greater than y_2, but the two are directed in opposite directions. Hence, their algebraic sum is $(y_1 - y_2)$, and this is plotted in the direction of y_1. The reverse of this case is depicted in Figure 9.1 (c).

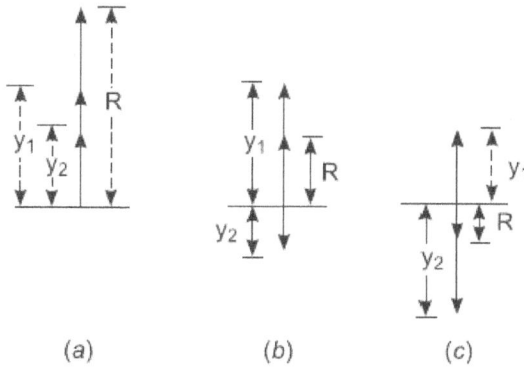

FIGURE 9.1. Illustrates the principle of the superposition of wave motions.

9.2 DISCOVERY OF INTERFERENCE OF LIGHT

Young's Experiment. Historically, the phenomenon of *interference of light* was first discovered by Thomas Young in 1801, when he announced an experiment capable of exhibiting an interference pattern due to the superposition of two beams of light. This experiment was regarded as a crucial one at that time, since it definitely established the wave nature of light. The corpuscular theory was found to be totally inadequate to explain the experimental results.

The apparatus is shown schematically in Figure 9.2.

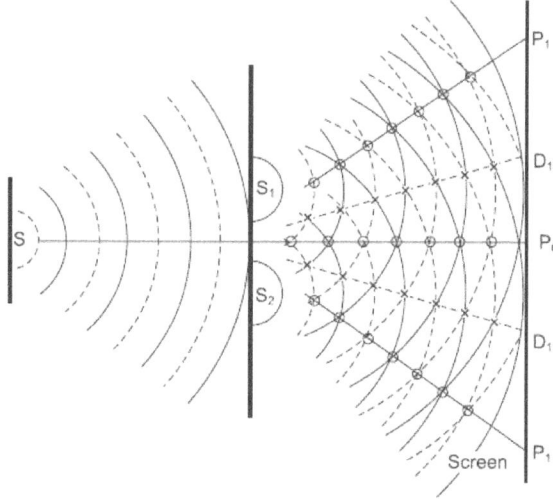

FIGURE 9.2. Young's double-slit experiment.

Young allowed the sunlight to pass through a pinhole S and then at some distance through two sufficiently close pinholes S_1 and S_2 in an opaque screen. Finally, the light was received on a screen on which he observed an uneven distribution of light intensity. Young found that the illumination on the screen consisted of many alternate bright and dark spots. In accordance with modern laboratory technique, narrow parallel slits replace pinholes, and the slit S is illuminated with the monochromatic light of wavelength λ. Light is received on a screen placed at a certain distance to the right and parallel to the plane containing the slits S_1 and S_2. According to *Huygens's principle*, cylindrical wavelets spread out from slit S and, as the path $SS_1 = SS_2$, the wavelets reach slits S_1 and S_2 at the same instant. A train of Huygens's wavelets, therefore, diverges to the right from both of these slits which have precisely equal phases at the start. Furthermore, their amplitude, wavelength, and velocity are also equal. Suppose in two dimensions, as in Figure 9.2, continuous circular arcs represent the wave crests while dotted circular arcs represent the wave troughs in each wave. At points marked by o's, a crest of one wave is superposed on a crest of the other or a trough of one is superposed on a trough of the other. In other words, at these points the two waves meet in the same phase. Therefore, according to the principle of superposition, at these points the resultant amplitude is twice that of each component wave. On the other hand, at points marked by x's, the crest of one wave is superposed on the trough of the other and vice versa; that is, the two waves meet in opposite phase. Hence, according to the principle of superposition, at these points they neutralize each other, and the resultant intensity is zero. The solid lines intersecting the screen at P_0 and P_1 connect the points marked o's, and therefore along these lines the resultant intensity is always maximum (*constructive interference*), while the dotted lines intersecting the screen at D_1 are the loci of points marked x's, and therefore along these lines the resultant intensity is zero (*destructive interference*). Actually, the loci of points marked x's or o's are confocal hyperbolae. Thus, on the screen a number of alternate bright and dark regions of equal width, called interference fringes, are observed parallel to the slits.

That the observed pattern is truly due to interference of two waves of light can be demonstrated by covering one of the double slits. The well-defined dark and bright fringes are replaced by a new pattern much coarser due to the diffraction of light by the uncovered single slit. A comparison of the two patterns indicates that a point on the screen, bright when only one slit is covered, changes to dark when both slits are exposed. A corpuscular theory is totally inadequate to account for this fact, which can be easily explained in terms of the wave nature of light.

The interference effects can be conveniently demonstrated in the laboratory with ripples on the surface of water. Two pins are attached to the same prong of an electrically maintained tuning fork and adjusted so that the pins just touch the surface of water placed underneath, while the prongs are capable of vibrating in a vertical plane. The periodic vibratory motion of the prong is also taken up by the attached pins, which are therefore always in equal phase. This periodic, vibratory, up-and-down motion of the pins on the surface of the water generates two waves or ripple systems of equal amplitude, equal wavelength, and equal velocity. The ripples spread out with their centers at the points of action of the two pins. In Figure 9.2, S_1 and S_2 may be now imagined as the points of action of two vibrating pins, and the continuous and dotted circular arcs are the traces of crests and troughs respectively of the ripple systems. It will be observed that along the dotted straight lines water remains permanently calm, while along the solid straight lines it permanently vibrates up and down with double the amplitude of the component waves.

9.3 THEORY OF INTERFERENCE

We shall now derive an expression for the resultant intensity an any point P of the screen due to superposition of two waves of light having the same amplitude, frequency, and wave number. Suppose that these waves are emitted from two point sources A and B separated by a distance $2d$, as in Figure 9.3, which is of the order of several wavelengths.

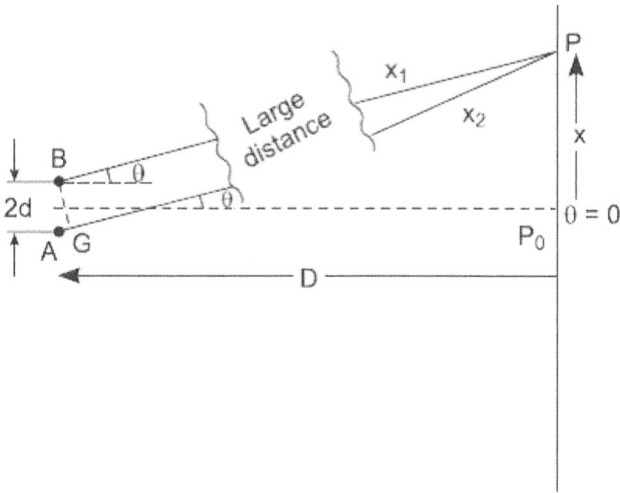

FIGURE 9.3. Interference effects.

The distribution of light intensity on a screen placed at a large distance from the sources A and B can then be determined as follows.

If the two point sources are locked in phase or have a constant phase difference, this intensity distribution will be independent of time. Further, if the distance D is large, the two waves reaching any field point P will have nearly the same amplitude, and the corresponding "rays" will be effectively parallel and inclined at the same angle θ to the axis $(\theta = 0)$ of the system. This amplitude, nevertheless, depends upon the average distance $x = \frac{1}{2}(x_1 + x_2)$, which the waves cover to reach the point P. With these reservations, we suppose that the waves reaching P are represented by

$$\phi_1 = ae^{-i(\omega t - kx_1)} \tag{9.3a}$$

and

$$\phi_2 = ae^{-i(\omega t - kx_2)} \tag{9.3b}$$

where the amplitude is $a = a(x)$. The resultant wave at P is then given by

$$\phi = \phi_1 + \phi_2$$

$$\phi = ae^{-i\omega t}\left\{e^{ikx_1} + e^{ikx_2}\right\} \tag{9.4}$$

Both the distances x_1 and x_2 may be expressed in terms of the average distance x. In this manner, Equation (9.4) may be written as

$$\phi = ae^{-i(\omega t - kx)}\left\{e^{ik(x_1 - x)} + e^{ik(x_2 - x)}\right\}$$

$$\phi = ae^{-i(\omega t - kx)}\left\{e^{\frac{1}{2}ik(x_1 - x_2)} + e^{-\frac{1}{2}ik(x_1 - x_2)}\right\} \tag{9.5}$$

The factor $(x_1 - x_2)$ occurring as the exponential power in Equation (9.5) is the path difference BG between the waves reaching P from the sources A and B. This is given by

$$\text{Path difference: } BG = 2d\sin\theta \tag{9.6}$$

Combining Equations (9.5) and (9.6), we find that the equation of the resultant disturbance at P is

$$\phi = ae^{-i(\omega t - kx)}\left\{e^{ikd\sin\theta} + e^{-ikd\sin\theta}\right\}$$

or

$$\phi = 2a\cos(kd\sin\theta)e^{-i(\omega t - kx)}$$

$$\phi = 2a\cos\left(\frac{2\pi}{\lambda}d\sin\theta\right)e^{-i(\omega t - kx)} \tag{9.7}$$

This equation of the resultant disturbance at any field point P is in fact true for all wave motions, provided we extract the real part from the R.H.S. It will then be equally applicable to acoustic waves or water waves under corresponding situations. Also, the treatment shall be valid even if the point sources A and B are replaced by line or slit sources of light. Thus

$$\phi = 2a\cos\left(\frac{2\pi}{\lambda}d\sin\theta\right)\cos(\omega t - kx) \tag{9.8}$$

is, in fact, the disturbance which leads to a definite intensity distribution pattern on the screen in the case of light waves. The amplitude of the resultant disturbance is now a function of both x and θ and may be written as

$$a(x,\theta) = 2a\cos\left(\frac{2\pi}{\lambda}d\sin\theta\right) \tag{9.9}$$

where twice the argument of the cosine is the difference of phase between the waves reaching P from A and B. The resultant intensity at P is proportional to the square of this amplitude, and hence we write

$$I\alpha\, 4a^2\cos^2\frac{1}{2}\delta \tag{9.10}$$

where

$$\delta = 2\left\{\left(\frac{2\pi}{\lambda}\right)d\sin\theta\right\}$$

With k as constant of proportionality, we can also write

$$I = 4ka^2\cos^2\frac{1}{2}\delta \tag{9.11}$$

$$I = 2ka^2\left(1 + \cos\delta\right) \tag{9.12}$$

which varies with the magnitude of δ.

9.3.1 Constructive Interference

When $\delta = 0, 2\pi, 4\pi$, and so on, then $\cos\delta = +1$, which is the maximum value of the cosine function. As a consequence, it follows from Equation (9.12) that

$$I = 4ka^2 \quad \text{[Maximum]} \tag{9.13}$$

which is, obviously, greater than what is to be expected from the separate intensities. The two waves reinforce each other completely, and the resultant intensity is thus maximum when

$$\delta = 2n\pi \quad [\text{Maxima}] \tag{9.14}$$

where $n = 0, 1, 2, 3, 4$, and so forth. Using this value of δ, we easily get the condition for maximum intensity in terms of the optical path difference, which is

$$AP - BP = 2n\left(\frac{\lambda}{2}\right) \quad [\text{Maxima}] \tag{9.15}$$

The two waves are said to be in the same phase at any point where their phases differ exactly by even multiples of π, and their interference is called *constructive interference*.

9.3.2 Destructive Interference

When $\delta = \pi, 3\pi, 5\pi, 7\pi$, and so on, then $\cos \delta = -1$, which is the least value of the cosine function. As a consequence, the Equation (9.12) yields

$$I = 0 \quad [\text{Minimum}] \tag{9.15}$$

which is obviously less than what is to be expected from the separate intensities. Two waves interfere destructively when

$$\delta = (2n + 1)\pi \quad [\text{Minima}] \tag{9.16}$$

where $n = 0, 1, 2, 3, 4$, and so forth. In terms of the optical path difference, the condition for minimum intensity is

$$AP - BP = (2n + 1)\left(\frac{\lambda}{2}\right) \quad [\text{Minima}] \tag{9.17}$$

where $n = 0, 1, 2, 3, 4$, etc.

Two waves are said to be in opposite phases at any point where their phases differ exactly by odd multiples of π, and their interference is called *destructive interference*.

It can be concluded that when two waves having a constant phase difference initially reunite after traversing difference optical paths, they interfere constructively or destructively according to how their phases, at the point of observation, differ exactly by even or odd multiples of π. This condition, it should be remarked, is the general one and applies for locating the regions of constructive and destructive interference in all experimental arrangements exhibiting interference effects.

9.3.3 Spacing of Interference Fringes—Fringe Width

In order to plot the intensity distribution in the interference pattern arising from the interference of light waves from two coherent sources S_1 and S_2, it is essential to investigate the spacing of these fringes on a screen placed parallel to the line joining S_1 and S_2. It will be recalled that at any point P on the screen, the resultant intensity is maximum or minimum according to its distance from the two coherent sources S_1 and S_2, which are always in equal phase and differ exactly by even or odd multiples of $\left(\dfrac{\lambda}{2}\right)$, where λ is the wavelength of the light emitted by the sources. In symbols the conditions are expressed as:

$$AP - BP = (2n)\left(\frac{\lambda}{2}\right) \quad [\text{Bright fingers}] \tag{9.18}$$

and

$$AP - BP = (2n+1)\left(\frac{\lambda}{2}\right) \quad [\text{Dark fingers}] \tag{9.19}$$

where $n = 0, 1, 2, 3, 4$, etc.

In Figure 9.3, in the plane of the diagram, let X be the distance of the point P from the central Point P_o, a point where the perpendicular bisector of AB intersects the screen. We can see that

$$(AP)^2 = D^2 + (X+d)^2 \tag{9.20}$$

which can be rearranged as

$$AP = D\sqrt{1+\left(\frac{X+d}{D}\right)^2} \tag{9.21}$$

On expanding the right side of the previous relation by the *binomial theorem* and neglecting powers of $\left(\dfrac{(x+d)}{D}\right)^2$ higher than one, which of course is permissible since, in general, D is some thousand times larger than d and x, we get

$$AP = D\left(1+\frac{1}{2}\left(\frac{X+d}{D}\right)^2\right) \tag{9.22}$$

Similarly,

$$(BP)^2 = D^2 + (X-d)^2 \tag{9.23}$$

from which, as previously, BP can be easily approximated,

$$BP = D\left(1 + \frac{1}{2}\left(\frac{X-d}{D}\right)^2\right) \tag{9.24}$$

Whence,

$$AP - BP = \frac{1}{2D}\left((X+d)^2 - (X-d)^2\right) = \frac{2Xd}{D} \tag{9.25}$$

This is the value of the path difference to be substituted in Equation (9.18) or Equation (9.19), and according to the point P, a bright or dark fringe is formed. Thus, we have for bright fringes,

$$\frac{2Xd}{D} = n\lambda$$

or

$$X = n\lambda\left(\frac{D}{2d}\right) \quad \text{[Bright fringes]} \tag{9.26}$$

and for dark fringes, we have the relation

$$\frac{2Xd}{D} = (2n+1)\left(\frac{\lambda}{2}\right)$$

or

$$X = (n+1)\lambda\left(\frac{D}{2d}\right) \quad \text{[Dark fringes]} \tag{9.27}$$

The separation on the screen between the nth and $(n+1)^{\text{th}}$ order bright fringes may now be easily computed, using Equation (9.26) as follows:

$$\bar{X} = X_{n+1} - X_n = \left((n+1)\lambda\frac{D}{2d}\right) - \left(n\lambda\frac{D}{2d}\right) = \frac{D\lambda}{2d} \tag{9.27}$$

which is also the separation between any two consecutive dark fringes. Since the separation between two consecutive bright or dark fringes is independent of n, for given values of λ, $2d$, and D the spacing of the fringes is constant. The distance between any two consecutive bright or dark fringes, denoted by $\bar{X}$, is called the *fringe width*, which varies directly with slit-screen separation D, inversely with the separation of slits $2d$, and directly with the wavelength λ of the light employed. These conclusions are in perfect agreement with the observed interference pattern on the screen.

9.3.4 Shape of the Interference Fringes

We now very easily form an idea regarding the shape of the interference fringes by deriving the equation of the locus of points having a given path difference from two slits S_1 and S_2. Let O, the middle point of S_1S_2, be chosen as the origin of coordinates of the axis of x along OX and the axis of z perpendicular to the plane of slits. Let P be any point with coordinates (z, x), and then from Figure 9.4, we get

$$\left(S_1P\right)^2 = z^2 + \left(x - d\right)^2 \quad \text{and} \quad \left(S_2P\right)^2 = z^2 + \left(x + d\right)^2 \tag{9.28}$$

where $2d$ is the separation of the slits.

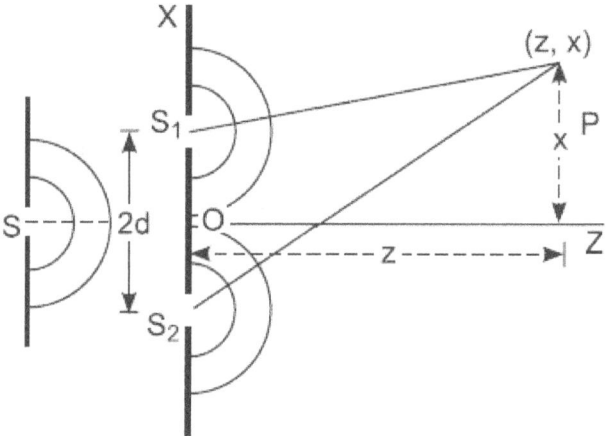

FIGURE 9.4. Shape of fringes.

The path difference Δ is,

$$\Delta = S_2P - S_1P = \sqrt{z^2 + \left(x + d\right)^2} - \sqrt{z^2 + \left(x - d\right)^2} \tag{9.29}$$

On rearranging,

$$\Delta + \sqrt{z^2 + \left(x - d\right)^2} = \sqrt{z^2 + \left(x + d\right)^2}$$

On squaring both sides,

$$\Delta^2 + 2\Delta\sqrt{z^2 + \left(x - d\right)^2} + z^2 + \left(x - d\right)^2 = z^2 + \left(x + d\right)^2$$

or

$$2\Delta\sqrt{z^2 + \left(x - d\right)^2} = 4xd - \Delta^2 \tag{9.30}$$

On squaring both sides, this equation easily reduces to

$$4\left(4d^2 - \Delta^2\right)x^2 - 4\Delta^2 z^2 = \Delta^2\left(4d^2 - \Delta^2\right)$$

or

$$\frac{x^2}{\left(\dfrac{\Delta^2}{4}\right)} - \frac{z^2}{\dfrac{\left(4d^2 - \Delta^2\right)}{4}} = 1 \tag{9.31}$$

This is the equation of *hyperbola* in a standard form with the foci S_1 and S_2 on the x axis. The loci of points of constant path difference in the XZ plane are thus *hyperbolae*. Furthermore, if instead of slits we have two coherent point sources S_1 and S_2, then the loci of points of constant path difference (hence fringes) are *concentric circles* in any plane parallel to the YZ plane. The loci of maxima and minima in space form a system of confocal hyperboloids.

It is left as an exercise for students to establish this statement analytically. The eccentricities of the hyperbolae are given by

$$e = \frac{\sqrt{\dfrac{\Delta^2}{4} + \dfrac{4d^2 - \Delta^2}{4}}}{\left(\dfrac{\Delta}{2}\right)} = \frac{2d}{\Delta} \tag{9.32}$$

In an optical experiment, the path difference Δ, corresponding to the condition of constructive or destructive interference, is of the order of 10^{-8} cm and $2d$ of the order of 10^{-2} cm. The eccentricity is, therefore, very large. As a consequence, hyperbolae are practically straight lines.

9.3.5 Interference Fringes on Screen

Figure 9.5 illustrates the interference fringes on screen. The conditions of constructive and destructive interference of two waves derived from a single source, it will be recalled, are respectively

$$\delta = 2n\pi \quad [\text{Maxima}] \tag{9.33}$$

$$\delta = \left(2n + 1\right)\pi \quad [\text{Minima}] \tag{9.34}$$

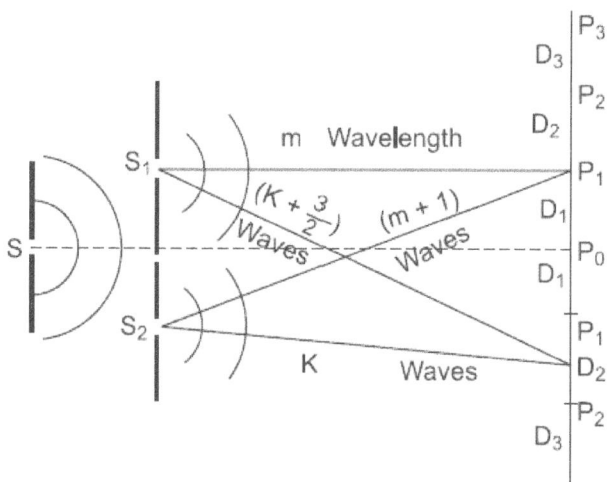

FIGURE 9.5. Interference fringes on screen.

Furthermore, it will be recalled that Huygens's wavelets from the source S reach slits S_1 and S_2 at the same instant, when both are equidistant from S. Therefore, the light sources (S_1 and S_2) will have precisely equal phases at all times. The phase difference between two waves on reaching any point P, therefore, depends only on the optical path difference (S_2P–S_1P) in this experimental arrangement. At the point P_o on the screen, equidistant from S_1 and S_2, the path difference and hence the phase difference δ between the two waves is zero. A bright fringe is, therefore, formed at P_o. This is the central fringe of the pattern, and since $\delta = 0$ can be obtained by putting $n = 0$ in Equation (9.33), this interference is spoken of as *zero order*. On either side of P_o on the screen, the optical path difference gradually increases, and we arrive at points D_1, for which the relation $\left(S_2D_1 - S_1D_1 = \dfrac{1}{2} \right)$ and hence ($\delta = \pi$) hold. The first dark fringe, on either side of the central bright fringe is, therefore, formed at D_1. At P_1 the optical path difference increases to $\left(\dfrac{2\lambda}{2} \right)$, and hence ($\delta = 2\pi$). A bright fringe is, therefore, formed at P_1. Since ($\delta = 2\pi$) can be obtained by putting ($n = 1$) in Equation (9.33), this bright fringe is spoken of as a *first order bright fringe*, and the interference is spoken of as *first order*. In this way it is possible to locate points P_0, P_1, P_2, P_3, and so forth on the screen in such a way that their distances from S_1 and S_2 differ exactly by $0, \dfrac{2\lambda}{2}, \dfrac{4\lambda}{2}, \dfrac{6\lambda}{2}, \dots$ in order, and the phases of the two waves at these points differ exactly by $0, 2\pi, 4\pi, 6\pi, ...$, in order. Therefore, at P_0, P_1, P_2, P_3, and so on are formed bright fringes of zero order, 1st order, 2nd order, 3rd order,….., respectively. In between these points are other points D_1, D_2, D_3, D_4, and so forth so that the distance from the two slits differs exactly by $\dfrac{\lambda}{2}, \dfrac{3\lambda}{2}, \dfrac{5\lambda}{2}, ...$ in order. Therefore at D_1, D_2, D_3, D_4, and so on, waves interfere destructively, thereby producing dark fringes at these points.

The interference fringes on the screen are practically straight, alternate bright and dark bands. The integer n, it should be emphasized, characterizes the order of interference and, therefore, the order of bright fringe. The bright fringes with $n = 0, 1, 2, 3, ...$ are of zero, first, second,…, order.

9.3.6 Intensity Distribution Curve

We are now in a position to plot the intensity distribution for interference fringes from coherent monochromatic beams with the knowledge that the intensity at any point of the screen is expressed by (with k taken as unity)

$$I = 2a^2 (1 + \cos \delta) \tag{9.35}$$

where δ is the phase difference between the two waves at P. If a bright fringe of nth order is formed at P, then $\delta = 2n\pi$. For the sustained fringes, the value of δ at P must at all times remain $\delta = 2n\pi$, with further additional knowledge that the spacing of the fringes on the screen is constant for given values of λ, $2d$, and D. Along a horizontal line, the values of δ, which are $0, \pi, 2\pi, 3\pi, ...$, and so on are marked at equidistant points, and the axis of I is taken along the normal line through $\delta = 0$. For $\delta = 0, \pm 2\pi, \pm 4\pi, \pm 6\pi$, and so on, $I = 4a^2$, and for $\delta = 0, \pm \pi, \pm 3\pi, \pm 5\pi$, and so forth, $I = 0$. In fact, as δ gradually increases from 0 to π, $\cos\delta$ gradually decreases from +1 through 0 to −1 and, as a consequence, I gradually diminishes from $4a^2$ to 0. Hence, the shape of the intensity distribution curve is as depicted in Figure 9.6.

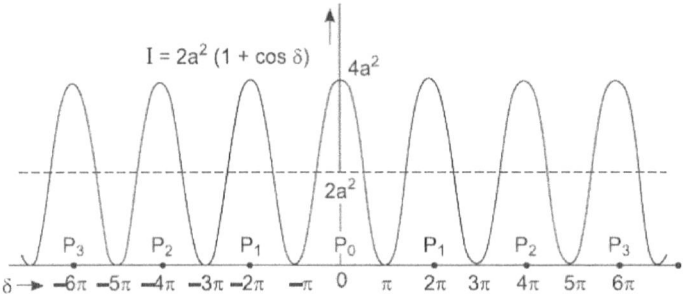

FIGURE 9.6. Intensity distribution for the interference fringes from two beams.

Before concluding our discussion, it should be emphasized that no destruction of light energy occurs in the phenomenon of the *interference of light*. Interference does not transform light energy into any other form of energy, so the total amount of light energy must remain constant. What has happened is merely a redistribution of energy; the energy which apparently disappears at the dark fringes is actually present in the bright fringes, where the energy is $4a^2$. However, the average value of the energy over any number of fringes is the same as if the interference effects were absent. For example, the average value of the intensity on the screen over the range $\delta = 0$ to $\delta = 2\pi$ is given by

$$I_{Average} = \frac{\int_0^{2\pi} I d\delta}{\int_0^{2\pi} d\delta} = 2a^2 \frac{\int_0^{2\pi} (1+\cos\delta) d\delta}{\int_0^{2\pi} d\delta} = 2a^2 \tag{9.36}$$

and this justifies the statement made previously as illustrated in Figure 9.7. There is no violation of the law of conservation of energy in the interference phenomenon.

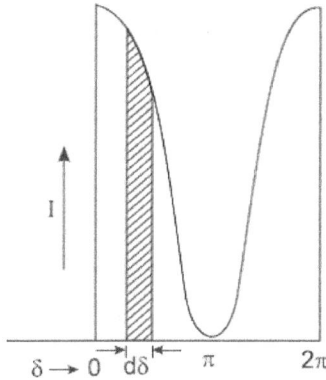

FIGURE 9.7. Average value of the intensity on the screen over the range $\delta = 0$ to $\delta = 2\pi$.

9.4 CONDITIONS FOR INTERFERENCE

In a well-defined interference pattern, the intensity at regions corresponding to destructive interference must remain zero, while at regions corresponding to constructive interference, it must remain maximum for all values of time. To accomplish this, it is essential that in every experimental arrangement, the following conditions are fulfilled:

A. The two beams of light which interfere must originally originate in the same source of light.

B. The waves must have the same period and wavelength. Also, their amplitudes must be equal or very nearly equal.

C. The original source must emit light of a single wavelength or be very nearly monochromatic. On the other hand, if the light source is heterogeneous, the optical path difference between the two interfering beams must be very small.

D. The two interfering waves must be propagated in almost the same directions or the two interfering wavefronts must intersect at a very small angle.

E. In addition to the previous conditions, the following condition must also be satisfied if we are dealing with the interference phenomenon produced by polarized light: The two interfering waves must be in the same state of polarization.

9.4.1 Condition A is Fundamental for the Production of Stationary Maxima and Minima

This can be seen from the following considerations. It will be recalled that the resultant intensity I at any point P due to the superposition of light waves from two sources is given by the expression in Equation (9.35). This total phase difference depends on two factors:

(i.) Initial phase difference δ_1 between the two sources.
(ii.) Phase difference due to optical path difference $(S_2P{-}S_1P)$, namely,

$$\delta_2 = \left(\frac{2\pi}{\lambda}\right)(S_2P - S_1P).$$

Obviously, δ_2 remains constant at P throughout the experiment. Hence, in order to keep δ constant at P, the initial phase difference δ_1 between the two vibrating sources must remain constant. To satisfy this requirement, the vibrating sources must be identical in all respects. That is, if there is any sudden phase change in one, there must simultaneously be a corresponding phase change in the other, with the result that the phase difference between the two interfering waves is not changing with time. Such two sources having a point-to-point phase relationship are said to be coherent. This is only possible when both become the sources of light under the influence of light waves originating from the same source.

Two coherent sources can be accomplished experimentally by either making one source image of the other by reflection (Lloyd's single mirror) or by dividing the light waves into two parts by reflection (Fresnel's double mirror) or by refraction (Fresnel's biprism) or by partial reflection (Newton's rings, Michelson's interferometer). In these circumstances, any change of phase which the original light wave undergoes is shared by its two parts instantaneously, with the result that the fringes remain stationary.

Sustained interference effects can never be accomplished with two independent sources due to the fact that any two arbitrary beams of light are always incoherent.

Condition B follows from the previous mathematical analysis. Even with nearly unequal amplitudes, the intensity in the region of destructive interference would be very small so as to render it dark as compared with the maximum intensity in the constructive interference region.

Condition C is essential in order to avoid the complete masking of interference patterns due to the presence of different wavelengths in the light emitted by light sources. We shall again refer to this point when discussing white light fringes.

The necessity of **Condition D** being satisfied can be clearly seen by a comparison of the two systems of plane wavefronts originating from the same source and traveling to the right, as shown in Figure 9.8. The solid lines in the figure represent the regions of maximum positive displacement, while dotted lines represent the regions of maximum negative displacement in these waves. As in Figure 9.2, the regions of constructive and destructive interferences are marked in the present figure as well. Then angle θ between the two wavefronts is small in Figure 9.8 (a) as compared with that in Figure 9.8 (b).

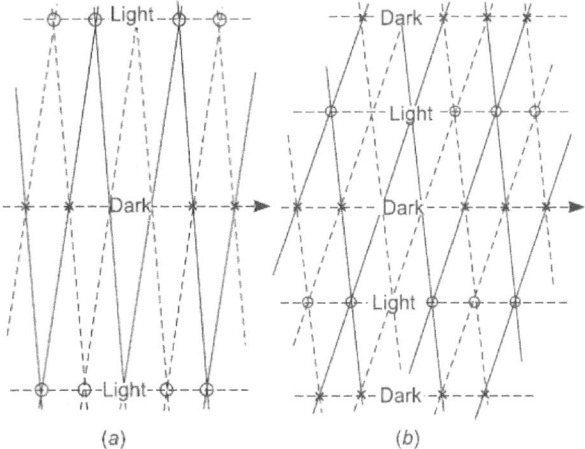

(a)   (b)

FIGURE 9.8. Superposition of two plane waves (a) at a small angle, (b) at a large angle.

A comparison of these figures clearly illustrates that the larger the angle θ, the smaller the spacing between interference fringes. Although, for a larger angle θ, interference effects are sustained, yet the fringes may become indistinguishable even under high magnification. Therefore, the condition that the interfering wavefronts must intersect at a small angle with each other, in mathematical language, $2d$, the separation between the two coherent sources must be small so as to produce widely spaced fringes, essentially applies to the observation rather than to the production of sustained interference effects.

9.5   CLASSIFICATION OF INTERFERENCE PHENOMENON

The experimental devices for producing two interfering beams of light from the same source may be conveniently classified under the following two main heads:

A.   *Division of Wavefront.* The devices which divide the incident wavefront into two parts by utilizing the phenomena of reflection, refraction, or diffraction in such a way that after traversing different optical paths they eventually reunite at a small angle to produce interference bands come under this class of interference phenomenon. In every experiment employing such devices it is absolutely essential to employ either a point source or a line source, for example, a narrow illuminated slit parallel to the line of division of the incident wavefront. The Fresnel biprism and mirrors and Lloyd's mirror described in this chapter are examples of this class. In all such devices, since limited portions of the wavefront are employed, diffraction effects will also be present along with the interference effects.

B.   *Division of Amplitude.* Devices which divide the amplitude of the incoming wave of light into two or more parts by partial reflection and refraction, and thereby give rise to two or more beams which are later made to reunite to produce the interference effects, come under this class. In the case of instruments employing such devices, it is not essential to employ a point source or a narrow line source, but a broad light source may be employed to produce brighter bands. Since a large section of the wavefront is employed, diffraction effects are minimized.

9.6 FRESNEL'S EXPERIMENTAL ARRANGEMENTS

Soon after Thomas Young announced his crucial double slit experiment on the interference of light and explained the results with reference to *wave nature*, critics raised the objection that the bright and dark fringes which he had observed were not due to true interference of two beams of lights. On the other hand, they were inclined to attribute them to some complicated modification of light, possibly owing to diffraction at the edges of the double slit (or pinholes). In order to overcome this objection, Fresnel designed several novel experimental arrangements, in which diffraction was either largely eliminated or distinctly separated. Thus, he demonstrated the interference of two beams of light to the satisfaction of all, and thereby placed wave theory on a firm footing. The first of these arrangements is known as *Fresnel's biprism experiment*.

9.6.1 Fresnel's Biprism Interference Fringes

The apparatus employed in *Fresnel's biprism experiment* is shown schematically in Figure 9.9.

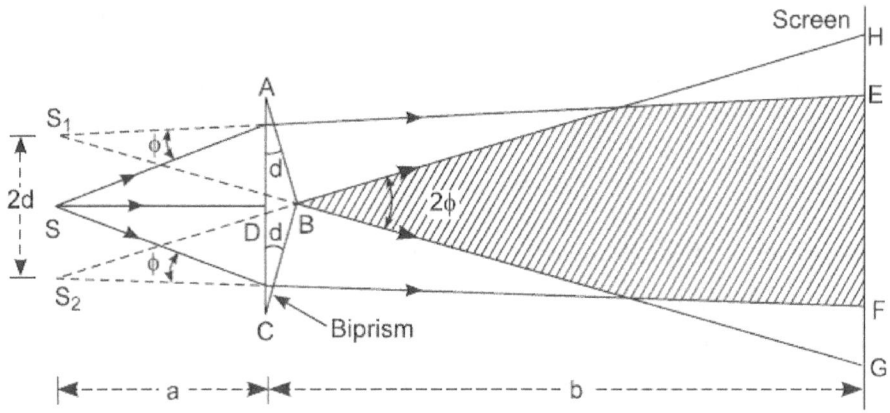

FIGURE 9.9. Fresnel biprism.

A biprism is essentially two prisms, each of very small refracting angle θ, placed base to base. In reality, the biprism is constructed from a single plate of glass by suitable grinding and polishing; the obtuse angle of the prism is only slightly less than 180°, and the other angles of the order of 30° are equal. In the experimental arrangement, described in sequence, this prism is so adjusted in relation to the source slit that the two halves of the incident wavefront suffer separate simultaneous refraction through the prism; hence, even this single prism is termed as a *biprism*. The essential idea is to divide the incident beam into two coherent interfering beams by utilizing the phenomenon of refraction.

Light from a narrow slit S, illuminated with monochromatic light of wavelength λ, is allowed to fall symmetrically on a biprism. The intersection of the two inclined faces forming the obtuse angle must be adjusted accurately parallel to the length of the slit. In Figure 9.9, the slit passes through S in a perpendicular direction to the plane of the diagram. Under this condition, this edge B divides the incident wavefront into two parts. First is the one which in passing through the upper half ABD of the biprism is deviated through a small angle toward the lower half of the diagram and appears to diverge from the virtual image S_1. Second is the one which in passing through the lower half CBD is deviated through a small angle toward the upper half of the diagram and appears to diverge from the virtual image S_2. The two emergent wavefronts, which intersect at a small angle, are derived from the same wavefront, and hence the fundamental condition of interference is satisfied. The virtual images S_1 and S_2, being the image of the slit S, obviously function as coherent sources in this experiment. Since the angles BAC and BCA are small and equal, these images are coplanar with, extremely close to, and equidistant from the source slit S, that is, $2d = S_1 S_2$ is extremely small. As a consequence, interference fringes are observed on the screen in the overlapping region EF of the two emergent beams of light. Interference patterns can also be seen through a powerful eyepiece in its principal focal plane. The fringes extend into space and are thus non-localized.

Determination of the Wavelength of Light

In order to determine the wavelength of monochromatic light with the help of biprism fringes, we employ the formula

$$\lambda = \left(\frac{2d}{D} \right) \bar{X} \qquad (9.37)$$

The value of fringe width $\bar{X}$, the distance $2d$ between the virtual coherent sources S_1 and S_2, and the normal distance D of the plane of observation of the fringes from the slit should be measured after making a few adjustments in the apparatus.

The experiment is performed on a heavy metallic optical bench about 2 meters in length and is supported on four leveling screws at the base. The bench is provided with a scale on one side, graduated in mm on corrodible material. The bench carries four uprights for supporting the adjustable slit, the biprism, a high-power micrometer Ramsden's eyepiece, and a convergent lens. These uprights are capable of movement along and also perpendicular to the length of the bench and may be adjusted to any desired height. The slit and the biprism may be rotated in their own plane with the help of tangent screws provided in the uprights. Each upright carries a vernier at the base, and thus its position on the bench can be accurately ascertained.

Adjustments. Before carrying out the measurement of the fringe-width, it is essential to obtain correct fringes in which the spacing is uniform on the entire field by carrying out the following adjustments in the apparatus:

1. The bed of the optical bench is first leveled with the help of a spirit level and leveling screws.

2. The eyepiece is focused on the cross-wires by moving the tube containing the lenses in the cross-wires tube until they are distinctly visible. One of the two wires in the cross is then made exactly vertical by observing a plumb line through the eyepiece and rotating the latter about its own axis until one wire exactly coincides with the image of the plumb line, which of course is vertical.

3. The slit and the eyepiece are adjusted to the same height above the bench. A real image of the illuminated slit is then formed in the plane of cross-wires by the help of a convex lens of small aperture. The slit is now rotated in its own plane by the help of a tangent screw until its image exactly coincides with the vertical wire in the eyepiece. The slit is then exactly vertical.

4. The biprism is mounted, keeping its refracting edge nearly vertical in its upright between the eyepiece and the slit, which is made quite narrow and illuminated with light whose wavelength is to be determined, as shown in Figure 9.10.

5. The edge formed by the intersection of inclined faces enclosing the obtuse angle in the biprism must be now adjusted exactly parallel to the vertical slit. To make this adjustment, two real images of the coherent sources S_1 and S_2 are formed in the focal plane of the eyepiece with the help of a convergent lens. By lateral movement of the prism, the images are made equally bright, that is, equally well focused and of equal height by rotation of the biprism in the vertical plane with the help of tangent screw. On removing the lens, this edge can now be made exactly parallel to the slit by giving finer rotation to the prism until the interference fringes become perfectly distinct and well defined.

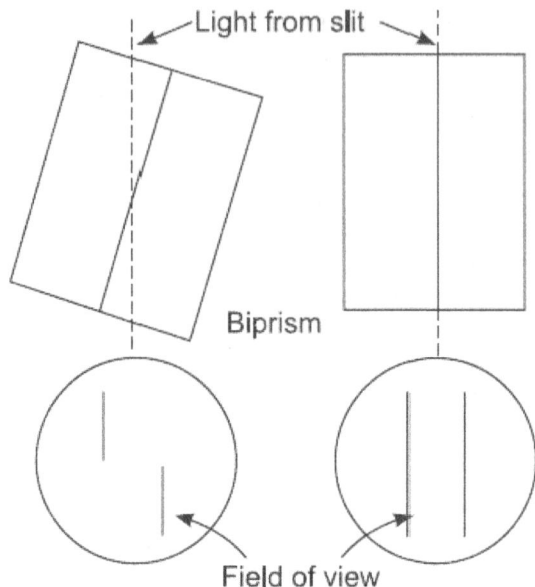

FIGURE 9.10. The biprism is mounted, keeping its refracting edge nearly vertical in its upright between the eyepiece and the slit.

6. **The axis of the experiment should be parallel to the length of the optical bench.** In other words, the line joining the obtuse edge of the biprism and the slit must be adjusted exactly parallel to the length of the bench.

It will be recalled that the expression for the fringe width $\bar{X}$, which is

$$\bar{X} = \left(\frac{D\lambda}{2d} \right) \tag{9.37}$$

was derived and shown uniform over the entire field for a given D, λ, and $2d$ on the supposition that the plane of observation of the fringes is exactly parallel to the plane containing the virtual coherent sources S_1 and S_2. It is absolutely essential that this condition should be satisfied in the experimental arrangement based on the division of the wavefront to obtain correct fringes. In our experiment, the plane of cross-wires is transverse to the length of the bench. Furthermore, a comparative study of diagrams sketched in Figure 9.11 leads us to the conclusion that the plane S_1S_2 will be transverse to the bench, as in the lower diagram, only when the line joining the obtuse edge of the prism and the slit is adjusted exactly parallel to the length of the bench.

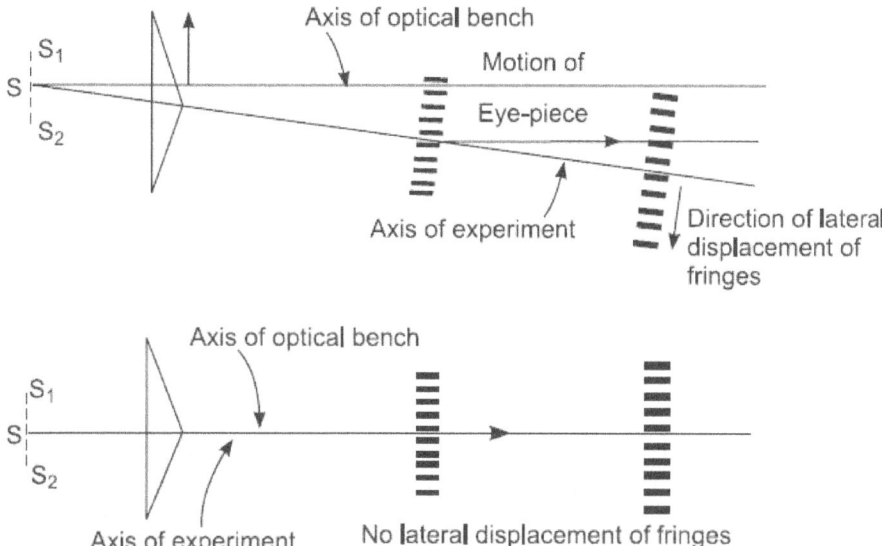

FIGURE 9.11. A Comparative study of diagrams.

If this adjustment is not perfect then, as shown in the upper diagram, the fringes would shift laterally relative to the cross-wires as the eyepiece is moved backward (or forward) along the bench. To remove this lateral shift, the biprism is moved to a small distance transversely to the bench in a direction opposite to the direction of the shift, and again the fringes are tested for lateral shift by taking back the eyepiece. When the position of the biprism relative to the slit is so adjusted as shown in the lower diagram in Figure 9.11, the lateral shift of the fringes would vanish.

Measurements. After making the previous adjustment in the apparatus, the following measurements are taken in sequence.

Measurement of $\overline{X}$. The eyepiece is fixed at a suitable distance from the biprism so that the fringes are fairly short and fairly wide in its principal focal plane. The cross-wire is set accurately at the center of the first bright fringe, and the reading of the micrometer screw is taken. The cross-wire is then moved in the same direction by the help of the micrometer screw, stopping at the center of every successive bright fringe, and the corresponding readings of the screw are noted. From these observations, the fringe width X may be easily deduced.

Measurement of D. The distance between the eyepiece and the slit can be directly found out by reading the positions of their uprights on the optical bench and taking the difference of the two readings. However, this distance is subject to the bench correction in order to obtain correct D. This is due to the fact that the slit and the cross-wires are not exactly at the zero mark of the vernier of their respective uprights, while D is the separation between the slit and the cross-wires. The bench correction should be determined after $2d$ has been measured.

Measurement of $2d$. A convex lens is introduced between the biprism and the eyepiece, and the latter is fixed at a distance from the slit which is greater than four times the focal length of the lens. As in the displacement method of measuring the focal length, the lens is adjusted to a position, marked 1st position in Figure 9.14, so that the magnified, distinct, and real images of the virtual coherent sources are formed in the plane of the cross-wires. The separation d_1 between these images is then measured with the help of cross-wires by giving it a lateral displacement from one image to the other. The lens is then moved to the conjugate position, marked 2nd position in Figure 9.12, so that distinct diminished images are formed in the plane of the cross-wires.

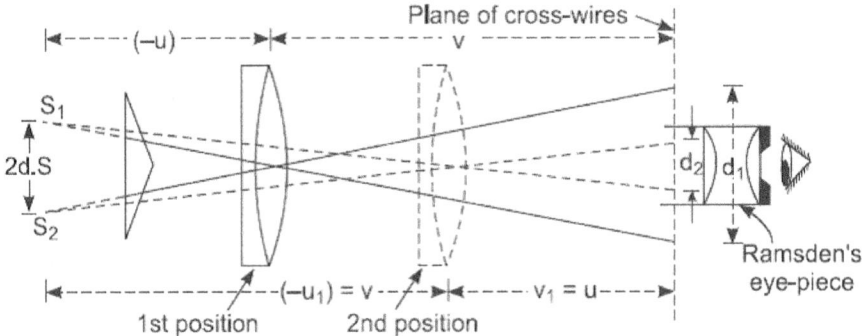

FIGURE 9.12. The lens moved to the conjugate position marked 2nd position.

The separation d_2 between these images is also measured. Since the magnification m_1 in the first position of the lens is just the inverse of the magnification m_2 in its conjugate position, we have

$$\left(\frac{d_1}{2d}\right) = \left(\frac{2d}{d_2}\right)$$

or

$$2d = \sqrt{d_1 d_2} \tag{9.38}$$

Evaluation of Bench Error. The biprism and the lens are removed. A rod of known length is held horizontally with its one end just touching the slit and the other in the plane of the cross-wires. The difference between the length of the rod and the separation between the two uprights is the bench error, which should be applied with proper sign to obtain correct D.

The measured values of X, $2d$, and D are substituted in Equation (9.37) to give the wavelength λ of the light employed in the experiment. The unit of wavelength is $Angstrom$. $Angstrom = 10^{-8}\,cm$.

9.6.2 Fresnel's Double Mirror

The second device of Fresnel for obtaining the sustained interference effects essentially consists of dividing the incident wavefront into two coherent interfering wavefronts by utilizing the phenomenon of reflection. It is illustrated schematically in Figure 9.13 and is known as the *Fresnel double-mirror experiment.*

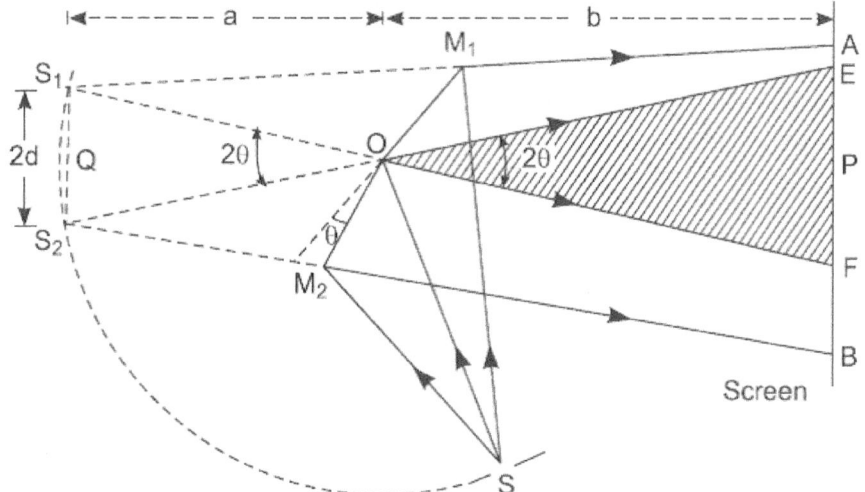

FIGURE 9.13. Fresnel double-mirror experiment.

Two optically plane mirrors OM_1 and OM_2, highly silvered on their front surfaces or blackened at the back to avoid multiple internal reflections, are mounted vertically inclined at a very small acute angle θ. Light from a narrow slit S, illuminated with monochromatic light of wavelength λ, is allowed to fall on the mirrors, whose line of intersection through O must be adjusted parallel to the length of the vertical slit through S. One half of the incident wavefront is reflected from OM_1 and appears to diverge from the point S_1, which is consequently the virtual image of the source slit S in OM_1. The other half of the incident wavefront is reflected from OM_2, giving rise to an image S_2 of the source slit S. Obviously, S_1 and S_2 are the virtual coherent sources in this experiment, since both are the images of the same source slit S. Thus, the fundamental condition of interference is satisfied. Furthermore, since the angle θ between the mirrors is very small, the separation $2d$ between the coherent sources S_1 and S_2, expressed by

$$2d = 2a\theta \qquad (9.39)$$

(where $OS = a$) is also very small. Thus, the condition for the observation of widely spaced interference fringes is also satisfied. The light on the screen AB appears to come from two virtual coherent sources S_1 and S_2, and in the region EF where the two beams overlap, interference fringes are observed parallel to the slit. That these bands were actually produced by interference and not by diffraction was demonstrated by Fresnel by covering one of the two mirrors, when all traces of equally spaced fringes disappeared. Therefore, this experiment gives a device evidence in favor of the wave theory of light. However, since the source slit S is employed at a position for almost grazing incidence of the light on the mirrors OM_1 and OM_2, the angle subtended by each mirror at the slit is very small. In other words, the incident wavefront is very narrow, and mirrors offer limitation to the wavefront. This results in the production of diffraction fringes, which have approximately the same spacing as the interference fringes. Therefore, the two systems are superposed on each other and cannot be easily distinguished.

9.7   INTERFERENCE WITH WHITE LIGHT

In Young's double slit and Fresnel's biprism and double mirror experiments, we have employed monochromatic light of a single wavelength to obtain a large number of bright and dark fringes. Let us now study the changes produced in the fringes when a monochromatic light source is replaced by white light in these experiments. White light consists of wavelengths varying between 4000 Å to 7500 Å, from the violet to the red end of the white light spectrum.

At any point P_o situated on the perpendicular bisector of coherent sources S_1 and S_2, the geometrical path difference is zero for the interfering light waves of all wavelengths. At this point, therefore, the condition of constructive interference is satisfied for all the wavelengths present in the white light.

Here we get the light of every color in exactly the same proportion as it exists in white light. As a consequence, the resultant illumination at P_0, due to the superposition of zero order (or central) bright fringes of all the wavelengths (color), is white.

The spacing between the consecutive bright or dark fringes,

$$\bar{X} = \frac{D\lambda}{2d} \tag{9.40}$$

is a function of the wavelength. Obviously, the smaller the wavelength, the closer will be corresponding fringes. The fringes of different wavelengths are, therefore, in step only at the central fringe but soon get out of step. On either side of the central white fringe, we shall have a dark fringe tinged with a violet color. Since λ, the wavelength of the violet end of the visible spectrum is the least, therefore the condition of constructive interference which is

Path difference = λ

will be first satisfied for the violet color and then for other colors, in the spectral order, as we move away from the central fringe. As a consequence, the bright fringe nearer to the central fringe shall have a strong violet tinge, followed by other bright fringes having a strong tinge of colors in the spectral order. In reality, no bright fringe is of saturated spectral color.

After eight or ten fringes, the path difference becomes so large that the condition of constructive interference, which is

Path difference = $n\lambda$

may be simultaneously satisfied for the number of wavelengths; for example, we may have the relation

Path difference $= 5\lambda_1 = 7\lambda_2 = 9\lambda_3$

and also at the same point the condition of destructive interference, which is,

Path difference $= (2n+1)\dfrac{\lambda}{2}$

may be simultaneously satisfied for many colors. As a consequence, for large optical path difference, the dark fringes of some wavelengths are completely masked by the bright fringes of other wavelengths.

Furthermore, at some points of the screen so many colors are present that the resultant illumination cannot be distinguished from white light with an unaided eye, and so the field appears essentially white. It should be noted that the interference is still occurring in this region but is not visible in ordinary circumstances. However, it can be clearly demonstrated by spectroscopic analysis of this region.

It can be concluded that for observable white light fringes, the optical path difference between the two interfering waves of light must be very small, and we observe eight or ten colored fringes on either side of the central white fringe.

9.8 DISPLACEMENT OF THE FRINGES

Let us now investigate the effect on the interference fringes of introducing a thin transparent plate in the path of one of the two interfering beams of monochromatic light. Let e be the thickness of the plate and μ its refractive index for the monochromatic light employed. From Figure 9.14 it is clear that a light wave in traveling from S_1 to P has to traverse a distance e in the plate while the rest $(S_1P - e)$ it traverses in air.

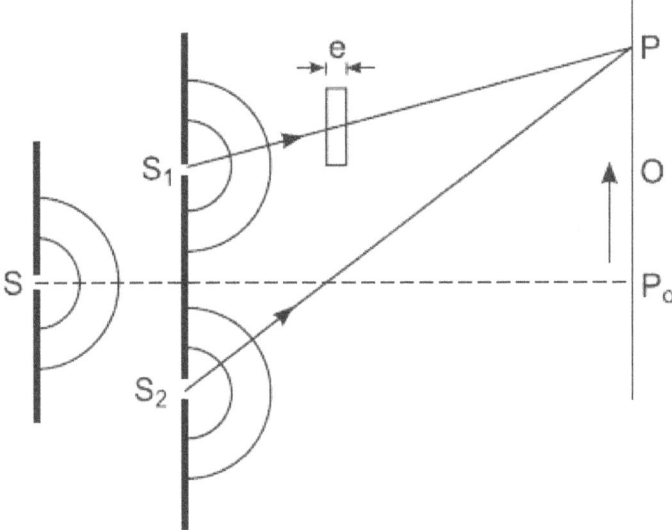

FIGURE 9.14. Displacement of fringes.

The time required for this journey is, therefore, given by

$$t = \frac{S_1P - e}{c} + \frac{e}{V_g} = \frac{1}{c}\left(S_1P - e + \frac{c}{V_g}e\right) = \frac{S_1P + (\mu - 1)e}{c} \tag{9.41}$$

The physical interpretation of the previous relation is that due to the introduction of the plate, the effective path from S_1 to P becomes $\left(S_1P + (\mu - 1)e\right)$ in air. Similarly, the effective path from S_1 to P_o, the point equidistant from S_1 and S_2, becomes $\left(S_1P_o + (\mu - 1)e\right)$ in air; and since

$$S_1P_o + (\mu - 1)e > S_2P_o \qquad (9.42)$$

the central bright fringe of zero order is not formed at P_o which, of course, is the normal position of the central fringe in the absence of the plate.

To locate the new position of the central fringe we should travel along the screen in such a direction that the left side of the previous inequality may decrease while the right side may increase, and eventually the two may become equal at some point. Suppose that in the presence of the plate, the central fringe of zero optical path difference is formed at O; we therefore write

$$S_2O = S_1O + (\mu - 1)e$$

or

$$S_2O - S_1O = (\mu - 1)e \qquad (9.43)$$

If $P_oO = x_o$, we can easily obtain the geometrical path difference

$$S_2O - S_1O = \frac{2x_o d}{D} \qquad (9.44)$$

This value of the path difference is to be substituted in Equation (9.43) to get the distance through which the fringe system has been displaced. Thus, we get

$$x_o = \frac{D}{2d}(\mu - 1)e = \frac{\overline{X}}{\lambda}(\mu - 1)e \qquad (9.45)$$

Furthermore, if the central fringe shifts to the nth bright fringe of the original system obtained without the plate, then

$$S_2O - S_1O = n\lambda \qquad (9.46)$$

Hence,

$$(\mu - 1)e = n\lambda \qquad (9.47)$$

It can be easily seen that the spacing of the interference fringes remains unaffected due to the introduction of the plate. If at the point P, in presence of the plate, bright fringe of the nth order is formed, then we must have the relation

$$S_2P - \left[S_1P + (\mu-1)e\right] = n\lambda$$

or

$$S_2P - S_1P = n\lambda + (\mu-1)e \tag{9.48}$$

If $P_oP = x_n$, then the geometrical path difference,

$$S_2P - S_1P = 2x_n\left(\frac{d}{D}\right) \tag{9.49}$$

where

$$x_n = \frac{D}{2d}\left(n\lambda + (\mu-1)e\right) \tag{9.50}$$

Similarly, for the $(n+1)^{th}$ bright fringe, we write

$$x_{n+1} = \frac{D}{2d}\left((n+1)\lambda + (\mu-1)e\right) \tag{9.51}$$

Hence,

$$\overline{X} = x_{n+1} - x_n = \frac{D\lambda}{2d} \tag{9.52}$$

a relation independent of the thickness of the thin plate and exactly similar to that which was derived in the absence of the plate.

On putting $n = 0$ in Equation (9.50) we may again get x_o, the distance through which fringes have been displaced. We conclude that the entire fringe system shifts laterally through a distance $\left(\dfrac{D(\mu-1)e}{2d}\right)$ toward the side on which the plate is placed, while the general shape and the spacing of the fringes remain unaffected provided, of course, the plate is thin.

By the measurement of the fringe shift, it is possible to employ Equation (9.45) to evaluate μ of the plate if we know its thickness or to evaluate its thickness if μ is known.

9.9   LLOYD'S SINGLE MIRROR

Lloyd (1837) has devised a very simple arrangement for satisfying the fundamental condition of interference, in which one part of the incident wavefront reaches the screen after suffering reflection while the other part reaches it directly. In addition to simplicity, it has another importance due to the fact that this experiment provides an experimental confirmation of the fact that a sudden phase change of π occurs on reflection in a rarer medium, that is, from the surface backed by a denser medium. Lloyd's arrangement is shown schematically in Figure 9.15.

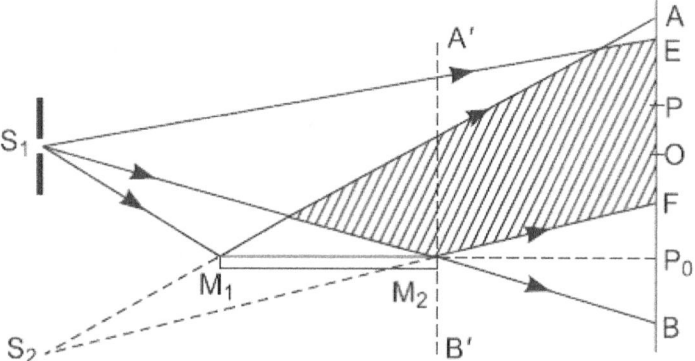

FIGURE 9.15. Lloyd's arrangement.

Light from a narrow slit S_1, illuminated with monochromatic light, is partly incident at a grazing angle on the surface of an optically plane front-surfaced mirror M_1M_2, while the rest reaches the screen directly. The reflected light appears to diverge from S_2, the virtual image of the source slit S_1, formed by reflection at the mirror. Accordingly, the virtual image S_2 acts as a coherent source with the slit S_1 itself in producing the interference fringes, and thus the fundamental condition of interference is satisfied. In the region of the overlapping of the direct and the reflected beams, shown shaded, interference occurs, and equally spaced interference fringes can be observed in the region EF of the screen. The mirror employed is one having silvering on its front surface to avoid multiple internal reflections, and its surface should be adjusted so as to make it exactly parallel to the vertical source slit S_1.

It will be observed that the point P of the screen, equidistant from S_1 and S_2, receives only the direct light from S_1 and not that reaching the screen after reflection. Therefore, the central fringe of zero geometrical path difference is not visible. Not only this, even less than half the interference pattern is visible with the arrangement sketched in Figure 9.15. However, half the central fringe can be brought into view by displacing the screen to the position $A'B'$ so that it just touches the edge M_2 of the mirror. When this is done, the edge M_2, which is equidistant from S_1 and S_2, passes through the center of the fringe of zero geometrical path difference. This fringe is dark instead of being a bright one. This can only occur at a point equidistant from S_1 and S_2, provided one of the two interfering beams suffers a sudden phase change of π. Since the direct beam could not suffer a sudden phase change, it must be the reflected beam which should suffer a sudden phase change of π, and the only possibility of this is on reflection from the surface backed by the denser medium. This may be interpreted as meaning that the coherent sources S_1 and S_2 differ in phase by π at all times instead of being in equal phase as in the double slit, Fresnel's biprism, and the double mirror experiments. As a consequence, the two waves on reaching any point P of the screen interfere constructively or destructively according to whether the geometrical path difference is an odd or even multiple of $\left(\dfrac{\lambda}{2}\right)$. In symbols, the conditions can be expressed as:

$$S_2 P - S_1 P = 2n\left(\frac{\lambda}{2}\right) \qquad \text{[Dark fringe]} \qquad (9.52)$$

$$S_2 P - S_1 P = 2(n+1)\left(\frac{\lambda}{2}\right) \qquad \text{[Bright fringe]} \qquad (9.53)$$

It will be observed that these conditions are just the reverse of the conditions of the constructive and destructive interference in the biprism arrangement.

Complete central fringe can also be brought into view by introducing a thin lamina of glass or mica in the path of the directly transmitted beam, keeping the screen fixed in the original position. Owing to the retardation of the directly transmitted waves, the complete fringe system shifts in the direction in which the plate is introduced, in Figure 9.15, in the upward direction. If the lamina is of suitable thickness, there can be one point O on the screen in the region of the overlapping of the two beams so that the condition

$$S_2 O = S_1 O + (\mu - 1)e \qquad (9.54)$$

is fulfilled. The central fringe of zero geometrical path difference will appear at O. But with monochromatic light, say sodium light, the fringes being alternately dark and bright, having one color it is not possible to distinguish the central fringe from the other bright fringes. Therefore, it is customary to employ a white light source when the fringes are to be displaced for locating the central fringe. With white light, the central fringe of zero geometrical path difference will be black while the other fringes will have other colors. At the central fringe, not only is the geometrical path difference zero, but also the phase difference is π for all the wavelengths due to an initial phase difference of π between the coherent sources S_1 and S_2. As a consequence, the central dark fringes corresponding to all the wavelengths are superposed at this point. This is also shown in Figure 9.16, in which the intensity distribution curves for the violet and the yellow colors are shown with ordinary and dotted curves respectively, while curves of other colors are not shown for the simplicity of the diagram.

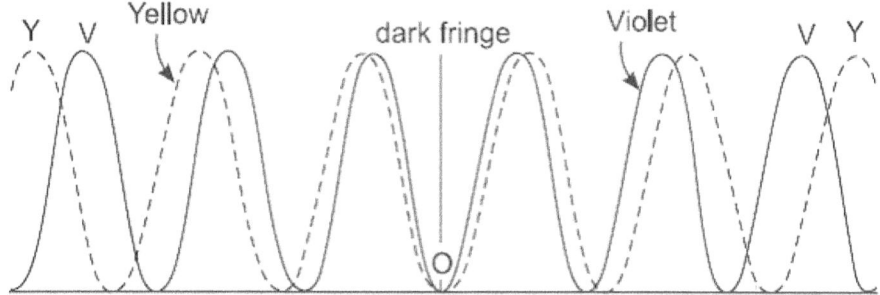

FIGURE 9.16. The origin of white light fringes with a dark fringe at the center.

It should be remarked that owing to grazing incidence of light on the mirror, the effective aperture on viewing from P is extremely small. The mirror, therefore, appears like a slit, giving rise to diffraction fringes due to the narrow slit. But, in this experiment, diffraction effects are of secondary importance.

9.9.1 Determination of the Wavelength

The wavelength of light can also be determined by the measurement of the fringe width X of the fringes produced by Lloyd's single mirror arrangement. For the validity of the formula

$$\lambda = \left(\frac{2d}{D} \right) \bar{X} \tag{9.55}$$

it should be again emphasized that the plane containing the coherent sources S_1 and S_2 must be exactly parallel to the plane of the observation of the fringes. Since the line joining the object and its image, formed due to reflection in a plane mirror, is always perpendicular to the reflecting surface, for the validity of the previous equation, this reflecting surface must be adjusted exactly perpendicular to the plane of cross-wires of the eyepiece. Furthermore, the reflecting surface must be adjusted exactly parallel to the length of the vertical source slit. We now give the following experimental details to achieve this end.

The experiment is performed on a leveled optical bench. Lloyd's mirror is mounted vertically on an upright, which is provided with screws to effect the rotation of the mirror about a vertical as well as about a horizontal axis. A vertical slit, illuminated with light whose wavelength is to be determined, is mounted on another upright and adjusted close to the line of the mirror so that the light may be reflected as far as possible at grazing incidence. The fringes are observed in the focal plane of the focused micrometer eyepiece.

It is absolutely essential, as explained previously, that M_1M_2 should be normal to AB. To achieve this, a convergent lens is introduced between the mirror and the eyepiece, and it is adjusted so that its center lies opposite the edge M_2 of the mirror and on a level with the center of the source slit S_1. Finally, the lens is fixed in a position so as to form the real images of S_1 and S_2 in the plane of cross-wires. Both images will be in focus in the plane of cross-wires for one setting of the lens, provided S_1S_2 is parallel to plane (AB) of the cross-wires. In other words, M_1M_2 is normal to AB. We therefore rotate M_1M_2 slightly about the vertical axis of rotation until the two images are exactly in focus simultaneously.

The reflecting surface of the mirror may be easily adjusted parallel to the vertical slit. On removing the lens, fringes are observed through the eyepiece. Mirror M_1M_2 is rotated about its horizontal axis to get the position when the fringes become well defined and brightest. The adjustment of M_1M_2 normal to the plane of cross-wires may be now tested by the absence of the lateral shift in the fringes as the eyepiece is moved backward (or forward) along the optical bench. Now the measurement of X, $2d$, and D may be carried out as in the biprism experiment.

9.10 ACHROMATIC INTERFERENCE FRINGES

An *achromatic fringe* is defined as one which shall result due to superposition of the same order fringes of light of all wavelengths emitted by the heterogeneous source. If it were possible to devise an experimental arrangement in which the geometrical path difference for the two coherent interfering waves from the source to a point of observation is different for different wavelengths, then the geometrical path difference can also be made proportional to the wavelength. As a consequence, the phase difference expressed as

$$\delta = \frac{2\pi}{\lambda} \times \text{Path difference}$$

becomes the same for all wavelengths at the point. In other words, the phase difference is independent of the wavelength over the whole region of the observation. The physical conception of this conclusion is that the first order fringes of all wavelengths are superposed on each other, and obviously the same is true for 2nd order fringes, 3rd order fringes, and so on; that is, the fringe width $\bar{X}$ becomes independent of the wavelength. We then have a system of achromatic fringes.

In *Young's experiment* and *Fresnel's biprism* and *double mirror experiments*, there is only one point in the plane of observation, which is the point equidistant from the coherent sources, where the phase difference is the same (zero) for all wavelengths. At this point zero order fringes of all wavelengths coincide, giving rise to an achromatic fringe. But the spacing between two consecutive bright or dark fringes is a function of the wavelength,

$$\bar{X} = \left(\frac{D\lambda}{2d} \right)$$

being greatest for the red and least for the violet end of the spectrum. As a consequence, the same order fringes of different wavelengths, excluding zero order, do not coincide. All three arrangements are thus perfectly achromatic only for the central fringe, provided we neglect the secondary effects due to finite slit width. This is also true for Lloyd's interference fringes but of course independent of the slit width. The fringe of zero geometrical path difference in Lloyd's mirror is not formed under ordinary arrangements. But the central fringe when brought into view by introducing a thin lamina of mica or glass in the path of direct light is found to be black even with a white light source. The other fringes are strongly colored, as in Figure 9.16. However, if by some device we render the fringe-width $\bar{X}$ the same for fringes of all wavelengths, even with white light, the bright fringes of a given order of different wavelengths will be perfectly superposed. Figure 9.17 illustrates the origin of white light fringes with central achromatic white fringe. The solid curve represents the intensity distribution for violet light and the dotted curve that for the yellow light, drawn with central fringes superposed on each other. In the region near the central fringe, it should be emphasized, the optical path difference between the two interfering waves is small. The same is also true for dark fringes, thereby giving rise to alternately white and dark fringes, according to our definition the achromatic fringes, illustrated in Figure 9.18.

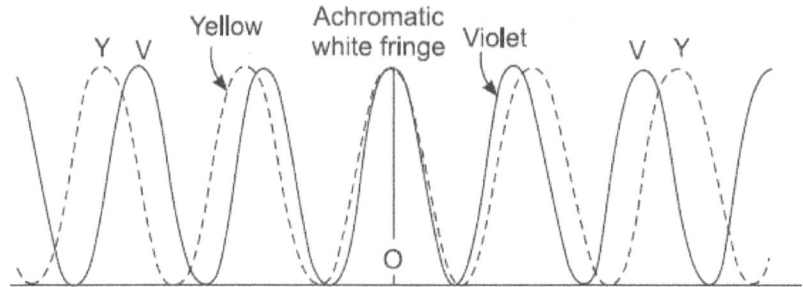

FIGURE 9.17. The origin of white light fringes with central achromatic white fringe.

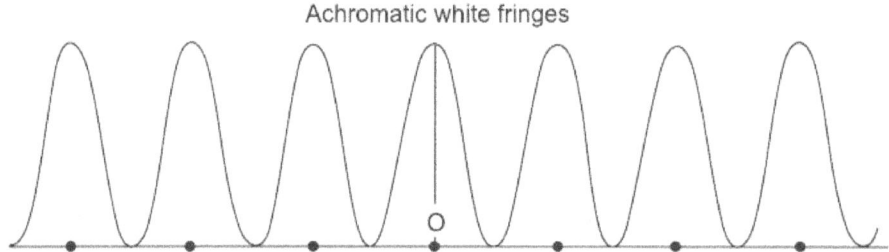

FIGURE 9.18. The origin of achromatic white fringes.

The inspection of the expression for the fringe width $\left(\overline{X} = \dfrac{D\lambda}{2d} \right)$ gives us a clue toward designing an experimental arrangement for achieving our object. It is obvious that D will be the same for all the wavelengths in any arrangement. Hence, for the constancy of $\overline{X}$, it follows from the previous expression for the fringe width that by some device the ratio $\left(\dfrac{\lambda}{2d} \right)$ must be made the same for all wavelengths. Physically, this means that the coherent sources for different wavelengths should be situated at different points in one plane and arranged in such a way that as λ increases, the separation $2d$ between the corresponding coherent sources should also increase in the same proportion. For example, we should satisfy the relation, like

$$\frac{2d_R}{2d_v} = \frac{\lambda_R}{\lambda_v} \tag{9.56}$$

for every possible pair of different wavelengths emitted by the white light source. Obviously, the fringe width $\overline{X}$ becomes the same for all wavelengths, and truly achromatic fringes may be obtained.

9.11 BILLET SPLIT LENS

In 1858, Billet obtained two coherent sources by forming two real images S_1 and S_2 of the illuminated slit S with the help of two halves of a lens, obtained by splitting a convergent lens into two parts along a diameter of its circular periphery as shown in Figure 9.19. The two halves of the lens can be separated or brought closer together by means of a micrometer screw, the motion being along the line perpendicular to the optical axis. The positions of the two halves can thus be very accurately regulated in relation to each other. In other words, the separation $2d$ between the coherent sources S_1 and S_2 can be suitably adjusted as seen with the naked eye. Fringes may be seen on the screen or with the aid of an eyepiece in the overlapping region EF of the two beams of monochromatic light.

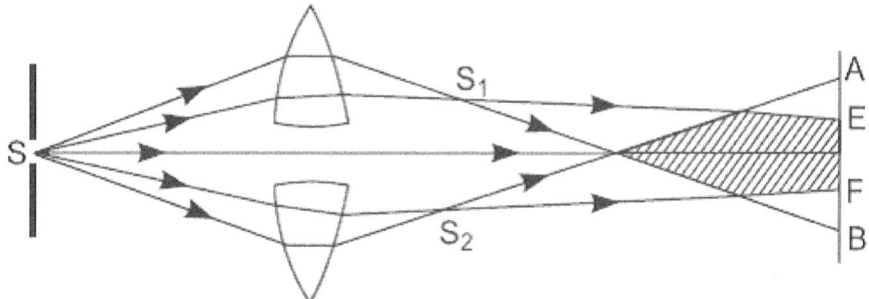

FIGURE 9.19. Billet's split lens for producing interference fringes.

We may also arrange the split lens with respect to the illuminated slit S so as to form two virtual images S_1 and S_1 which shall function as two coherent sources for the production of interference fringes.

9.12 PRODUCTION OF CIRCULAR INTERFERENCE FRINGES— MESLIN'S SPLIT LENS

It will be recalled that the loci of constant path difference in space, from two coherent monochromatic point sources S_1 and S_2, are hyperboloids of revolution surfaces generated by revolving hyperbolae about S_1S_2 as an axis. The intersections of these hyperboloids with a screen arranged perpendicularly to S_1S_2 are, therefore, concentric circular fringes with a common center on the prolongation of the line joining the coherent sources S_1 and S_2. The real point sources emit light in all directions; hence, we expect full concentric circular fringes on the screen in the previous stated position.

But it is impossible to have two real point coherent sources. However, Meslin (1893), by an ingenious modification of the Billet split lens, practically demonstrated alternate bright and dark semicircular concentric fringes on the screen. His arrangement is sketched in Figure 9.20.

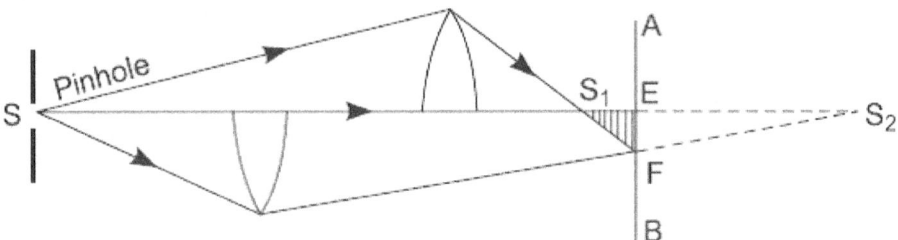

FIGURE 9.20. Meslin's split lens for producing circular fringes.

The two halves of the Billet split lens are placed a short distance apart on the optical axis. Two real point images S_1 and S_2 of a pinhole illuminated with monochromatic light are then formed on the optical axis. The real images function as coherent point sources which are not real point sources, because the images do not radiate light in all directions but only in the shape of a narrow cone. In the region of the overlapping of the two beams shown shaded in Figure 9.20, interference effects occur. A system of fringes, concentric semicircular arcs in shape, may be then obtained on a screen arranged between S_1 and S_2, normally to the optic axis.

In Meslin's arrangement, interference is in reality between the waves radiating from a source S_1 with waves converging to a similar source S_2. It should be emphasized that here circular fringes are obtained by the division of the wave-front in contrast to Newton's rings, which arise as a result of an interference phenomenon due to the division of amplitude.

Example 9.1

In an experiment with a biprism, the readings on the optical bench of the position of the eyepiece and the two positions of the lens were respectively 100.00, 67.00, and 34.00 cm. The distances between the two images for the two positions of the lens were respectively 0.3000 and 1.2000 mm, and the width of 10 fringes was 9.720 mm. Assuming that there is no index error, calculate (*a*) the distance between the focal plane of the eyepiece and the plane of interfering sources and (*b*) the wavelength of the light used.

Solution:

(*a*)

The distance between the eyepiece and the second position of the lens is given by

$$v_1 = 100 - 67 = 33 \text{ cm} = u$$

Hence,

(1st position of lens) $-u = 34 - 33 = 1$ cm.

Therefore, the distance between the focal plane of the eyepiece and the plane of interfering sources is

$$D = 100 - 1 = 99 \text{ cm}$$

(*b*)

$$2d = \sqrt{d_1 \times d_2} = \sqrt{1.2 \times 0.3} = 0.6 \text{ mm} = 0.06 \text{ cm}$$

$$\bar{X} = \frac{9.720}{10} \text{ mm} = 0.09720 \text{ cm}$$

Now,

$$\lambda = \frac{\bar{X} 2d}{D} = \frac{0.0972 \times 0.06}{99} = 5891 \times 10^{-8} \text{ cm}.$$

Example 9.2

Using sodium light with a Fresnel's biprism, the fringes were found to have a width of 0.0196 cm when observed at a distance of 100.0 cm from the slit. When a convex lens was placed between the biprism and the observer to give an image of the source at 100.0 cm from the slit, the distance apart of the images was found to be 0.70 cm; calculate the wavelength. The given distance from slit to lens was 30.0 cm.

Solution:

In the given problem, $X = 0.0196$ cm and $D = 100$ cm. To evaluate $2d$, we observe that the distance of the lens from the slit is 30 cm, and the images of the coherent sources are formed by the lens in a plane at $(100 - 30) = 70$ cm from the lens. Distance apart of the images of coherent sources is 0.70 cm. For lens in air $\mu_1 = \mu_2 = 1$ for magnification, we have

$$\frac{(-0.70)}{2d} = \frac{70}{(-30)} \to 2d = 0.30 \text{ cm}$$

Hence,

$$\lambda = \frac{\bar{X} 2d}{D} = \frac{0.0196 \times 0.30}{100} = 5880 \times 10^{-8} \text{ cm}$$

Example 9.3

When a thin monochromatic source of light was placed at a distance of 50 cm from a Fresnel biprism of $\mu = 1.5$, the distance between two consecutive bands formed on a screen placed at a distance of 100 cm from biprism was found to be 0.012 cm. If the wavelength of light was 5893×10^{-8} cm, find the magnitude of the obtuse angle of the biprism.

Solution:

We know

$$\bar{X} = \frac{(a+b)\lambda}{2a(\mu-1)\alpha}$$ where, it should be emphasized, α is the refracting angle

(acute angle) of the biprism in radian measure.

In the given problem, $a = 50$ cm, $b = 100$ cm, $\mu = 1.5$, $\bar{X} = 0.012$ cm, and $\lambda = 5893 \times 10^{-8}$ cm.

Hence,

$$\alpha = \frac{(50+100) \times 5893 \times 10^{-8}}{2 \times 50 \times (1.5-1) \times 0.012} \text{ radian}$$

But, π radian $= 180°$. Thus, α in degree measure is given by

$$\alpha = \frac{(50) \times 5893 \times 10^{-8} \times 180}{2 \times 50 \times (1.5-1) \times 0.012 \times \pi} = 0°.844$$

Therefore, the obtuse angle of the biprism $= 180 - 2\alpha = 178°31$.

Example 9.4

Fringes are produced with monochromatic light of wavelength 5.45×10^{-5} cm. A thin plate of glass of refractive index 1.5 is placed normally in the path of one of the interfering beams, and the central bright band of the fringes system is found to move to a position occupied by the third band from the center. Calculate the thickness of the glass plate.

Solution:

The central fringe shifts to the 3rd bright fringe from the center of the original system obtained without the plate. Hence, $(\mu - 1)e = n\lambda$. We have $(1.5 - 1)e = 3 \times 5.45 \times 10^{-5}$. Thus, $e = 3.27$.

9.13 EXERCISES

1. Consider the superposition of two simple harmonic disturbances and show that the resultant intensity is not just the sum of the intensities due to the separate disturbances. On the basis of this result, explain why coherent disturbances interfere and incoherent disturbances do not.

2. Discuss the important conditions for the interference of light.

3. Describe, giving experimental details, Fresnel's biprism method for determining the wavelength of light. Derive the formula used.

4. Describe an experimental arrangement for the observation and measurement of Lloyd's mirror fringes. Show how you would use it to determine the wavelength of monochromatic light, and how would you modify the arrangement to obtain achromatic fringes using a white light source.

5. Calculate the displacement of fringes when a thin transparent lamina is introduced in the path of one of the interfering beams in a biprism. Show how this method is used for finding the thickness of a mica sheet.

6. What are coherent sources? Discuss why two independent sources of light of the same wavelength cannot produce interference fringes. Give diagrams showing clearly how coherent sources are produced in (*i*) Newton's rings arrangement, (*ii*) biprism arrangement. Derive the formula for fringe width in the biprism experiment.

7. In an experiment with Fresnel's biprism bands, 0.196 mm in width are observed at a distance of 100 cm from the slit. A convex lens is then put between the observer and the biprism so as to give an image of the sources at a distance of 100 cm from the slit. The distance apart of

the images is found to be 0.70 cm, the lens being 30 cm from the slit. Calculate the wavelength of light used.

8. A thin sheet of glass $(\mu = 1.5)$ of 6 microns thickness introduced in the path of one of the interfering beams in a biprism arrangement shifts the central fringe to a position normally occupied by the fifth. Find the wavelength of light use.

9. The central fringe of the interference pattern produced by light of wavelength 6000 A.U. is shifted to the position of the fifth bright fringe by introducing a thin glass plate $(\mu = 1.5)$. What is the thickness of the plate?

10. Interference fringes are formed using a biprism having base angles of $4°$ each and refractive index 1.5. The slit is kept at a distance of 10 cm from the biprism and is illuminated with light of wavelength 5890 A.U. Calculate the fringe width at 100 cm from the biprism.

11. Interference fringes are being formed by a source of light $\lambda = 6000$ A.U. It is found that by introducing a transparent plate in one of the paths, the central fringe is shifted to a position occupied by the 6th bright fringe. Find the refractive index of the plate if its thickness is 6 microns. Establish the formula used.

12. Two straight and narrow parallel slits 3 mm apart are illuminated with a monochromatic source $(\lambda = 5.9 \times 10^{-5}$ cm$)$. Fringes are obtained at a distance of 30 cm from the slits. Find the width of fringes.

13. The distance between the slit and the biprism and that between the biprism and the screen are each 50 cm. The angle of the biprism is $179°$ and its refractive index is 1.5. If the distance between the successive bright fringes is 0.0135 cm, calculate the wavelength of light.

14. The inclined faces of a biprism of refractive index 1.5 make angles of $2°$ with its base. A slit illuminated by a monochromatic light is placed at a distance of 10 cm from the biprism. If the distance between two dark fringes, observed at a distance of 1 meter from the biprism, is 0.18 mm, find the wavelength of the light used. Derive the formula used.

15. Fringes are produced by a Fresnel's biprism in the focal plane of a reading microscope, which is 100 cm from the slit. A lens inserted between the biprism and microscope gives two images of the slit in two positions. In one case the two images of the slit are 4.05 mm and in the

other are 2.9 mm apart. If sodium light ($\lambda = 5893\,\text{Å}$.) is used, find the distance between the interfering bands.

16. Interference fringes of yellow light ($\lambda = 5800\,\text{Å}$) are formed by a Billet split lens. The distance from the source S to the lens L is 25 cm. The focal length of the lens is 15 cm. The lens halves are separated by 0.08 mm, and the source to screen distance is 200 cm. Find the fringe separation.

17. A Lloyd's mirror of length 5 cm is illuminated with monochromatic light, $\lambda = 5460\,\text{Å}$, from a narrow slit 0.1 cm from its plane, and 5 cm, measured in that plane, from its near edge. Find the separation of the fringes at a distance of 120 cm from the slit, and the total width of the pattern produced there.

10

DIFFRACTION OF LIGHT

Chapter Outline

10.0 INTRODUCTION

The wave theory of light in its original form as proposed by Huygens was not successful in explaining the observed phenomenon that light appears to travel in straight lines. The phenomenon is easily borne out by corpuscular theory. A minute investigation of the fact, however, reveals that light suffers some deviation from its straight path in passing close to the edges of opaque obstacles and narrow slits. Some of the light does bend into the region of geometrical shadow, and its intensity falls off rapidly. This deviation, moreover, is extremely small when the wavelength of the waves of light is small in comparison to the dimensions of the obstacle or aperture. But the deviation becomes much more pronounced, though never of the same order as the bending of sound waves around corners, when the dimensions of the aperture or the boundary of the obstacle are comparable with the wavelength of light. Even then, only a careful examination of the screen on which the geometrical shadow of the obstacle is received reveals this deviation of light from its rectilinear path.

For example, let us suppose, with reference to Figure 10.1, that waves of light diverging from a narrow slit S, which acts as a secondary source under the influence of light from the monochromatic source O, pass an obstacle AB with a straight edge A parallel to the slit. If the shadow of the obstacle is received at C on a screen, it is observed that the boundary of the shadow is never sharp. A small portion of light has "bent" around the edge into the geometrical shadow; outside the shadow parallel to its edge are observed several bright and comparatively dark bands.

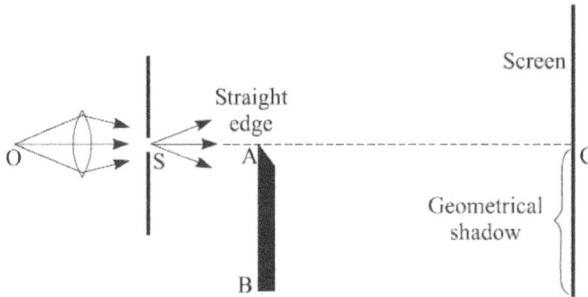

FIGURE 10.1. Geometrical shadow of a straight edge.

The term *diffraction* is applied to all phenomena like this which are produced whenever there is any limitation on the width of the beam of light. Since in most diffraction patterns some light penetrates within the region of geometrical shadow, diffraction is sometimes defined as *"the bending of light around an obstacle."* This phenomenon was first discovered in 1665 by an Italian scientist named Grimaldi, and was later on studied by Newton. In the days when corpuscular theory was prevalent, attempts were made by Newton to interpret diffraction effects as due to attractive or repulsive forces exerted by the edges of obstacles on flying corpuscles, so as to deflect them from their rectilinear path. In terms of wave theory, the first attempt to explain the diffraction phenomenon was made by Dr. Young, who attributed it to the interference between the direct light waves which pass near the edge of the obstacle and the wave of light reflected at grazing incidence from the edge. This explanation could scarcely be applied to explain the penetration of light within the region of geometrical shadow. Moreover, the maxima and minima regions are not equally spaced as demanded by the mode of interference imagined by Young. Even then if we accept Young's explanation to be correct, the intensity distribution on the screen should depend upon (a) the sharpness of the edge, (b) its degree of polish, and (c) its material. But in 1818 Fresnel, after a series of experiments, proved that the details of the diffraction pattern were independent of the previously mentioned factors so long as the material of the obstacle was opaque.

We owe the correct interpretation of diffraction to the brilliant work of Fresnel, who attributed this phenomenon to the mutual interference of secondary wavelets originating from various points of a wavefront which were not blocked off by an obstacle or were allowed to pass by a slit. That is, instead of finding the new wavefront by constructing the envelope of these secondary wavelets, we must find, by the principle of superposition, their resultant at every point of the screen, taking due account of their relative amplitudes and phases. Fresnel thus applied Huygens's principle of secondary wavelets in conjunction with the *principle of interference* and calculated the position of fringes in general agreement with the observed diffraction pattern.

It should be emphasized, however, that the process responsible for the production of the diffraction phenomenon is going on continuously during the propagation of every wavefront. But the diffraction effects are observed only when a portion of the wavefront is cut off by some obstacle. Every optical instrument, in fact, makes use of only a limited portion of the wavefront. A telescope or a microscope, for example, utilizes only a limited portion of a wavefront transmitted by the objective lens. Therefore, some diffraction is always present in the image. Hence, a clear comprehension of the nature of diffraction is of fundamental importance for a complete understanding of practically all the optical phenomena.

10.1 FRESNEL'S EXPLANATION OF THE RECTILINEAR PROPAGATION OF LIGHT

One of the greatest difficulties encountered by the supporters of the wave theory of light in its early days was the explanation of the observed fact that light appears to be propagated in straight lines. The correct interpretation in terms of wave theory was given by Fresnel in 1815 by applying principles of interference in conjunction with Huygens's principle of secondary wavelets, but at the same time he showed that the rectilinear propagation of light is only approximate. To explain this let us first discuss Fresnel's method of finding the effect of a plane wavefront at a point ahead of it.

10.1.1 Fresnel's Half-Period Zones

In Figure 10.2, let $ABCD$ represent a plane wavefront of monochromatic light of wavelength λ, traveling from left to right. According to Huygens's principle, every point in this plane may be regarded as the origin of the secondary wavelets, and at any given instant, every one of these secondary wavelets passes through the point P.

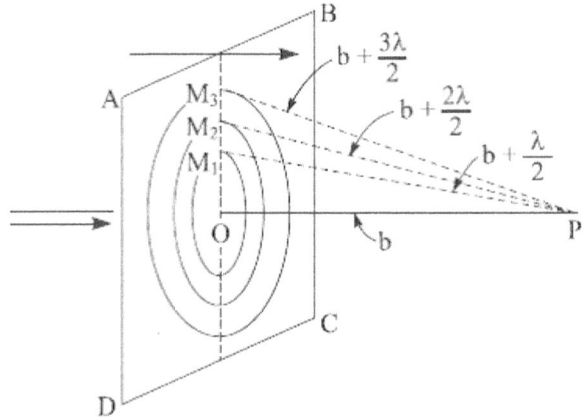

FIGURE 10.2. Half-period zones on a plane wavefront.

The resultant effect at P due to the whole wavefront will simply be equal to the resultant of all these secondary wavelets. To find this resultant, we divide the entire wavefront into concentric zones by the following construction. We drop a perpendicular from P on $ABCD$. The point O, the foot of perpendicular from P, is called the pole of the wave with respect to P. Let $PO = b$ and λ be the wavelength of the light waves. With P as center and radius $\left(b + \dfrac{\lambda}{2} \right)$, we construct a sphere which intersects the wavefront in a circle M_1. Then

$$PM_1 = b + \left(\frac{\lambda}{2} \right) \text{ and } PM_1 - PO = \left(\frac{\lambda}{2} \right) \tag{10.1}$$

This simply means that the secondary wavelets originating from O and from the points on the circumference of the circle M_1, on reaching P simultaneously, will differ in their phase by

$$\frac{2\pi}{\lambda}(PM_1 - PO) = \pi \text{ radians} \qquad (10.2)$$

But π radians is equivalent to the phase difference of $\left(\dfrac{T}{2}\right)$. Consequently, the area enclosed by the M_1 is called Fresnel's first half-period zone. Similarly, we construct other spheres of radii

$$b + \left(\frac{2\lambda}{2}\right), b + \left(\frac{3\lambda}{2}\right), b + \left(\frac{4\lambda}{2}\right), \text{ etc.} \qquad (10.3)$$

which intersect the wavefront in circles M_2, M_3, M_4, and so on. This construction divides the entire wavefront into a number of Fresnel's half-period zones. These zones are so situated that the distance of the point P increases by $\left(\dfrac{\lambda}{2}\right)$ as we pass from the inner to the outer boundary of each zone. The first half-period zone, as explained previously, is a circle, while the second half-period zone is an annular space between the circles M_1 and M_2 and so on.

The area A_n, of the nth zone is from Figure 10.3,

$$A_n = \pi\left(r_n^2 - r_{n-1}^2\right) = \pi\left\{\left(PM_n^2 - b^2\right) - \left(PM_{n-1}^2 - b^2\right)\right\} \qquad (10.4)$$

$$A_n = \pi\left\{\left(b + \frac{n\lambda}{2}\right)^2 - \left(b + (n-1)\left(\frac{\lambda}{2}\right)\right)^2\right\} \qquad (10.5)$$

$$A_n = \pi b\lambda + \pi(2n-1)\left(\frac{\lambda^2}{4}\right) \qquad (10.6)$$

On neglecting the term in λ^2, as b is large in comparison with λ, we get

$$A_n \cong \pi b\lambda \qquad (10.7)$$

The approximate value of A_n is independent of n, and hence all zones are approximately equal in area. Actually, the area will increase with the order n of the zone according to Equation (10.6).

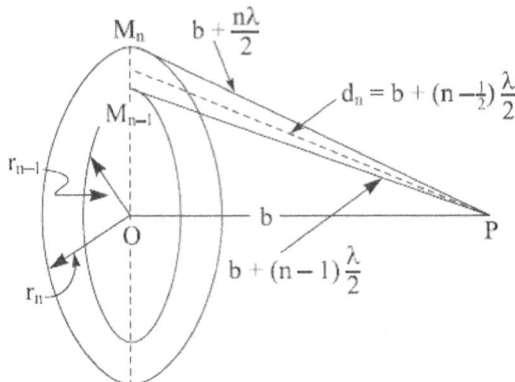

FIGURE 10.3. Geometry of the nth half-period zone.

10.1.2 Phase of the Resultant of Secondary Wavelets from Each Zone

Consider an instant at which the phase at P of the secondary wavelet from O is zero. At this instant, the phase at P of the wavelets from the points on the outer edge of the first zone will be π, since $PM_1 - PO = \left(\dfrac{\lambda}{2}\right)$, and the wavelets start from the wavefront in the same phase. Consequently, the phase of the resultant at P of all the wavelets from the first zone, at this instant, will be approximately the mean of the phases of these wavelets and may be taken to be $\left(\dfrac{\pi}{2}\right)$. At this same instant, the phases at P of the secondary wavelets from the inner and outer boundaries of the second zone are π and 2π respectively, so the mean phase of the resultant due to this zone is $\left(\dfrac{3\pi}{2}\right)$. Thus, the mean phases of the resultants of wavelets from the successive outer zones are $\left(\dfrac{\pi}{2}\right), \left(\dfrac{3\pi}{2}\right), \left(\dfrac{5\pi}{2}\right), \left(\dfrac{7\pi}{2}\right), \left(\dfrac{9\pi}{2}\right)$, and so on. Let $R_1, R_2, R_3, R_4, ..., R_n$ be the resultant amplitudes at P of wavelets reaching P from the first, second, third . . . and nth zones. Calling the resultant amplitude due to the entire wavefront as R, this may now be expressed, by the principle of superposition, as the sum of the series,

$$R = R_1 - R_2 + R_3 - R_4\left(-1\right) + \cdots + \left(-1\right)^{n-1} R_n \qquad (10.8)$$

10.1.3 Factors Governing the Magnitude of R_n

The magnitudes of successive terms of the previous series depend upon the following three factors:

(i.) The area of each zone determines the number of secondary wavelets reaching P from that zone, and the magnitude of R_n depends on the number of wavelets from the nth zone.

$$R_n \alpha A_n \qquad (10.9)$$

(ii.) The magnitude of R_n is inversely proportional to the average distance d_n from the point P of the nth zone.

(iii.) Last, $R_n \alpha (1 + \cos\theta_n)$, where θ_n is the angle which the direction of P from the nth zone makes with OP. It is assumed that the obliquity factor varies slowly and that it may be regarded as constant over a single half-period zone. But

$$\frac{A_n}{d_n} = \frac{\pi\lambda\left(b + (2n-1)\frac{1}{4}\lambda\right)}{\frac{1}{2}\left(b + \frac{1}{2}n\lambda + b + (n-1)\frac{1}{2}\lambda\right)} = \pi\lambda \qquad (10.10)$$

The amplitudes of successive zones, therefore, depend on the obliquity factor $(1 + \cos\theta)$. As θ increases from zero, $\cos\theta$ decreases very slowly at first, but more rapidly for larger values of θ. Therefore, the successive amplitudes R_1, R_2, R_3, R_4... at first decrease slowly but then more rapidly for higher values of n.

10.1.4 Summing up of the Series

We have seen previously that the successive terms in the series

$$R = R_1 - R_2 + R_3 - R_4 + \cdots + (-1)^{n-1} R_n \qquad (10.11)$$

are in decreasing order of magnitude. With this knowledge we will try to sum the previous series by a method given by *Schuster*. Let n be an odd integer. The terms in the previous series can then be grouped in the following two alternate ways:

$$R = \frac{1}{2}R_1 + \left(\frac{1}{2}R_1 - R_2 + \frac{1}{2}R_3\right) + \left(\frac{1}{2}R_3 - R_4 + \frac{1}{2}R_5\right) + \cdots + \frac{1}{2}R_n \qquad (10.12)$$

or

$$R = R - \frac{1}{2}R_2 - \left(\frac{1}{2}R_2 - R_3 + \frac{1}{2}R_4\right)$$
$$- \left(\frac{1}{2}R_4 - R_5 + \frac{1}{2}R_6\right) - \cdots - \frac{1}{2}R_{n-1} + R_n \tag{10.13}$$

Suppose first that each term of the original series is greater than the arithmetic mean of two adjacent terms, that is,

$$R_2 > \frac{1}{2}(R_1 + R_3), R_3 > \frac{1}{2}(R_2 + R_4), R_4 > \frac{1}{2}(R_3 + R_5) \tag{10.14}$$

then the bracketed terms are all negative. Therefore, the previous equations may be written as

$$R = \frac{1}{2}(R_1 + R_n) - \alpha \tag{10.15}$$

and

$$R = R_1 - \frac{1}{2}(R_2 + R_{n-1}) + R_n + \beta \tag{10.16}$$

where α and β represent the sum of all terms within brackets. It is now obvious that the following inequalities must hold

$$\frac{1}{2}(R_1 + R_n) > R > R_1 - \frac{1}{2}(R_2 + R_{n-1}) + R_n \tag{10.17}$$

If the original series is such that each term is less than the mean of two adjacent terms, the bracketed terms are all positive. Now the following inequalities must hold,

$$R_1 - \frac{1}{2}(R_2 + R_{n-1}) + R_n > R > \frac{1}{2}(R_1 + R_n) \tag{10.18}$$

But, $R_1 = R_2$ and $R_n = R_{n-1}$, approximately. Making these substitutions in the previous inequalities, we get the equality

$$\frac{1}{2}R_1 + \frac{1}{2}R_n = R = \frac{1}{2}R_1 + \frac{1}{2}R_n \tag{10.19}$$

Similarly, when n is even, we get

$$\frac{1}{2}R_1 - \frac{1}{2}R_n = R = \frac{1}{2}R_1 - \frac{1}{2}R_n \tag{10.20}$$

If n becomes very large, the amplitude due to the nth zone, on account of the obliquity factor $(1 + \cos\theta_n)$, becomes negligibly small. The resultant amplitude at P due to the whole wave, whether n is odd or even, is therefore merely $\frac{1}{2}R_1$, that is, one half of that which could be produced by the first Fresnel zone alone.

10.1.5 Rectilinear Propagation of Light

Let us now apply the previous theory to explain the rectilinear propagation of light. Let us suppose a plane wavefront of monochromatic light to be incident normally on a square aperture $EFGH$ in a screen as illustrated in Figure 10.4 (a). Let A be the pole of the wavefront with respect to the point P, at which it is desired to find the resultant intensity due to the entire wavefront. The pole A is within the aperture, sufficiently far away from its edges. Suppose the wavefront is now divided into zones around A. By the time these intersect the edges, their number will be fairly large, and all the effective zones will be exposed. The resultant amplitude at P, according to the previous theory, will be $\frac{1}{2}R_1$, that is, it will be just the same as with the screen with the aperture removed as in Figure 10.4 (b).

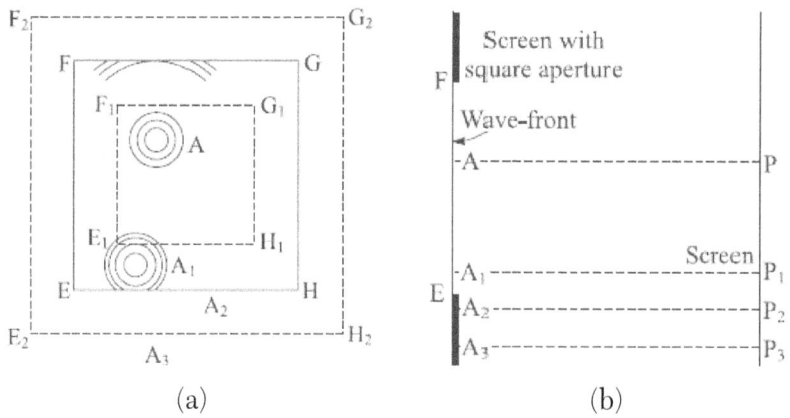

FIGURE 10.4. (a) Explanation of rectilinear (approximate) propagation, (b) Screen with aperture.

In the same way, if the point P_3 is so far inside the geometrical shadow of the aperture that the pole A_3 corresponding to it is outside the aperture, far from its edges, the effect at P_3 reduces to that of one-half of the Fresnel first zone around A_3, which is obviously zero. The point P_3 will be just as dark as if light traveled strictly in straight lines. It is only when the pole is in the region enclosed by dotted squares that the previous theory does not apply. This is simply due to the reason that for points like P_1 and P_2, only some of the effective zones are exposed and the rest blocked. Therefore, Fresnel's theory becomes inadequate in deciding the intensity at these points. The wavelength of light being extremely small, the sides of the two dotted squares lie extremely close to the edges of the aperture.

We have thus arrived at the law of rectilinear propagation for all points whose poles lie within the dotted square $E_1F_1G_1H_1$ and for all points whose poles lie outside $E_2F_2G_2H_2$. The law fails in the case of those points whose poles lie between the two dotted squares—in fact when light passes close to the edges of obstacles, there is some deviation from its rectilinear path. Thus, the propagation of light is approximately linear.

We can see in another way how rectilinear propagation (approximate) follows from *Fresnel's theory* of half-period zones. Let us first find out the radius of the first zone of the plane wave for a point, P, 20 cm ahead of it, taking λ as 5000×10^{-8} cm.

Radius of first zone $= \sqrt{b\lambda} = \sqrt{20 \times 5000 \times 10^{-8}} = 0.0316$ cm, and

Radius of 10,000th zone

$= \sqrt{10,000 \times b\lambda} = \sqrt{10,000 \times 20 \times 5000 \times 10^{-8}} = 3.16$ cm.

The diameter of the first zone is therefore less than a mm and that of the 10,000th zone is less than 10 cm, which is quite a normal width of the wavefront. Therefore, the outermost zone for this case will be of an order even greater than 10,000, and its amplitude R_n at P will be quite negligible. The resultant amplitude at P is simply $\dfrac{R_1}{2}$; that is, the portion of the wavefront within about 0.3 mm (radius of first zone) around the point O, the pole of the wavefront with respect to P. In other words, only secondary wavelets from the first half-period zone are effective at P; the remainder of the wavefront simply destroys at P half the amplitude of the first zone so that light travels to P from a region 0.3 mm in radius around O. Thus, it travels approximately in a straight line.

10.2 ELIMINATION OF REVERSE WAVE IN HUYGENS'S PRINCIPLE

An early objection to Huygens's principle of wave propagation, it will be recalled, lies in the fact that it also gave the propagation of a wave in the backward direction toward the source. This principle, of course, correctly postulated the propagation of a wave in the forward direction. By the help of Fresnel's idea of interference of Huygens's secondary wavelets in conjunction with the inclination factor $(1 + cos\ \theta)$ governing the amplitudes of secondary wavelets which was introduced *ad hoc* by Fresnel, it is possible to give a simple explanation of the absence of a backward wave. The method of dividing the wavefront into Fresnel half-period zones with respect to a point ahead of it may be also applied to divide the wavefront into zones with respect to a point behind it. An analysis on the lines shows that the amplitude of a reverse wave at any given point is half of the amplitude which would be produced by the secondary wavelets from Fresnel's first zone alone. The inclination θ for the wave traveling in the backward direction for the first zone is π and, accordingly, the inclination factor $(1 + cos\ \theta)$ of the amplitude for this wave is zero. Hence, the amplitude of the reverse wave is zero.

We may also explain the absence of a reverse wave on the basis of the electromagnetic theory of light. According to this theory, light is a propagation of the vibration of an electric vector $\vec{E}$ coupled with the propagation of the vibration of a magnetic vector $\vec{H}$, with both vectors in synchronism to always be in equal phase, mutually at right angles, and at right angles to the direction of the velocity vector $\vec{V}$. Conversely, if the instantaneous values of $\vec{E}$ and $\vec{H}$ are known, the direction of the velocity $\vec{V}$ of the wave motion can be uniquely determined analytically. This analysis always gives the velocity in the forward direction, and thus there is no possibility of a reverse wave.

10.3 TWO CLASSES OF DIFFRACTION PHENOMENA

Diffraction phenomena, which arise as a result of some limitation on the width of a wavefront, are divided into two classes known for historical reasons as Fraunhofer diffraction and Fresnel diffraction. In the former class, the source of the light and the screen are effectively at infinite distance from the obstacle of aperture causing the diffraction. This means that the wavefront incident on the obstacle or aperture is plane and the secondary wavelets, which we consider as originating from the unblocked portion of the wavefront at the moment it just touches the aperture, all start at the same time. In other words, the phase of secondary wavelets is the same at every point of the aperture. The screen is at infinity. Therefore, we have to consider the interference as taking place between parallel diffracted rays which can be brought into focus by placing a convergent lens behind the aperture. Thus, it is very conveniently observed by employing two convergent lenses: one to render the light from the source parallel before its incidence on the aperture, and the other to unite parallel diffracted rays in focus on the screen—an arrangement which effectively removes the source and screen at infinity.

In the Fresnel class of diffraction, the source of light or the screen or both are at finite distances from the obstacle or aperture, but no lenses are employed for rendering the rays parallel or convergent. Therefore, the incident wavefront is spherical or cylindrical instead of being plane. As a consequence of this, the phase of secondary wavelets is not the same at all points in the plane of aperture causing the diffraction. The resultant amplitude at any point of the screen, however, is obtained by the mutual interference of secondary wavelets from different elements of the unblocked portion of the wavefront.

10.4 DIVISION OF A CYLINDRICAL WAVEFRONT INTO HALF-PERIOD STRIPS

The envelope of secondary wavelets diverging from a long narrow slit SS', illuminated by the monochromatic light of wavelength λ, is a cylindrical surface WW' as in Figure 10.5.

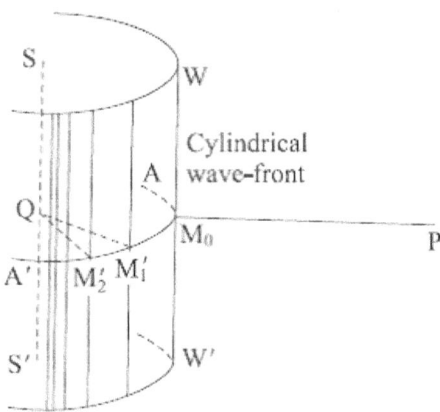

FIGURE 10.5. Half-period strips on a cylindrical wavefront.

It is essential first to find its effect at a point P situated ahead of the wavefront, by dividing it into a number of elementary half-period strips by a construction closely similar to that of the division of a plane wavefront into zones. We draw a perpendicular PQ from P on SS'. In Figure 10.5, the point of intersection, M_0, of PQ with the wavefront is the pole of the wave with respect to P.

Let us now consider an equatorial section AA' through the line PM_0Q of the wave surface as in Figure 10.6. Let $QM = a$ and $M_0P = b$. With P as center and radii

$$b+\frac{\lambda}{2}, b+\frac{2\lambda}{2}, b+\frac{3\lambda}{2}, \text{ etc.} \tag{10.21}$$

we draw arcs so as to cut the section AA' into points (M_1, M_1'), (M_2, M_2'), (M_3, M_3'),…, and so on.

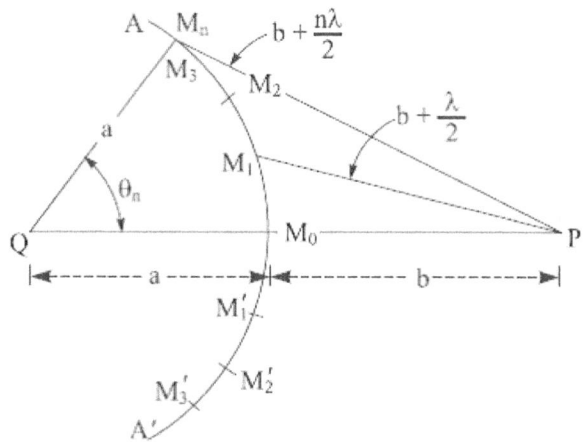

FIGURE 10.6. Equatorial section of cylindrical wavefront.

Let $PM_n = b + \frac{1}{2}n\lambda$, and then with the triangle PM_nQ, we have by the *law of cosine*,

$$PM_n^2 = QM_n^2 + PQ^2 - 2PQ \times QM_n \cos\theta_n \tag{10.22}$$

$$\left(b+\frac{1}{2}n\lambda\right)^2 = a^2 + (a+b)^2 - 2(a+b)a\cos\theta_n \tag{10.23}$$

As λ is small, the term $\left(\frac{n^2\lambda^2}{4}\right)$ in the previous equation can be neglected in comparison with other quantities. Also, if θ_n is small, we consider only the first two terms in the expression

$$\cos\theta_n = 1 - \frac{\theta_n^2}{2!} + \frac{\theta_n^4}{4!} - \frac{\theta_n^6}{6!} + \cdots \tag{10.24}$$

With these considerations, Equation (10.23) reduces to

$$bn\lambda = a(a+b)\theta_n^2 \tag{10.25}$$

$$\theta_n = \sqrt{\frac{nb\lambda}{a(a+b)}} = K\sqrt{n} \qquad (10.26)$$

where K is a constant, as a, b and λ are constants for a given arrangement. Hence

Replacing n by 1, 2, 3, 4, *etc.* we have respectively,

arc $M_0M_1 = aK$

$$\left.\begin{array}{l} \text{arc } M_0M_2 = aK\left(\sqrt{2}-1\right)=0.414\ aK \\[2mm] \text{arc } M_2M_3 = aK\left(\sqrt{3}-\sqrt{2}\right)=0.318\ aK \\[2mm] \text{arc } M_3M_4 = aK\left(\sqrt{4}-\sqrt{3}\right)=0.268\ aK \end{array}\right\} \text{ Rapid change}$$

$$\left.\begin{array}{l} \text{arc } M_{15}M_{16} = aK\left(\sqrt{16}-\sqrt{15}\right)=0.127\ aK \\[2mm] \text{arc } M_{16}M_{17} = aK\left(\sqrt{17}-\sqrt{16}\right)=0.123\ aK \end{array}\right\} \text{ Small change}$$

Therefore, the lengths of arcs M_0M_1, M_1M_2, M_2M_3, and so on decrease rapidly at first but rather slowly afterward. Through the points M_0, M_1, M'_1, M_2, M'_2, and so forth, lines are drawn parallel to the slit, thus dividing the wavefront into strips of equal lengths. The widths of these strips decrease rapidly at first but more slowly later on as one considers these in either direction from the central line through M_0. The areas of strips are, therefore, in decreasing order, and the lower order strips have larger areas as compared to higher order ones. Such strips are called *half-period zones* or *elements*.

The resultant effect of the whole wavefront at P is therefore the sum of the resultants of individual strips. Consider one such strip EE' which is at an *average* distance c from P. This strip can be divided into *half-period zones* by drawing lines, as shown in Figure 10.7, so that the middle points N_1, N_2, N_3, and so forth of these lines are respectively at distances

$$c+\frac{1}{2}\lambda, c+\lambda, c+\frac{3\lambda}{2}, \ etc. \qquad (10.27)$$

From P. In the right triangle PON_m, we have $\left(c+\frac{1}{2}m\right)^2 = c^2 + ON_m^2$

and neglecting $\frac{1}{4}m^2\lambda^2$, the previous equation reduces to $ON_m^2 = cm\lambda$, or

$$ON_m = \sqrt{cm\lambda} = K'\sqrt{m} \qquad (10.28)$$

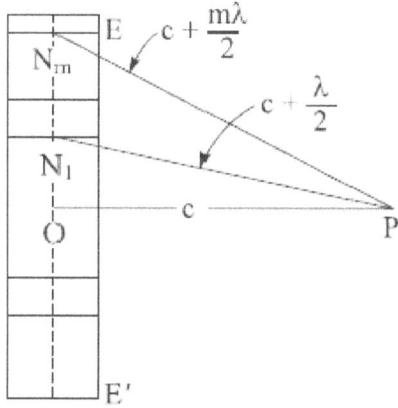

FIGURE 10.7. Half-period elements on a single strip.

where K' is a constant. The same analysis as before will show that the widths and, therefore, the areas of these half-period elements decrease at first rapidly but more slowly afterward. Consequently, the resultants at P of the higher order elements, being alternately in and out of phase with the first element, annul each other's effect. Thus, the resultant effect of the whole strip at P is due to a few half-period elements of lower order. These conclusions can be applied to all strips of the wavefront. Therefore, the resultant effect at P of the entire cylindrical wavefront is merely due to a narrow equatorial band surrounding the section AM_0A', which is divided into half-period elements by the strips. This is shown in Figure 10.8, which represents the equatorial band spread out in a plane.

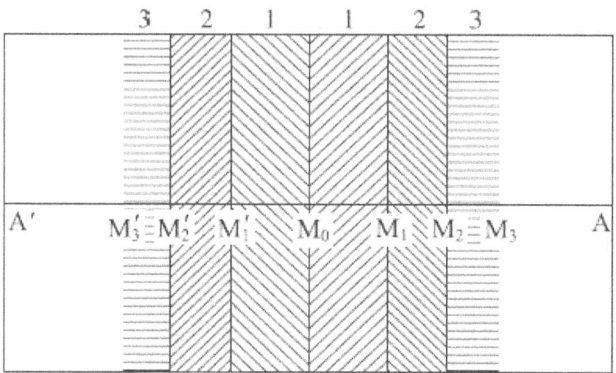

FIGURE 10.8. Equatorial band of cylindrical wavefront spread out in a plane.

Let R_1, R_2, R_3,...,R_n be the resultant amplitudes at the point P due to secondary wavelets reaching P from the first, second, third, . . . and nth zone on either side of the central line through M_0. Therefore, the resultant amplitude R at P due to either half of the wave, by the principle of superposition, is

$$R = R_1 - R_2 + R_3 - R_4 + ...+ \left(-1\right)^{n-1} R_n \tag{10.29}$$

where the alternate plus and minus signs occur because the resultants of successive zones are alternately out of and in phase. The summation of this series is given in Section 10.1, and the result is

$$R = \frac{R_1}{2} / \pm \frac{R_n}{2} \tag{10.30}$$

The plus sign corresponds to n odd while the negative sign corresponds to n even. And when n is large, then

$$R = \frac{R_1}{2} \tag{10.31}$$

Obviously, the resultant amplitude at P of the entire wavefront is R_1.

10.5 DIFFRACTION AT A STRAIGHT EDGE

Let A be the sharp straight edge of an obstacle placed perpendicular to paper and exactly parallel to the long narrow slit S, which is illuminated with a monochromatic light source. A train of cylindrical waves is incident on the edge A from the left. A careful examination of the screen reveals that immediately above C there are series of maxima and minima regions, parallel to the straight edge, called diffraction fringes. The intensity of maxima gradually decreases while that of minima increases as we move above the point C along the screen. At a short distance from this point, the fringes merge into a uniform illumination. It should be emphasized that minima regions are not absolute. The fringes, moreover, are not equidistant but gradually become closer as we move away from C. Some of the light also penetrates within the geometrical shadow, and the illumination falls off continuously and rapidly to zero, without showing any maxima and minima.

To explain these characteristics of a diffraction pattern due to a straight edge, we divide the incident cylindrical wavefront into *Fresnel's half-period elements*. In Figure 10.9 WW' represents the section of the wavefront, when it just touches the edge A, by the plane of paper.

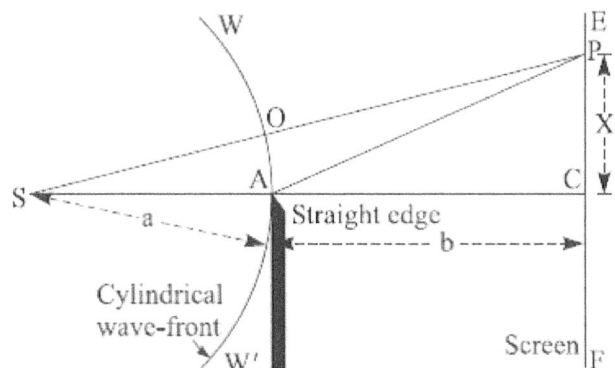

FIGURE 10.9. Fresnel diffraction of a cylindrical wavefront by a straight edge.

Let P be any point on the screen. A straight line joining P to S intersects the wavefront at O, its pole with respect to P. With F as the center and radii

$$PO + \frac{1}{2}\lambda, PO + \lambda, PO + \frac{3}{2}\lambda, \text{ etc.} \qquad (10.32)$$

we draw circles intersecting the section WW' in different points. By drawing lines parallel to the straight edge through these points, the wavefront is divided into half-period zones. Then the resultant effect at P due to whole wavefront is equal to the sum of the effect of the upper half OW and that of the lower half OW' of the wavefront. The resultant amplitude at P due to each half of the wave is given by

$$R = R_1 - R_2 + R_3 - R_4 + \cdots = \frac{1}{2}R_1$$

where R_1 is the amplitude at P due to first zone in either half of the wave and so on.

10.5.1 Uniform Illumination at Points Far above C

Let us suppose that the point P is sufficiently far above C so that the portion OA contains all the effective half-period zones of the lower half of the wave. In other words, the obstacle AB imposes no limitation on the zones effectively contributing to the resultant amplitude at P. The intensity at P will be, therefore,

$$\frac{1}{2}R_1 + \frac{1}{2}R_1 = R$$

i.e., as if the entire wave is unobstructed. This explains the uniform illumination on the screen beyond the diffraction fringes.

10.5.2 Maxima and Minima Diffraction Fringes on the Screen

Let us consider a point P_1 on the screen for which the pole of the wavefront is O_1, as shown in Figure 10.10(a), in such a position that the following relation

$$P_1A - P_1O_1 = \frac{1}{2}\lambda \qquad (10.33)$$

is satisfied. O_1A contains only the first half-period zone of the lower half of the wave. Therefore, the resultant amplitude D_1 at P_1 will simply be equal to the sum of the resultant amplitude due to the upper half of the wave and that of the first zone of the lower half; that is,

$$D_1 = \left(R_1 - R_2 + R_3 - R_4 + \cdots\right) + R_1 = \frac{1}{2}R_1 + R_1 = \frac{3}{2}R_1 \qquad (10.34)$$

and the intensity at P_1 is proportional to $\left(\dfrac{9}{4}R_1^{\ 2}\right)$, which is 2.25 times as great as that which would be produced by the unobstructed wave.

As we move above P_1, we come across a point P_2 as shown in Figure 10.10 (b) so that

$$P_2A - P_2O_2 = \lambda \qquad (10.35)$$

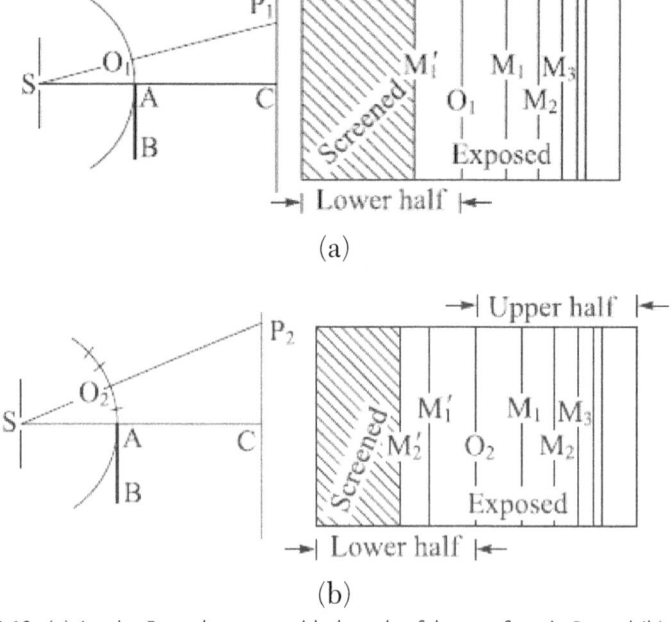

(a)

(b)

FIGURE 10.10. (a) A point P_1 on the screen with the pole of the wavefront is O_1, and (b) a point P_2 on the screen with the pole of the wavefront is O_2.

Then O_2A contains, for this point, the first and the second half-period zones of the lower half of the wave. The resultant amplitude at P_2 is, therefore,

$$D_2 = \frac{1}{2}R_1 + R_1 - R_2 = \frac{3}{2}R_1 - R_2 \tag{10.36}$$

This is less than the magnitude of R_1, so that bright band just outside the shadow is followed by a darker one. However, the minima is not absolute, as D_2 has some positive value.

Further above, we come across a point P_3 so that

$$P_3A - P_3O_3 = \frac{3}{2}\lambda \tag{10.37}$$

Then O_3A exposes the first three half-period zones. Therefore, the resultant amplitude at P_3 is

$$D_3 = \frac{1}{2}R_1 + R_1 - R_2 + R_3 \tag{10.38}$$

For a point P_4 such that $P_4A - P_4O_4 = 2\lambda$, O_4A exposes the first four half-period zones, and the resultant amplitude is

$$D_4 = \frac{1}{2}R_1 + R_1 - R_2 + R_3 - R_4 \tag{10.39}$$

We can form some idea about the relative magnitudes of D_1, D_2, D_3, D_4, and so forth by writing in the previous expressions the approximate relations

$$R_2 = \frac{1}{2}\left(R_1 + R_3\right), \ R_4 = \frac{1}{2}\left(R_3 + R_5\right), \quad etc.$$

Then we get,

$$D_1 = \frac{3}{2}R_1$$

$$D_2 = R_1 - \frac{1}{2}R_3$$

$$D_3 = R_1 + \frac{1}{2}R_3$$

$$D_4 = R_1 - \frac{1}{2}R_5$$

$$D_5 = R_1 + \frac{1}{2}R_5$$

Since R_1, R_3, R_5, R_7, and so on are in decreasing order of magnitude, hence $D_3 > D_2$ and $D_3 > D_4$; $D_5 > D_4$ and $D_5 > D_6$; and so forth. Thus, D_1, D_3, D_5, D_7, and so on correspond to regions of maximum illumination, while D_2, D_4, D_6, and so on correspond to regions of minima illumination of the screen. Moreover, it is obvious that $D_1 > D_3 > D_5 > D_7$ and $D_2 < D_4 < D_6$, and so forth. Thus, the amplitude and therefore the intensity of the maxima decreases gradually while that of the minimum increases gradually as we move along the screen above the point C, until finally the fringes merge into a uniform illumination.

From the previous analysis, it should be obvious that the intensity at any point P of the screen is maximum or minimum as OA exposes an odd or even number of zones of the lower half of the wave. Mathematically, the condition for maxima and minima can be expressed as

$$\text{Maximum Brightness: } PA - PO = \frac{(2n+1)\lambda}{2} \tag{10.40}$$

and

$$\text{Minimum Brightness: } PA - PO = \frac{2n\lambda}{2} \tag{10.41}$$

The positions of the maxima and minima on the screen can be easily calculated. Let $SA = a$, $AC = b$ and $CP = x$, then

$$PA^2 = b^2 + x^2$$

$$PA = b\sqrt{1+\left(\frac{x}{b}\right)^2} \cong b + \frac{1}{2}\left(\frac{x^2}{b}\right)$$

Similarly,

$$PS = b\sqrt{(a+b)^2 + x^2} \cong (a+b) + \frac{1}{2}\left(\frac{x^2}{a+b}\right)$$

Hence,

$$PO = PS - SO = (a+b) + \frac{1}{2}\left(\frac{x^2}{a+b}\right) - a = b + \frac{1}{2}\left(\frac{x^2}{a+b}\right) \tag{10.42}$$

Therefore,

$$PA - PO = \frac{1}{2}\left(\frac{x^2}{b}\right) - \frac{1}{2}\left(\frac{x^2}{a+b}\right) = \left(\frac{ax^2}{2b(a+b)}\right) \tag{10.43}$$

From Equations (10.40) and (10.43), we have for maxima,

$$\frac{ax_{max}^2}{2b(a+b)} = \frac{(2n+1)\lambda}{2}$$

or

$$x_{max} = K\sqrt{2n+1} \tag{10.44}$$

where $K = \dfrac{b(a+b)\lambda}{a}$, a constant for a given experimental arrangement and wavelength. Similarly, from Equations (10.41) and (10.43), we have for minima,

$$x_{min} = \sqrt{\dfrac{b(a+b)2n\lambda}{a}} = K'\sqrt{n} \qquad (10.45)$$

where $K' = \sqrt{\dfrac{b(a+b)2n\lambda}{a}}$, a constant.

For 1st Max at P_1, $n = 0$ in Equation (10.44), therefore, $x_1 = K$

For 2nd Max at P_3, $n = 1$ in Equation (10.44), therefore, $x_2 = K\sqrt{3}$

For 3nd Max at P_5, $n = 2$ in Equation (10.44), therefore, $x_3 = K\sqrt{5}$

But x_1, x_2, x_3, x_4, and so on are the distances from the point C of the first, second, third, and fourth maxima. Hence, the separation between the first and second maxima, and between second and third maxima, and so forth, is as follows

$$x_2 - x_1 = K\sqrt{3} - K = 0.732\ K$$

$$x_3 - x_2 = K\sqrt{5} - K\sqrt{3} = 0.504\ K$$

$$x_4 - x_3 = K\sqrt{7} - K\sqrt{5} = 0.430\ K$$

Thus, the diffraction fringes are not equidistant as in the case of two small coherent sources of light but get closer and closer as we move above the edge of the geometrical shadow. Also, these diffraction fringes are parallel to the edge of the geometrical shadow.

10.5.3 Intensity at the Edge of the Geometrical Shadow

The edge A of the obstacle itself is the pole of the wave for the point C. Thus, the upper half of the wave is only exposed while the lower half is completely obstructed by the obstacle. The resultant amplitude at C is, therefore,

$$C = R_1 - R_2 + R_3 - R_4 \cdots$$

and the intensity at C is proportional to $\dfrac{R_1^{\ 2}}{4}$, which is simply one-fourth of the intensity produced by the full wave. Also, the intensity at C is less than that at the first maxima point P_1.

10.5.4 Penetration of Light within the Geometrical Shadow

Having explained the diffraction pattern above the point C, let us now explain the penetration of light to a certain distance within the geometrical shadow of the obstacle as shown in Figure 10.11.

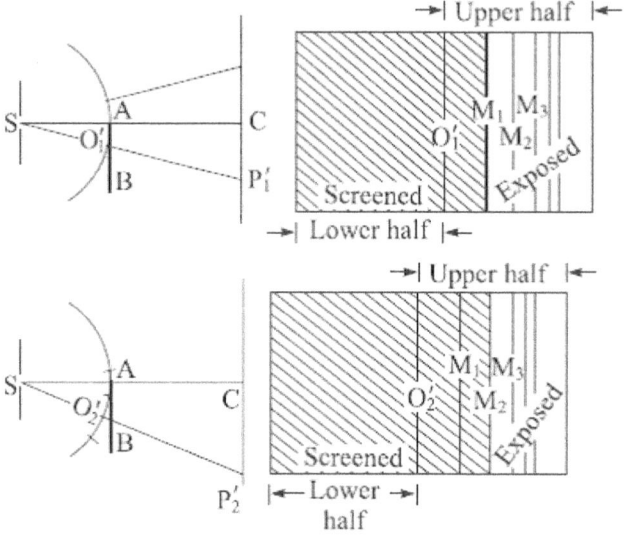

FIGURE 10.11. Penetration of light within the geometrical shadow.

Let us suppose a point P_1' on the screen below C so that

$$P_1'A - P_1'O_1' = \frac{\lambda}{2}$$

Then for this point, the obstacle not only cuts off completely the lower half of the wave but also the first zone of the upper half. Therefore, the resultant amplitude at P_1' due to the exposed portion of the upper half is

$$D_1' = R_2 - R_3 + R_4 - R_5 + \cdots = \frac{R_2}{2}$$

and the intensity at P_1' is proportional to $\frac{R_2^2}{4}$, which is evidently less than the intensity at C.

For the point P_2' below C so that

$$P_2'A - P_2'O_2' = \lambda$$

the obstacle obstructs, in addition to the lower half, the first two zones of the upper half of the wave also. The resultant amplitude at P_2' is

$$D_2' = R_3 - R_4 + R_5 - R_6 + \cdots = \frac{R_3}{2}$$

and the intensity is proportional to $\dfrac{P_3^2}{4}$, which is evidently less than the intensities at the points P_1' and C. Thus, as the positions below C are taken to be farther and farther into the geometrical shadow, the number of zones contributing to the resultant amplitude on the screen will be smaller. Therefore, the intensity falls off continuously but rapidly without showing any maxima or minima, till all the effective zones are cut off and the higher order ones in the upper half AW of the wavefront simply annul the effects of one another. Thus, the resultant intensity becomes zero.

The graphs of amplitude and intensity of illumination on the screen against the distance from the edge of the geometrical shadow can now be plotted. These graphs are depicted in Figures 10.12 and 10.13 respectively.

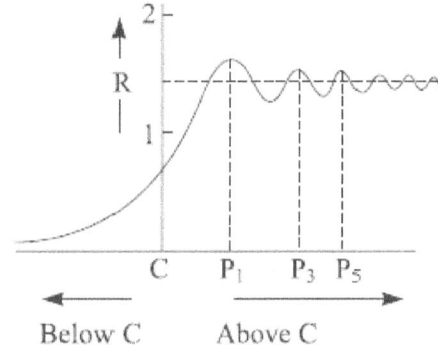

FIGURE 10.12. Amplitude contour.

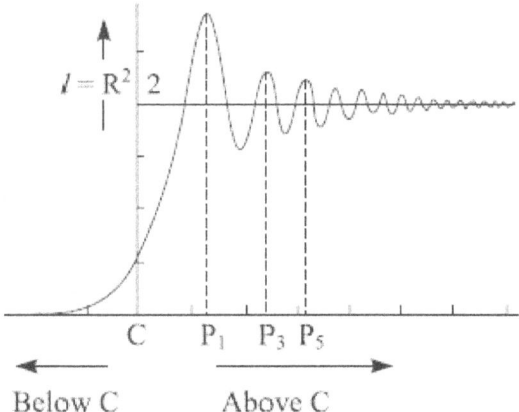

FIGURE 10.13. Intensity contour.

10.5.5 Measurement of Wavelength

To carry out the measurement of wavelength of light by diffraction at a straight edge, a narrow slit is mounted vertically in one of the uprights of the optical bench. In another upright a tin plate is mounted, with its straight edge parallel to the slit, which is illuminated with monochromatic light. The parallelism of the straight edge and the slit is adjusted first with the naked eye. The finer adjustment is, however, done by observing the diffraction bands through the micrometer traveling eyepiece and rotating the edge in its own plane by the help of the tangent screw of the holder until the bands are most distinct. Now the distance between the first or second bright band and one of the most distant that can be seen distinctly is measured by the help of the micrometer eyepiece. If the former be mth and the latter nth, we have by the use of Equation (10.44)

$$x_n = \sqrt{\frac{b\lambda(2n-1)(a+b)}{a}} \text{ and } x_m = \sqrt{\frac{b\lambda(2m-1)(a+b)}{a}}$$

Hence,

$$x_n - x_m = \sqrt{\frac{b\lambda(a+b)}{a}}\left(\sqrt{2n-1} - \sqrt{2m-1}\right) \tag{10.46}$$

10.6 DIFFRACTION AT A NARROW WIRE

Let AB represent, in Figure 10.14, a long narrow wire perpendicular to the plane of the paper and placed exactly parallel to a long narrow slit S_1, which is illuminated by a source of monochromatic light. The cylindrical waves diverging from the slit are incident on the wire from the left. The lines joining the slit S with the extremities of the wire when produced meet the screen EF at points C and C'. Therefore, CC' is the region, on the screen, of the geometrical shadow of the wire. The wire offers some limitation to the incident wave. Therefore, what is actually observed on the screen on both sides of the geometrical shadow CC' is a diffraction pattern characterized by diffraction fringes exactly similar to those in the case of the straight edge. These fringes gradually get closer and closer, independent of the width of the wire, finally merging into uniform illumination. Within the geometrical shadow are found some dark and bright fringes which are equidistant and narrow, their width being inversely proportional to the width of the wire.

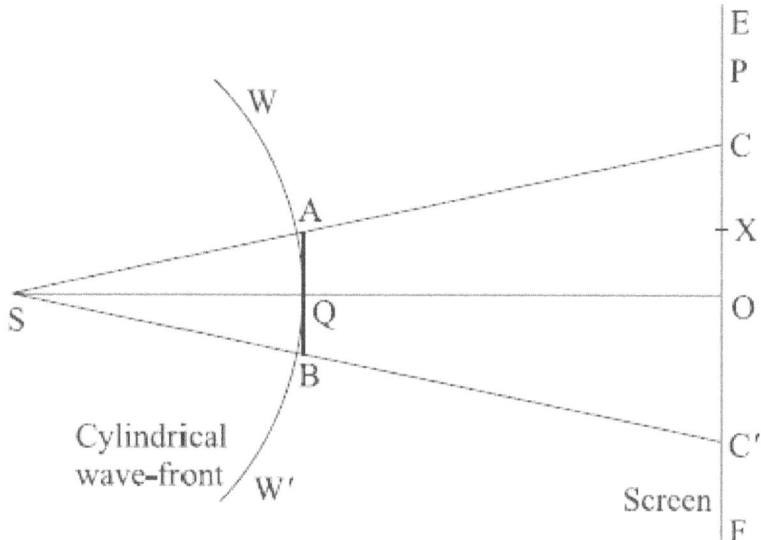

FIGURE 10.14. Fresnel diffraction of a cylindrical wavefront by a narrow wire.

To explain the origin of this diffraction pattern, we should consider separately the portions CE, CC', and $C'F$ of the screen. The intensity at any point P in the portion CE of the screen is the resultant effect of half-period zones in one-half of the wave above the pole O' and the exposed half-period zones between O' and A of the lower half of the wave. It is assumed that the width AB of the wire is such that the portion BW' of the wave contains, for the point P', higher order half-period zones which mutually cancel each other's effect at P and hence are quite ineffective. Thus, the diffraction fringes in the portion CE of the screen arise due to diffraction of the portion AW of the incident wave by the extremity A of the wire, and fringes are independent of the portion BW' of the wave. Exactly in a similar way, the diffraction fringes in the portion $C'F$ arise on account of the diffraction of the portion BW' by the extremity B of the wire, and these are independent of the portion AW.

It now remains to interpret the system of equidistant fringes within the geometrical shadow of the wire. Suppose X is a point within the shadow. A straight line joining S and X cuts BA at some point between A and B. The wire cuts off a few half-period zones from the upper and the lower half of the wave for this point. The resultant intensity at X is, therefore, due to the combined effect of the resultant of the half-period zones contained in AW and the resultant of zones in the portion BW' of the wave. The resultant amplitude at X of the portion AW is simply equal to half the amplitude of the zone adjacent to A. Therefore, the phase of this resultant will be almost the same as that of the optical disturbance from the zone adjacent to A. Similarly, the phase of the resultant at X of the portion BW' will almost be the same as that of disturbances from the zone adjacent to B. The figure on both sides of the line SO is symmetrical in the plane of the paper. Hence, it is obvious that the disturbances start from A and B in the same phase. Therefore, on reaching X they will mutually interfere constructively or destructively, according to whether the difference of their paths BX and AX is an even or odd multiple of $\frac{1}{2}\lambda$. This set of fringes is, therefore, simply the interference fringes arising as if waves of light from two coherent sources, situated at A and B, interfere in the region of the geometrical shadow of the wire. Let the diameter AB of the wire be $2d$ and the perpendicular distance between the wire and the screen be D. The fringe width of the interference fringes is given by

$$\bar{X} = \frac{D\lambda}{2d} = \lambda \frac{OQ}{AB} \qquad (10.47)$$

where λ is the wavelength of the light waves incident on the wire. The fringe width is the same for all internal fringes and is inversely proportional to the diameter of the wire.

The explanation of the origin of internal fringes was experimentally verified by Young. He showed that when the light which passed through one edge was intercepted by an opaque screen, fringes within the geometrical shadow as well as diffraction fringes on the side of the opaque screen vanished. Therefore, it is clear that the internal fringes arise on account of the combined action of the portions AW and BW', according to the principles of interference of light.

10.7 DIFFRACTION AT A CIRCULAR APERTURE

Fresnel diffraction at a circular aperture is characterized by alternate concentric circular maxima and minima as shown in Figure 10.15. The intensity at minima regions, however, is not absolutely zero. Moreover, as the screen is moved toward or away from the aperture, the intensity of the central ring alternately becomes maximum and minimum.

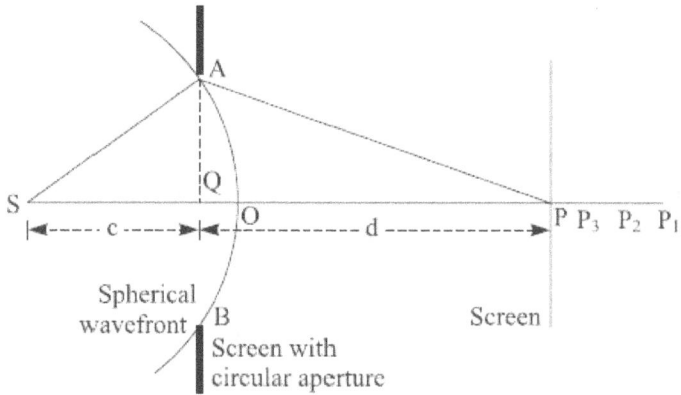

FIGURE 10.15. Fresnel diffraction of a spherical wavefront by a circular aperture.

The same effect is observable by increasing or decreasing the diameter of the aperture. We shall now explain the origin of this diffraction pattern with the aid of *Fresnel's half-period zones*.

Let *AB* represent a section, in the plane of paper, of the spherical wavefront diverging from a point source *S* situated on the axis of the circular aperture. Let λ be the wavelength of the light waves. *P* is an axial point on the observing screen. The amplitude at *P* is the resultant of a large number of secondary wavelets, which we consider as originating from various points of the spherical wavefront at the instant they just touch the periphery of the circular aperture. To find this resultant, we divide the spherical wavefront into *Fresnel's half-period zones* around the point *O*, the pole of the wave with respect to the point *P*. The Fresnel zones are shown in Figure 10.16 where

$$PO = b, PM_1 = b + \left(\frac{\lambda}{2}\right), \; PM_2 = b + \left(\frac{2\lambda}{2}\right)$$

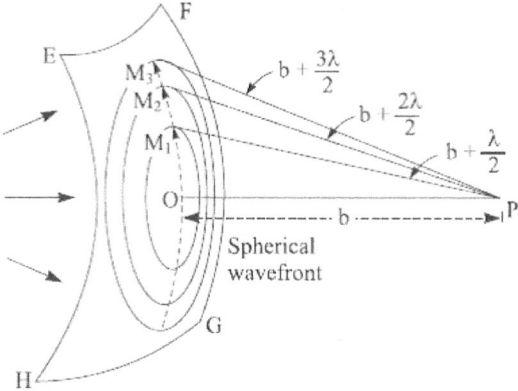

FIGURE 10.16. Half-period zones on a spherical wavefront.

The segment of the spherical wavefront by the first circle is the first half-period zone, the annular space between circles M_1 and M_2 is the second half-period zone, and so on. The *area of Fresnel zones*, that is, of rings between successive circles, can be very easily obtained. Let the radius of the spherical wavefront at a moment it just touches the periphery of the aperture be a, and let us consider a segment of the wavefront which subtends an angle 2α at the point source S. Then, from the geometry of Figure 10.17, $EG = a \sin \alpha$. The width of the annular ring, shown in Figure 10.17, is $EF = a\, d\alpha$. Hence, the area of the elementary ring is given by

$$dA = 2\pi a \sin \alpha \times a\, d\alpha = 2\pi a^2 \sin \alpha\, d\alpha \qquad (10.48)$$

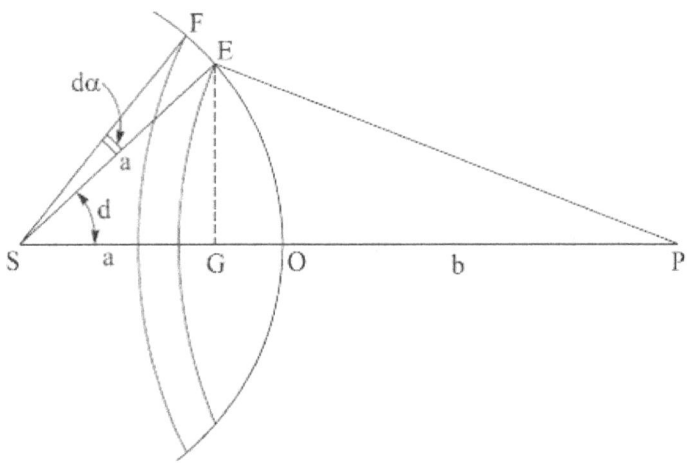

FIGURE 10.17. Illustrating the geometry for finding the area of Fresnel zones.

Integrating this elementary area between the limits 0 and α, we get the area of the segment which subtends 2α at S,

$$A = 2\pi a^2 \int_0^\alpha \sin \alpha d\alpha = 2\pi a^2 (1 - \cos \alpha) \qquad (10.49)$$

The angles α_n for the boundary of the zones are determined by the condition

$$PM_n = b + \left(\frac{n\lambda}{2} \right) \qquad (10.50)$$

If the point E is on the boundary of nth zone, then

$$EP = PM_n = b + \left(\frac{n\lambda}{2} \right) \text{ and } \alpha = \alpha_n$$

Hence,

$$\cos\alpha_n = \frac{(a+b)^2 + a^2 - \left(b + \frac{n\lambda}{2}\right)^2}{2a(a+b)}$$

Substituting this value in Equation (10.49), we get

$$\text{Area of first } n \text{ zones} = 2\pi a^2 - \frac{\pi a}{(a+b)}\left(2a^2 + 2ab - bn\lambda - \frac{n^2\lambda^2}{4}\right)$$

Replacing n by $(n-1)$ in the previous expression, we have

Area of first $(n-1)$ zones

$$= 2\pi a^2 - \frac{\pi a}{(a+b)}\left(2a^2 + 2ab - b(n-1)\lambda - \frac{(n-1)^2\lambda^2}{4}\right)$$

To obtain the area A_n of the nth zone, we subtract the area of the first $(n-1)$ zones from that of the first n zones. Thus, we get

$$A_n = \frac{\pi a}{(a+b)}\left(b\lambda + (2n-1)\frac{\lambda^2}{4}\right) \tag{10.51}$$

In practice, b is always large as compared to the wavelength λ. Therefore, the term $\dfrac{(2n-1)\lambda^2}{4}$ may be neglected. Hence

$$A_n = \frac{a}{(a+b)}(\pi b\lambda) \tag{10.52}$$

This area is independent of n, and hence all zones are approximately equal in area. To be more accurate, the areas will increase slowly with n according to Equation (10.51).

The average distance d_n of the nth zone from P is

$$d_n = \frac{1}{2}\left(b + \left(\frac{n\lambda}{2}\right) + b + \frac{(n-1)\lambda}{2}\right) = b + \frac{(2n-1)\lambda}{4}$$

Hence,

$$\frac{A_n}{d_n} = \frac{\pi a\lambda}{(a+b)}$$

This ratio, being independent of n, is the same for all zones. The resultant amplitudes of successive zones starting from the first, owing to obliquity factor $(1 + \cos\theta)$, will decrease slowly at first but more rapidly afterward as θ increases. The resultant amplitude at P is given by

$$R = R_1 - R_2 + R_3 - R_4 + R_5 \cdots + (-1)^{n-1} R_n$$

where the alternate plus and minus signs occur, because at P the effects of the zones are alternately out of and in phase.

10.7.1 Maxima and Minima Illumination of the Central Ring

Let the screen be at such a distance from the aperture that for the axial point P_1, the aperture exposes only the first Fresnel zone. The amplitude at P_1 is simply R_1, which is twice the amplitude if the entire wavefront were exposed. The intensity at P_1 is, therefore, four times as great as the intensity obtainable if the entire wave were exposed. Hence, the point P_1 is extremely bright.

For the light of given wavelength, the area of Fresnel zone

$$A_n = \frac{\pi a \lambda}{1 + \left(\dfrac{a}{b}\right)}$$

decreases with a decrease in the value of b. Hence, the aperture exposes a greater number of zones as the screen is moved toward it. Let the screen be moved until it arrives at the point P_2 for which the aperture exposes the first and the second *Fresnel zones*. The amplitude at P_2 is therefore $R_1 - R_2$, which since R_1 and R are nearly equal, is very small. The intensity is approximately zero. Thus, the center of the diffraction pattern is practically dark in contrast to the *Fraunhofer* diffraction pattern where the central disc is always bright. If the screen is moved still nearer to the aperture, it arrives at P_3, for which the first three zones are exposed. The amplitude at P_3 becomes

$$R_1 - R_2 + R_3 \cong \frac{1}{2}(R_1 + R_3)$$

which is again large. Thus, the axial point alternately becomes bright and dark as the screen is moved toward the aperture.

If the position of the screen is kept fixed while the aperture is gradually increased, it successively exposes one, two, three, or four zones or more. Therefore, again, the intensity at the axial point is alternately maximum and minimum. The illumination at the axial point P of the screen is, therefore, maximum or minimum according to whether the aperture exposes odd or even numbers of Fresnel zones.

We can now easily calculate the position of the screen for the maximum or minimum brightness of the axial point P. Let $SQ = c$, $QP = d$, $AQ = r$ where Q is the center of the circular aperture.

$SA^2 = c^2 + r^2$

Hence,

$$SA = c\sqrt{1 + \frac{r^2}{c^2}} = c + \frac{1}{2}\left(\frac{r^2}{c}\right)$$

on expanding by binomial theorem and neglecting powers of $\frac{r^2}{c^2}$, since r is very small in comparison with c and d.

Similarly,

$$AP = d + \left(\frac{r^2}{2d}\right)$$

Hence,

$$SA + AP = (c + d) + \frac{r^2}{2}\left(\frac{1}{c} + \frac{1}{d}\right)$$

Therefore, the path difference between the extreme ray SAP relative to the axial ray SQP is

$$SAP - SQP = (c + d) + \frac{r^2}{2}\left(\frac{1}{c} + \frac{1}{d}\right) - (c + d) = \frac{r^2}{2}\left(\frac{1}{c} + \frac{1}{d}\right)$$

For the maximum at P, the aperture contains an odd number of Fresnel zones. Therefore, the path difference $(SAP - SQP)$ must be an odd multiple of $\frac{\lambda}{2}$, that is,

$$\text{Maxima:} \quad \frac{r^2}{2}\left(\frac{1}{c} + \frac{1}{d}\right) = (2n + 1)\frac{\lambda}{2} \tag{10.53}$$

Similarly, for the minimum at P, we have

$$\text{Minima:} \quad \frac{r^2}{2}\left(\frac{1}{c} + \frac{1}{d}\right) = (2n)\frac{\lambda}{2} \tag{10.54}$$

Equation (10.53) is a general equation connecting the distance c of the source and d of the axial point P, both measured from Q, the center of the aperture, when intensity is maximum at P.

10.7.2 Intensity at Non-Axial Points

The mathematical treatment for finding the intensity at points a short distance off the axis is more complicated, but we can form an idea about the result by a simplified treatment. For concreteness, let us suppose that the aperture is of such a size as to expose three zones for an axial point P_3, so the center of the pattern is bright.

The Fresnel zones that should be used to find the amplitude at non-axial point P' have their center on the point O', the intersection of the wavefront with the line joining S and P'. As seen from P', the aperture and the half-period zones appear as in Figure 10.18, where it would be seen that the first and the second zones are still entirely exposed, together with about half of the third and fourth zones. The resultant amplitude at P' due to exposed zones is equal to

$$R_1 - R_2 + \frac{1}{2}R_3 - \frac{1}{2}R_4 = R_1 - \frac{1}{2}(R_1 + R_3) + \frac{1}{2}R_3 - \frac{1}{2}R_4 \cong \frac{1}{2}(R_1 - R_4)$$

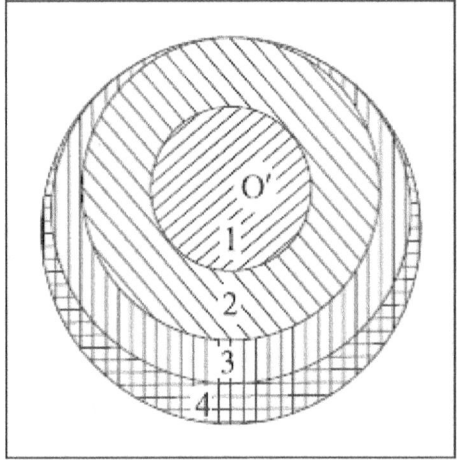

FIGURE 10.18. Fresnel zones exposed by a circular aperture for a non-axial point.

which is a minimum value as compared to the amplitude at the axial point P_3. The intensity at P' is, therefore, less than that at the center. The aperture is perfectly symmetrical about its axis. Therefore, the central bright spot in this case is surrounded with an approximately dark ring. In a similar way it can be shown that there are other bright and dark rings surrounding the central spot.

10.7.3 Penetration of Light within the Geometrical Shadow

Let us consider a point on the screen just above the edge of the geometrical shadow of the aperture. The exposed zones are shown in Figure 10.19.

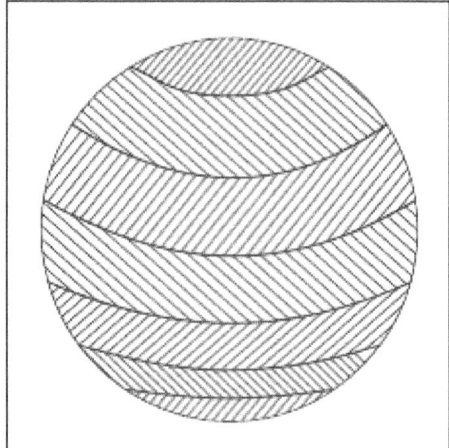

FIGURE 10.19. Fresnel zones exposed by a circular aperture for a point within the geometrical shadow.

About one-third of the first zone, a somewhat larger portion of the second, and varying parts of many more zones are exposed. There is, therefore, a small but definite amount of intensity within the geometrical shadow. But if the point is so well within the geometrical shadow that the first few zones and the last few zones are entirely cut off, then the contributions of the exposed portions of different zones mutually cancel, with the result that the intensity is zero.

10.8 DIFFRACTION AT AN OPAQUE DISC

A surprising result was deduced by Poisson in 1815, that is, that a bright spot should always be formed at the center of the shadow of a small circular disc. He concluded that this seemingly absurd prediction was so far at variance with the recognized properties of light that it disproved wave theory. The experiment was tried by Argo, who employed a circular disc of 2 mm in diameter. He showed that the geometrical shadow had a central bright spot. The complete diffraction pattern, besides the central spot and faint rings in the shadow, consists of bright circular fringes bordering the outside of the shadow. We shall show, by the help of Fresnel's half-period zones, that the center of the shadow should always be bright, irrespective of distance of the screen from the disc.

Suppose a spherical wavefront WW of monochromatic light is incident on the circular disc AB, from the left as shown in Figure 10.20. We divide this wavefront into Fresnel zones for an axial point in the usual manner. The areas of these zones depend on the distance b of the axial point from the disc. Let us suppose that for the point P_1, the disc covers up the first zone. The second zone is then the first exposed zone, sending light to P_1. Therefore, to find the resultant amplitude at P_1 due to the entire exposed zones, we should perform the summation beginning with R_2.

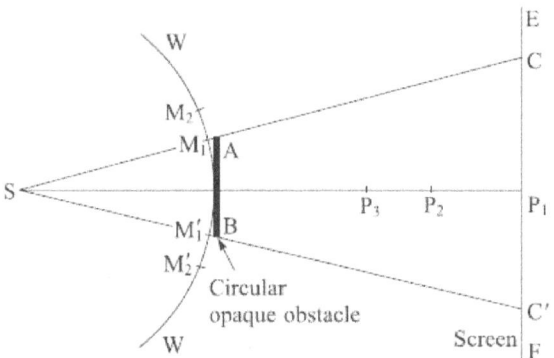

FIGURE 10.20. Fresnel diffraction of a spherical wavefront by a circular opaque obstacle.

We get for the amplitude at P_1,

$$D_1 = R_2 - R_3 + R_4 - R_5 + \cdots = \frac{R_2}{2}$$

and the intensity at P_1 is proportional to $\dfrac{R_2^{\,2}}{4}$, which is approximately the same as $\dfrac{R_1^{\,2}}{4}$, the intensity obtainable if the entire wave were exposed. The center of shadow, therefore, is nearly as bright as when the disc was absent.

When the screen is moved toward the disc, the area of each zone diminishes and the disc, therefore, covers up a large number of zones. Let us suppose that the disc covers up the first two zones when the screen is at P_2, a point nearer to the disc than P_1. The resultant amplitude at P_2 is therefore given by

$$D_2 = R_3 - R_4 + R_5 - R_6 + \cdots = \frac{R_3}{2}$$

and the intensity at P_2 is proportional to $\left(\dfrac{R_3}{2}\right)^2$. Similarly, the intensity at any point P is proportional to $\left(\dfrac{R_4}{2}\right)^2$, $\left(\dfrac{R_5}{2}\right)^2$, $\left(\dfrac{R_6}{2}\right)^2$, and so on, according to when the first three, four, five . . . zones are intercepted. Thus, the center of the geometrical shadow is always bright. The central spot becomes smaller in diameter as the disc becomes large.

The rings in the shadow are similar in origin to the interference fringes within the shadow of the narrow wire, while the bright and comparatively dark fringes bordering the outside of the shadow are similar in origin to the diffraction fringes due to a straight edge.

10.9 ZONE PLATE

A remarkable experimental confirmation of Fresnel's theory of half-period zones is provided by an optical device known as a *zone plate*. The zone plate is so called because it is constructed following the Fresnel zone law. It is simply a plane parallel glass plate, having concentric circles of radii accurately proportional to the square roots of the consecutive natural numbers 1, 2, 3, and so on. The even or odd order annular spaces between the circles are completely dark. Let us suppose that a plane monochromatic light wave, for which the plate is constructed, is incident normally on it. On the other side on its axis we shall encounter a series of maxima points having intensities in increasing order, as the distance of the points from the plate increases. The zone plate acts, therefore, like a convergent lens but, unlike the lens, it has a series of foci and associated focal lengths. There are also series of virtual foci on the side of the source. In other words, the zone plate is capable of acting as a divergent lens also. The theory of the zone plate and its properties are discussed as follows.

Let *AB* in Figure 10.21 represent a plane parallel transparent screen perpendicular to the plane of paper, and let *S* be a point source of monochromatic light of wavelength λ.

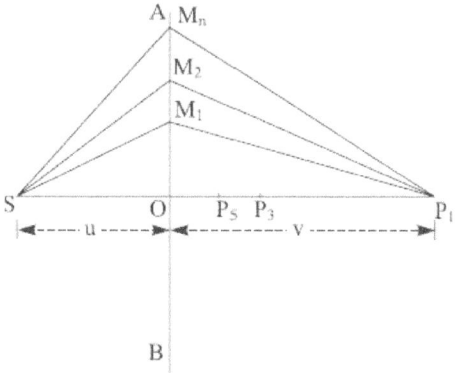

FIGURE 10.21. Theory of the zone plate.

We have to find the resultant amplitude at a point P_1 due to secondary wavelets diverging from the various points of the screen under the influence of spherical waves diverging from S. We must, of course, consider that the screen is not a wave surface, and so the virtual secondary sources are not in phase. The resultant amplitude can be determined by dividing the entire screen into half-period zones. We mark off points M_1, M_2, and so forth on the screen so that

$$SM_1P_1 - SOP_1 = \frac{1}{2}\lambda$$

$$SM_2P_1 - SOP_1 = \lambda$$

$$\cdots \cdots \cdots \cdots$$

$$SM_nP_1 - SOP_1 = \frac{1}{2}n\lambda$$

A circle of radius OM_1 drawn on the transparent screen constitutes what is known as Fresnel's first half-period zone; the annular space between the concentric circles of radii OM_1 and OM_2 constitutes the second half-period zone, and so on. In this way the entire screen can be divided into a large number of half-period zones.

Let r_n be the radius OM_n of the outer boundary of the nth zone, and let $OS = u$ and $OP_1 = v$. Then, from the right-angled triangle SOM_n, we have

$$SM_n{}^2 = SO^2 + OM_n^2 = u^2 + r_n^2$$

Hence,

$$SM_n = u\sqrt{1 + \frac{r_n^2}{u^2}} = u + \frac{1}{2}\frac{r_n^2}{u}$$

since r_n is small in comparison with u. Similarly

$$P_1M_n = v + \frac{1}{2}\frac{r_n^2}{v}$$

$$SM_n + M_nP_1 = (u+v) + \frac{r_n^2}{2}\left(\frac{1}{v} + \frac{1}{u}\right)$$

$$SM_nP_1 = SOP_1 + \frac{r_n^2}{2}\left(\frac{1}{v} + \frac{1}{u}\right)$$

But by construction $SM_nP_1 - SOP_1 = \dfrac{n\lambda}{2}$
Hence,

$$\frac{r_n^2}{2}\left(\frac{1}{v} + \frac{1}{u}\right) = \frac{n\lambda}{2}$$

$$r_n^2 = \frac{nuv\lambda}{(u+v)} \tag{10.55}$$

$$r_n^2 = nr_1^2 \tag{10.56}$$

The value of r for the first, second, . . . zones can be obtained by substituting 1, 2, 3, . . . for n in Equation (10.55). Thus, the external radii of Fresnel's half-period zones, for given values of u, v, and λ, are proportional to square roots of natural numbers.

The area of the nth half-period zone is given by

$$A_n = \frac{\pi v\lambda}{\left(1 + \dfrac{v}{u}\right)}$$

This being independent of n, all zones are equal in area (approximately). If u becomes infinite, we shall have plane waves falling on the screen. The area of each half-period zone reduces to $\pi v\lambda$.

Let R_1, R_2, R_3,..., R_n be the resultant amplitudes at P_1 due to secondary wavelets from the 1st, 2nd, 3rd, . . . and the nth zone. Owing to the obliquity factor, the magnitudes of R_1, R_2, R_3, and so on are in decreasing order. Therefore, the resultant amplitude at P_1 due to entire zones is given by

$$D = R_1 - R_2 + R_3 - R_4 + \cdots + (-1)^{n-1} R_n$$

$$D = \frac{1}{2}(R_1 \pm R_n)$$

where n is finite and small. But when n is very large, then

$$D = \frac{1}{2}R_1$$

The resultant amplitude at P_1 due to either the odd-numbered zones or due to the even-numbered zones alone would be clearly much greater than that of the entire wave. This may by tested experimentally by blocking off by some means the even numbered zones, for example. The resultant amplitude at P_1 in this case becomes

$$D_1 = R_1 + R_3 + R_5 + R_7 + \cdots \cong \frac{1}{2}NR_1 \tag{10.57}$$

where N is the total number of zones in the screen and $\dfrac{N}{2}$ zones are exposed.

The intensity at P_1 is, therefore, proportional to $\dfrac{1}{4}N^2R_1^2$, which is N^2 times greater than the resultant intensity $\dfrac{R_1^2}{4}$ due to all zones. For example, if the only alternate exposed zones are 10 in number, that is $N = 20$, even then the intensity is 400 times as great as that obtained when all zones are exposed. Under this condition, P_1 will be a point of maximum intensity. In other words, most of the light from S will be concentrated at P_1. If odd zones are blocked, the resultant amplitude at P_1 becomes

$$D_1' = R_2 + R_4 + R_6 + \cdots$$

and again P_1 will be a point of maximum intensity. We can therefore say that under these conditions, P_1 is the image of S, and its distance from the zone plate satisfies the relation Equation (10.55), which can be expressed in the form

$$\frac{1}{v} + \frac{1}{u} = \frac{n\lambda}{r_n^2} \tag{10.58}$$

a result similar to the lens equation.

A zone plate can be very conveniently prepared by the following method:

On a sheet of white paper, we first draw concentric circles of radii accurately proportional to the square roots of consecutive integers 1, 2, 3, 4, and so on. The alternate zones are then blackened, and a much-reduced copy of the same is obtained by photographing it on a plane parallel glass negative. When the plate has been fixed, the negative is a zone plate. Because of the difficulty of drawing a large number of circles accurately, another method for constructing the zone plate may be employed. Dark rings of a Newton rings interference pattern, it will be recalled, have radii proportional to the square roots of consecutive integers 1, 2, 3, 4, and so forth. The dark and bright rings have equal areas when parallel monochromatic light is employed. Thus, it is possible to construct a zone plate by photographing Newton's rings as seen by the reflected light.

The zone plate acts as a convergent lens not only for axial points but also for small extended objects. Let us suppose that a small object PQ is situated at a distance u from the plate on its axis. Let $PQ = x$ and $P'Q' = y$. It is easy to see from Figure 10.22 that

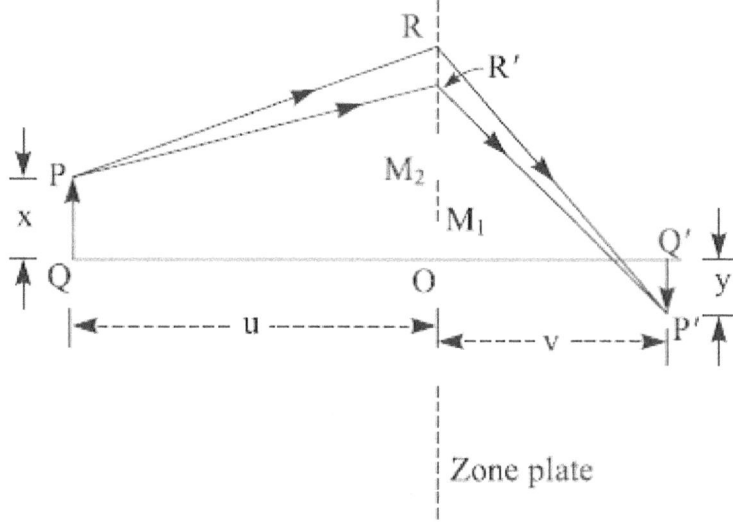

FIGURE 10.22. Illustration of image formation by zone plate.

$$PR = \sqrt{u^2 + (OR - x)^2} = u + \frac{1}{2}\frac{(OR - x)^2}{u}$$

$$P'R = \sqrt{v^2 + (OR + y)^2} = v + \frac{1}{2}\frac{(OR + y)^2}{v}$$

Hence,

$$PRP' = (u + v) + \frac{1}{2}\frac{(OR - x)^2}{u} + \frac{1}{2}\frac{(OR + y)^2}{v}$$

Similarly,

$$PR'P' = (u + v) + \frac{1}{2}\frac{(OR' - x)^2}{u} + \frac{1}{2}\frac{(OR' + y)^2}{v}$$

Thus,

$$PRP' - PR'P' = \frac{1}{2}(OR^2 - OR'^2)\left(\frac{1}{v} + \frac{1}{u}\right) - RR'\left(\frac{x}{u} - \frac{y}{v}\right) \tag{10.59}$$

Let R and R' be the two corresponding points on two successive transparent zones. For concreteness, let R be on the outer boundary of the nth zone and R' on that of the $(n-2)$th zone, the $(n-1)$th zone being opaque. Therefore, by the help of Equation (10.55), we have

$$OR^2 - OR'^2 = \frac{uv\lambda}{(u + v)}(n - (n - 2)) = \frac{2uv\lambda}{(u + v)}$$

Hence, Equation (10.59) assumes the form

$$PRP' - PR'P' = \lambda - RR'\left(\frac{x}{u} - \frac{y}{v}\right)$$

If the path difference between the waves through corresponding points of the zone plate is λ, the waves reinforce each other at P'. Then, the second term of the previous equation vanishes and gives the relation

$$\frac{y}{x} = \frac{v}{u}$$

Under these conditions P' is the image of P. The zone plate, therefore, also forms well-defined images of small extended objects, and the magnification $\frac{y}{x}$ follows the same law as in the case of a lens.

10.9.1 Multiple Foci of a Zone Plate

The focal length of the zone plate can be easily obtained by making u infinite in Equation (10.58)

$$f_n = \frac{r_n^2}{n\lambda} \tag{10.60}$$

For $n = 1$,

$$f_1 = \frac{r_1^2}{\lambda} \tag{10.61}$$

The point P_1 given by the previous relation is called the *primary focus* or the first *order focal point* and is the most intense. Here one Fresnel zone coincides with each actual zone of the plate. f_1 is known as the primary focal length of the plate.

There are a series of foci, unlike in a convergent lens, of diminishing intensity as we go along the axis towards the plate. The existence of these foci is simply because the area of each zone diminishes as points nearer to the plate are considered. Thus, for a point P_3 on the axis at one-third the distance from the plate to the first focus, the actual first zone contains 1st, 2nd, and 3rd half-period zones. The second black zone intercepts the 4th, 5th, and 6th new Fresnel zones, while the wavelets from the 7th, 8th, and 9th new zones, contained in the original 3rd zone, are transmitted, and so on.

Therefore, the resultant amplitude at P_3 is given by

$$D_3 = (R_1 - R_2 + R_3) + (R_7 - R_8 + R_9) + (R_{13} - R_{14} + R_{15}) + \cdots$$

$$D_3 = \left(R_1 - \frac{1}{2}(R_1 + R_3) + R_3\right) + \left(R_7 - \frac{1}{2}(R_7 + R_9) + R_9\right) + \cdots$$

$$D_3 \cong R_1 + R_3 + R_7 + R_9 + \cdots$$

The intensity at P_3 is a maximum which is, therefore, the second focal point, but it is much less intense than first one. The light rays transmitted through each successive transparent zone have a path difference of 3λ. Hence, P_3 is also called the third-order focal point. The first original zone exposes three new zones, all of equal area. Therefore, we have the relation

$$3\pi a_1^2 = \pi r_1^2$$

$$a_1 = \frac{r_1}{\sqrt{3}}$$

where a_1 is the radius of the new first zone. Therefore, replacing r_1 by a_1 in Equation (10.61), the focal length associated with the second focal point P_3 is obtained as

$$f_3 = \frac{a_1^2}{\lambda} = \frac{r_1^2}{3\lambda} = \frac{1}{3}f_1$$

The fifth-order focal point P_5 occurs at a position in which Fresnel zones are alternately passed and blocked off in sets of five. The resultant amplitude at P_5 is given by

$$D_5 = (R_1 - R_2 + R_3 - R_4 + R_5) + (R_{11} - R_{12} + R_{13} - R_{14} + R_{15}) + \cdots$$

$$D_5 \cong \frac{1}{2}(R_1 + R_5 + R_{11} + R_{15} + \cdots)$$

Thus, P_5 is still a less bright focus, and the corresponding focal length is given by

$$f_5 = \frac{b_1^2}{\lambda} = \frac{r_1^2}{5\lambda} = \frac{1}{5}f_1$$

where b_1 is the radius of the first new zone.

A zone plate has multiple foci at $f_1, \frac{1}{3}f_1, \frac{1}{5}f_1, \frac{1}{7}f_1$, and so on, and their intensity is in decreasing order. Generalizing the result we can say that if, for any point on the axis, an odd number of Fresnel zones are alternately passed and blocked off, the point under consideration is a focal point. The focal length associated with various foci can be obtained by putting $m = 0, 1, 2, 3$, and so forth in the expression

$$f_{2m+1} = \frac{r_1^2}{(2m+1)\lambda}$$

Similar reasoning shows that there should be no even-order focal points. However, all the zone plates tested had a second-order focal point at approximately $\frac{1}{2}f_1$. This was caused partly due to unequal areas of opaque and transparent zones. Owing to this reason, the two Fresnel zones, which should coincide with each actual zone of the plate at the second-order focal point, do not completely cancel each other but a partial reinforcement of light occurs.

The zone plate also acts simultaneously as a concave lens but, unlike the lens, it has multiple virtual foci on the side of the source.

Example 10.1

A very narrow vertical slit illuminated with sodium light of wavelength 5.9×10^{-5} cm casts a shadow of a vertical copper wire 1 mm in diameter and 2 meters away, on a screen 3 meters away from the wire as shown in Figure 10.23. Calculate the distance apart of the bands inside the shadow and the total number of bands which can be seen.

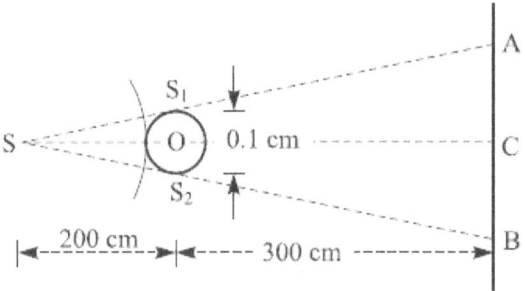

FIGURE 10.23. Example 10.1.

Solution:

In Figure 10.23, AB represents the region of geometrical shadow of the wire O. In this region of the screen, equidistant bright and dark bands are observed. Their distance apart is expressed by

$$\bar{X} = \frac{D\lambda}{2d}$$

In the given problem, $D = OC = 300$ cm, $2d = S_1 S_2 = 0.1$ cm, and $\lambda = 5.9 \times 10^{-5}$ cm

Hence,

$$\bar{X} = \frac{300 \times 5.9 \times 10^{-5}}{0.1} = 0.177 \text{ cm}$$

Also, from similar triangles SS_1S_2 and SAB, we have the relation

$$\frac{AB}{SC} = \frac{S_1S_2}{SO}$$

With the given data, $AB = \dfrac{0.1 \times 500}{200} = 0.25 \text{ cm}$

At the point C is equidistant from S_1 and S_2, a bright fringe will be observed. The separation between the two bright interference fringes, each of the order 1, would be 2×0.177 cm, that is, 0.354 cm (if formed). This separation exceeds the width of the geometrical shadow AB. Hence, only 2 dark bands will be seen within the geometrical shadow.

Example 10.2

A strong parallel beam of monochromatic light is incident normally on a thin plate having a small circular hole of diameter 1 mm. If the screen is moved through a distance of 12.5 cm from the first position where the center is black to the second similar position, what is the wavelength of light used?

Solution:

From the theory of Fresnel diffraction due to circular aperture, it follows that for the first position of the screen where the center of diffraction is black (not absolute), the aperture should expose two Fresnel's half-period zones of the wavefront. Accordingly, the distance of the center of the diffraction pattern from the edge of the aperture must be $(x+\lambda)$, where x is the separation between the screen and the aperture. It is now easy to write the relation

$$(x + \lambda)^2 = x^2 + r^2$$

Neglecting the term λ^2, being extremely small as compared with the other terms, the previous equation reduces to

$$2x\lambda = r^2$$

The second similar position of the screen is closer to the aperture. For this setting of the screen, the aperture should expose four half-period zones of the wavefront. Accordingly, the distance of the center of the diffraction pattern from the edge of the aperture would be $(x - 12.5 + 2\lambda)$. It is now easy to write the relation

$$(x-12.5+2\lambda)^2 = (x-12.5)^2 + r^2$$

Neglecting the term involving λ^2, the previous equation reduces to the form

$$4x\lambda - 50\lambda = r^2$$

Hence,

$$\lambda = \frac{r^2}{50} = \frac{0.05 \times 0.05}{50} = 5 \times 10^{-5} \text{ cm.}$$

Example 10.3

A zone plate is made by arranging that the radii of the circles which define the zones are the same as the radii of Newton's rings formed between a plane surface and surface whose radii of curvature is 2 meters. Find the primary focal length of the zone plate.

Solution:

The square of the radii of dark Newton's rings for the case of air film ($\mu=1$) reduces to

$$r_n^2 = n\lambda R$$

where $n=1, 2, 3, 4$, and so on for the first, second, third . . . dark rings respectively, and R is the radius of curvature of the curved surface of the film.

According to the condition of the problem, the previous equation also expresses the square of the radii of the circles which define the zones of the zone plate. The primary focal length of the zone plate, expressed by Equation (10.61), therefore reduces to

$$f_1 = R = 200 \text{ cm.}$$

10.10 EXERCISES

1. (a) Explain the phenomenon of the diffraction of light.
   (b) Explain the difference between the Fresnel and Fraunhofer classes of diffraction phenomena.
   (c) Distinguish clearly between interference and diffraction.

2. What are half-period zones? Show that the amplitude due to a large wavefront at a point in front is just half that due to the first half-period zone acting alone. Hence, explain the rectilinear propagation of light on the basis of wave theory.

3. What is a zone plate? How is it constructed? Show that a zone plate has multiple foci. Compare the zone plate with a convex lens.

4. Describe giving necessary theory the Fresnel type of diffraction produced by a straight edge. Explain how you would obtain the wavelength of light in this case, stating precisely what observations you would make. Draw a diagram to indicate the intensity distribution of light in the diffraction pattern.

5. Describe and explain the diffraction pattern formed by a narrow wire illuminated by a monochromatic light from a narrow slit parallel to the wire. How would you use the pattern to measure the thickness of the wire?

6. Give Fresnel's theory of diffraction and explain with its help why the center of the shadow of a small disc is bright.

7. A narrow circular aperture is held on one side of a fine hole and a screen is placed at some distance on the opposite side. Explain the illumination observed on the screen when a beam of monochromatic light is made to pass normally.

8. What is the radius of the first zone in a zone plate of focal length 20 cm for light of wavelength 5000 A.U.?

9. A zone plate has the radius of the first ring 0.05 cm. If plane waves ($\lambda = 5000$ A.U.) fall on the plate, where should the screen be placed so that light is focused to a bright spot?

10. A zone plate gives a series of images of a point source on its axis. If the strongest and the second strongest images are at 30 cm and 6 cm respectively from the plate, both on the same side remote from the source, calculate the distance of the source from the zone plate.

11. If the diameter of a wire is 0.2 mm, calculate the separation between the fringes formed on a screen placed at 50 cm from the wire. The light source used has a wavelength of 5000 A.U. and is placed at a finite distance from the wire.

12. A narrow slit illuminated by light of wavelength 6.4×10^{-5} cm is placed at a distance of 3 meters from a straight edge. If the distance between the straight edge and the screen is 6 meters, calculate the distance between the first and the fourth dark bands.

13. Light of wavelength 6×10^{-5} cm passes through a narrow circular aperture of radius 0.09 cm. At what distance along the axis will the first maximum intensity be observed?

14. A circular aperture of 1.2 mm diameter is illuminated by plane waves of monochromatic light. The diffracted light is received on a distant screen which is gradually moved toward the aperture. The center of the circular path of light first becomes dark when the screen is 30 cm from the aperture. Find the wavelength of light used.

15. A monochromatic beam of light, on passing through a slit 1.6 mm wide, falls on a screen held close to the slit. The screen is then gradually moved away, and the middle of the patch of light on it becomes dark when the screen is 50 cm from the slit. Calculate the wavelength of light used.

16. A very narrow vertical slit is illuminated with monochromatic light of wavelength 5461 A.U. A vertical wire of 0.1 cm diameter is placed in front of the slit, parallel to it. A screen is placed 3 meters from the wire. Calculate the bandwidth within the geometrical shadow.

17. The diameter of the central zone of the zone plate is 0.23 cm. If a point source emitting light of wavelength 6000 A.U. is placed 600 cm from the zone plate, find the position of the primary image and that of other weaker images.

CHAPTER 11

POLARIZATION OF LIGHT

Chapter Outline

11.0 INTRODUCTION

The successful interpretation of the phenomena of interference and diffraction of light on the basis of wave theory established conclusively that, like sound, light is also a wave motion. It was, however, not at all essential at that stage to raise the question whether light waves are longitudinal or transverse, owing to the fact that these phenomena occur for all types of waves. With the help of experiments illustrating these phenomena, we are able to measure accurately the wavelength of light, but we fail to obtain any information on the nature of light waves. In this connection we may ask whether light waves are transverse or longitudinal, or whether the vibrations are linear, circular, elliptical, or torsional. In order to get correct answers to these questions, we have to take recourse to another group of experiments which bring into evidence another property of light called the polarization of light. This phenomenon demands for its explanation that light must be a transverse wave motion in contradistinction to Huygens's conception of longitudinal wave motion. Before discussing the phenomenon of the polarization of light, we will first try to explain the meaning of the phrase "polarization of a wave."

Plane polarized light represents a special and relatively simple type of polarization. The light vector varies harmonically along a fixed straight line perpendicularly to the direction of propagation. By the superposition of two plane polarized beams of monochromatic light under suitable conditions, the resultant light vector so produced rotates in a plane perpendicular to the direction of propagation, and if its magnitude remains constant, the resulting light is spoken of as circularly polarized. If, however, the magnitude of the resultant light vector varies periodically between its maximum and minimum values during its rotation, its tip would trace out an ellipse, and the resulting light is known as elliptically polarized.

11.1 POLARIZATION OF A WAVE

In a longitudinal wave, the vibrations of the medium are parallel to the direction of the propagation. As a consequence, vibrations are exactly similar in all planes drawn through the direction of propagation. To be more precise, the appearance of a longitudinal wave is the same when viewed from different angles around this direction. This is expressed by saying that the longitudinal wave is perfectly symmetrical about the direction of propagation. On the other hand, in the case of a transverse wave, particles of the medium execute simple periodic vibrations in a direction at right angles to the direction of propagation. For example, in Figure 11.1 in a transverse wave progressing along the X-axis, the vibrations of the medium are confined to the XY plane. There are no vibrations in the XZ plane.

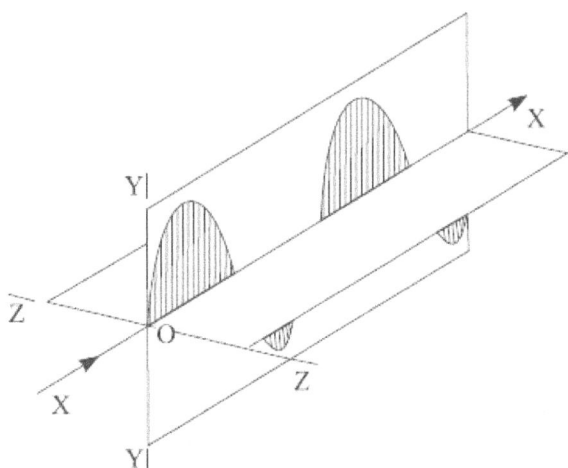

FIGURE 11.1. Linearly polarized wave.

As a consequence, the transverse wave would appear quite differently when viewed from different angles around the direction of propagation; that is, this wave is not symmetrical around the axis of X. This want of symmetry is the distinguishing feature of every transverse wave and is spoken of as the polarization of the wave. However, it is only so if we observe a limited portion of the wave or observe it momentarily, for it may happen that the displacement of the medium may be changing its direction continuously, although always remaining perpendicular to the direction of propagation to preserve its transverse character. Besides this, these changes may become too rapid to be perceived by the eye or photographic plate, both of which are capable of registering only the average effect. As a consequence, the resultant effect perceived by the eye may show it to be perfectly symmetrical about its direction of propagation, although there is no doubt about its transverse character. Such a wave is said to be unpolarized.

Ordinary or natural light behaves in this manner and is said to be unpolarized. To elaborate this statement we may add that the electric vector, transverse to the direction of propagation in the ordinary light wave, undergoes such rapid and random changes of direction that if it were possible to view the vibrations physically, then due to the persistence of vision the eye would observe the wave to be quite symmetrical about the direction of propagation. This is due to the fact that ordinary light contains millions of transverse waves about a meter in length and confined in all possible planes through the direction of propagation. Each wave is radiated out for 10^{-8} sec from one of the millions of excited atoms or molecules in random orientations in the source of light, and it is impossible for the electric vector to have all possible orientations simultaneously at any moment. It is assumed that the constituent transverse waves follow one another in such rapid succession that, during the minimum period required by the eye or photographic plate to register the effect, all possible orientations of the electric vector are equally represented. In other words, vibrations in any one direction cannot be isolated in the analysis of ordinary light. This accounts for the symmetrical character of ordinary or natural light around the direction of propagation.

To this change in the direction of vibration in the transverse waves we add another possibility of change in the character of vibration, namely, it may somehow undergo many random changes of shape, say from linear to circular and from circular to elliptical. Accordingly, we define the polarized wave as follows: a transverse wave is said to be polarized when the vibrations of the medium do not change in direction or form.

By suitable devices described later, the electric vector may be constrained to describe by the same locus linear, circular, or elliptical rays of fixed orientation transverse to the direction of propagation of the wave. Accordingly, we shall have linearly polarized, circularly polarized, or elliptically polarized waves.

11.2 MECHANICAL EXPERIMENT TO DEMONSTRATE THE POLARIZATION OF A WAVE

A very simple mechanical experiment may be arranged to demonstrate the polarization of a wave. Suppose a narrow loosely stretched string AB passes through two narrow rectangular slits N_1 and N_2, each a little wider than the diameter of the string. Its one end B is fixed while transverse waves are generated in it by shaking its free end A and changing the direction of shaking in a random fashion from one instant to the next. In the portion AN_1, the vibration of every element of the string is changing arbitrarily; accordingly, in this portion we have an unpolarized wave. The slit N_1, being only slightly wider than the string, will allow only those transverse waves to pass through it freely whose vibration planes are parallel to its length. The transmitted vibrations are thus confined to a single fixed plane; therefore, the wave between two slits is symmetrical and is spoken of as a plane polarized or linearly polarized wave, while the slit N_1 is spoken of as the polarizer in this experiment. This polarized wave will be transmitted by the second slit N_2 only if the lengths of N_1 and N_2 are mutually parallel as in Figure 11.2 (a), but it will be completely stopped if N_2 is rotated to a position perpendicular to the slit N_1 as shown in Figure 11.2 (b). The string beyond N_2 in the latter case will show no vibrations at all. The slit N_2 which functions as the analyzer or detector of polarization of incoming waves is spoken of as an analyzer.

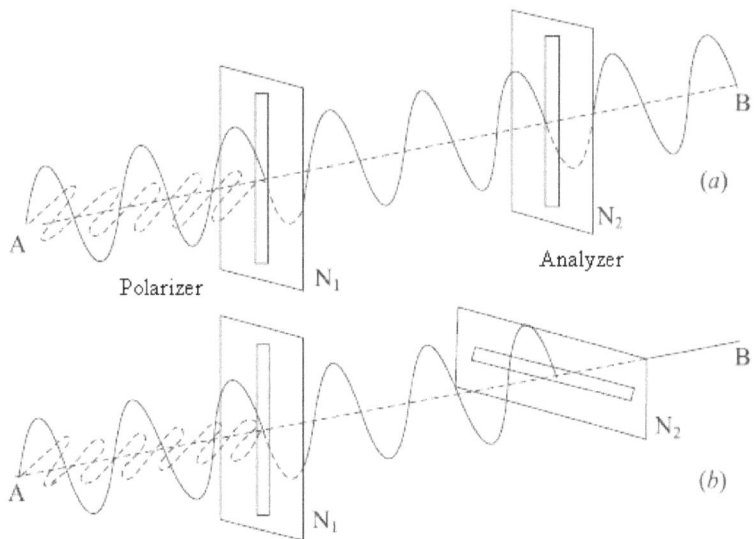

FIGURE 11.2. Mechanical experiment to demonstrate polarization of a wave.

If, however, the string were replaced by a spiral spring, longitudinal waves created in it would pass freely from one end to the other, unaffected by the presence of slits or their relative orientation. Polarization is thus the characteristic of transverse waves only. Figure 11.3 illustrates that the vertical disturbance in a rubber tube is transmitted freely through a vertical slit, in a horizontal disturbance it is not transmitted, and in a disturbance in an intermediate plane it is partially transmitted.

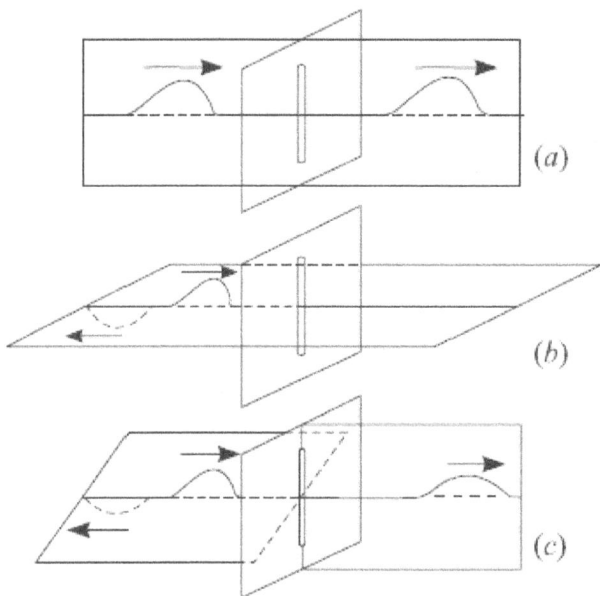

FIGURE 11.3. (a) A vertical disturbance in a rubber tube is transmitted freely through a vertical slit; (b) a horizontal disturbance is not transmitted; (c) a disturbance in an intermediate plane is partially transmitted.

11.3 OPTICAL EXPERIMENT TO ILLUSTRATE THE POLARIZATION OF A WAVE

A similar experiment in optics throws considerable light on the nature of vibrations in light. Tourmaline is a naturally occurring crystal, originally obtained from Tourmali in Ceylon. When a beam of ordinary light is allowed to pass through a plate N_1 of this crystal, cut with faces parallel to its vertical axis, only a part of the incident light is transmitted and is found to be plane polarized. The phrase plane polarized means that the electric vector in the beam is confined in one definite plane or parallel to it, through the direction of propagation. But to an eye which is incapable of detecting polarization of light, the transmitted beam appears to be only slightly colored due to the natural color of the crystal. If this beam is further allowed to pass through a second similar crystal plate N_2, placed with its axis parallel to N_1 as depicted in Figure 11.4 (a), the only observable effect is increased coloration of the emergent light, owing to two crystals in the path of the incoming light. But if N_2 is rotated with respect to N_1 in its own plane, the light transmitted through the combination gradually becomes dimmer and dimmer as in Figure 11.4 (b) and is finally cut off after a rotation of 90°—the axes of N_1 and N_2 are now crossed as in Figure 11.4 (c). On further rotation of N_2 through 90°, the transmitted light gradually attains full brightness; the axes of N_1 and N_2 are now again mutually parallel. Thus, brightness and darkness continue to alternate at settings of N_2 which are 90° apart.

This experiment clearly shows that the light wave emerging from N_1 is asymmetric, for had it been symmetrical, it would have passed freely through N_2 in all its orientations. Obviously, tourmaline crystals possess some property analogous to that of slits. This simple experiment, which illustrates the polarization of light, demands for its explanation, as first advanced by Fresnel, that the light wave must be transverse. For if it had been a longitudinal wave, it would have passed freely through the crystal plates irrespective of their mutual orientation. Accordingly, the intensity of transmitted light would have remained constant. After traversing N_1, the vibrations in the light wave are executed in one single direction, parallel to the crystallographic axis of N_1. Thus, the emergent light is plane polarized. The crystal plate N_1 is spoken of as the polarizer and N_2 as the analyzer because it enables us to observe the polarization of light. The plane of polarization is defined as the plane perpendicular to the plane of vibration in the light wave.

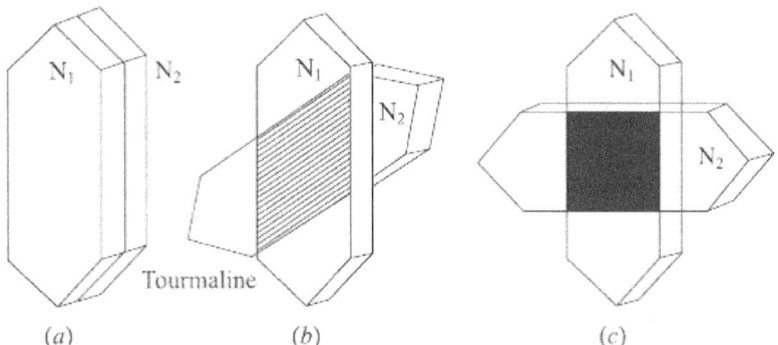

FIGURE 11.4. Optical experiment to demonstrate polarization of light.

11.4 DISCOVERY OF POLARIZATION OF LIGHT

The history of polarization began in 1669 with the discovery of the phenomenon of double refraction by a Dutch philosopher, Erasmus Bertholinus, while experimenting with a calcite crystal. This phenomenon was further studied by Huygens in 1690 who found that calcite forms two images of equal brightness, except when light traversed it in a direction parallel to the crystallographic axis. Furthermore, he discovered that each of the emergent beams from the calcite can be further subdivided, by its passage through a second crystal, into two beams of equal or unequal intensities or not decomposed at all, depending upon the orientation of the second crystal with respect to the first. This experiment clearly established that natural light is perfectly symmetrical, while the emergent beams from calcite are asymmetrical and therefore polarized. Although Huygens became the discoverer of polarization of light, he remained ignorant of its true nature and could not explain this phenomenon according to the prevalent longitudinal conception of light waves. Newton recognized the essential features of polarization when he said that a ray of light might have sides.

In 1808, Malus discovered that light reflected from glass possessed partly the characteristics (asymmetry) of beams of light emerging from calcite, but at one particular angle of reflection, it is completely polarized. He could not explain this asymmetry on the basis of Young's longitudinal vibration theory. Accordingly, he suggested that this phenomenon must be caused by some induced property of light corpuscles like that of the magnetic pole or an electric charge, giving rise to definite bias or polarity in the light beam. This suggestion gave birth to the title polarization for this new phenomenon. The interpretation of polarization in terms of the transverse nature of light waves was, however, given by Fresnel.

11.5 PICTORIAL REPRESENTATION OF LIGHT

For further study of polarization of light, we shall have to represent light waves on paper. Accordingly, we should adopt some convention for their proper representation. According to the electromagnetic theory of light, it will be recalled that light is nothing but the propagation of mutually perpendicular vibrating electric and magnetic vectors, and the electric vector functions as the light vector. A beam of ordinary or natural light, therefore, consists of millions of electromagnetic waves, the light vector of each having its own plane of vibration in arbitrary orientation due to random orientations of excited atoms or molecules in the source. Vectors of light are, therefore, arranged symmetrically about the direction of propagation, in all planes with equal probability, and its end on view is represented by Figure 11.5.

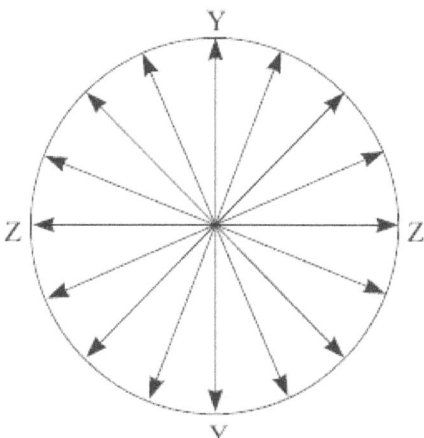

FIGURE 11.5. End on view of unpolarized beam of light.

The mode of vibration of a light vector in any one light wave may be represented as shown in Figures 11.6 (a) and 11.6 (c), depending upon the plane of vibration. Figure 11.6 (a) is a representation of a wave of light traveling from left to right with vibrations confined to the plane of paper or parallel to it. The vibrations are represented by short vertical lines. Accordingly, this is the representation of a plane polarized light beam with its plane of polarization perpendicular to the plane of the page. Figure 11.6 (c) is a representation of a light wave in which vibrations are executed perpendicular to the plane of the page; accordingly, they are represented by dots on the direction of propagation. End on views of the same waves are shown respectively in Figures 11.6 (b) and 11.6 (d).

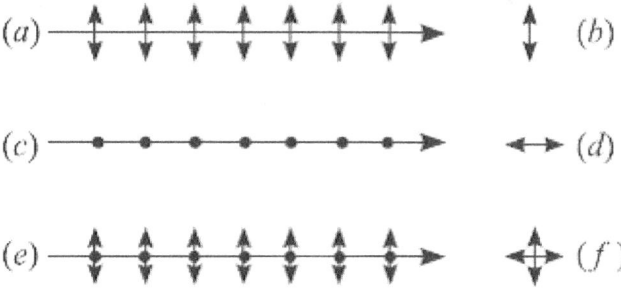

FIGURE 11.6. Pictorial representations of plane polarized and ordinary light beams.

In ordinary light, light vectors are in random orientations and all the amplitudes of light vectors may be resolved into components along any two transverse and mutually perpendicular directions, say in the Y and Z directions.

Figure 11.7 shows components OAY and OAZ of the light vector OA of one constituent wave, where

$$OAZ = OA\cos\theta \text{ and } OAY = OA\sin\theta \qquad (11.1)$$

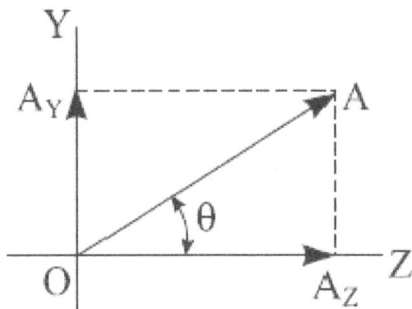

FIGURE 11.7. Components OAY and OAZ of the light vector OA of one constituent wave.

If this resolution of amplitude is continued for all light vectors, then owing to all vibration directions being equally probable, the average of the amplitude components along the axis of Y will be just equal to the average along the Z axis. Consequently, for theoretical discussion, we may replace an ordinary beam of light by two waves of equal amplitudes, vibrating in mutually perpendicular planes, of course, having no permanent phase relations to each other, due to the arbitrary phase changes in waves constituting the ordinary light. Figures 11.6 (e) and 11.6 (f) may be taken as the theoretical representation of ordinary light.

11.6 PRODUCTION OF PLANE POLARIZED LIGHT

The extreme importance of the discovery of the polarization of light lies in the fact that the transverse nature of the waves constituting the light was thereby established, in contradistinction to Huygens's conception of longitudinal waves. The common methods for the production and analysis of plane polarized light may be grouped under the following heads: polarization by (a) reflection, (b) refraction, (c) double refraction, (d) selective absorption, and (e) scattering.

11.7 POLARIZATION BY REFLECTION

In 1808, the French physicist Malus discovered that when a beam of ordinary light is incident at a particular angle of about 57° on a glass plate, the reflected light possesses properties similar to those found in light obtained by transmission of natural light through tourmaline or calcite; that is, reflected light is plane polarized light. The phrase plane polarized means that the light vector in the reflected light is vibrating transversely to the direction of transmission, in a fixed plane through this direction. Pursuing the enquiry further, he found that the same phenomenon occurs when light is reflected from water and other transparent substances.

The polarization of reflected light can be easily demonstrated by an experimental arrangement sketched in Figure 11.8 in which N_1 and N_2 are two plane parallel glass plates blackened on their back surfaces so as to absorb the refracted beam.

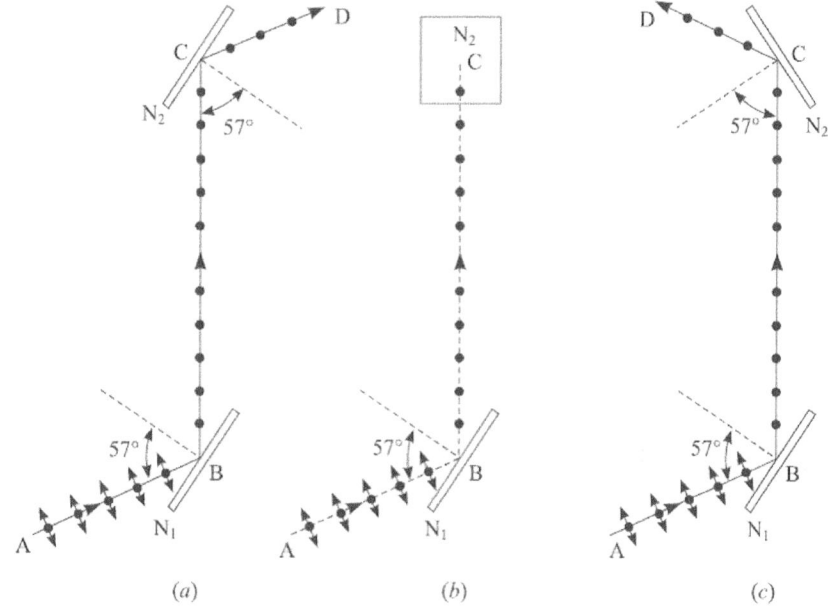

FIGURE 11.8. Polarization by reflection from glass surface.

A beam of unpolarized light AB consisting of parallel rays is incident at an angle $57°$ on N_1 at B, and the reflected beam BC is again reflected at the same angle from the plate N_2, placed exactly parallel to N_1. The plate N_2 is now rotated about BC as an axis through one complete rotation; this ensures the angle of incidence on N_2 to be the same in all its orientations. The intensity of twice the reflected beam is found to be maximum when N_1 and N_2 are parallel as in Figure 11.8 (a) or antiparallel as in Figure 11.8 (c), that is the planes of incidence are mutually parallel. The intensity is zero when they are crossed as in Figure 11.8 (b), that is, the planes of incidence are mutually perpendicular;

and in passing from the parallel to crossed position, the intensity diminishes from maximum to zero value. The variation in intensity of twice the reflected beam due to the rotation of N_2 clearly proves that when light is reflected at 57°, the reflected beam consists of waves in which vibrations are confined to a certain definite direction transverse to the ray; that is, it is a plane polarized beam. The lower plate N_1 is called the polarizer while the upper plate N_2 is called the *analyzer*.

If we rotate the lower plate N_1 about the incident beam as an axis, we get the reflected beam BC of the same intensity, thereby showing that the incident light is perfectly symmetrical about the direction of propagation.

If, however, the angle of incidence on the upper or lower plate is not 57°, the twice-reflected beam will exhibit maximum and minimum intensities as before, but the minimum intensity will not be zero. In other words, there will always be a reflected beam from N_2. At angles other than 57°, the reflected beam BC is not completely plane polarized, while at 57° the polarization is most complete.

The angle of reflection i_p at which the polarization of reflected light is most complete is called the *polarizing angle*, and it varies with the nature of glass and also with the wavelength of light. Malus defined the plane of polarization as the plane of incidence of polarized light when the reflected light was of maximum intensity. We shall explain in Section 11.9 that the vibration in the reflected plane polarized light BC is perpendicular to this plane.

11.8 BREWSTER'S LAW

Brewster made a remarkable discovery that at the polarizing angle, the reflected and refracted rays are just 90° apart. The angle of refraction r_p is, therefore, equal $\left(90 - i_p\right)$, and we have for the refractive index,

$$\mu_g^a = \frac{\sin i_p}{\sin r_p} = \tan i_p \qquad (11.2)$$

The tangent of the angle of polarization is equal to the refractive index of the reflecting substance; this is called *Brewster's law*. As a consequence of this law, it follows that the angle of polarization depends upon the wavelength λ of light employed.

Brewster's law also applies to reflection at the second surface of the glass plate in the denser medium. In this case, it states that the refractive index is equal to the cotangent of the polarizing angle. For a polished metallic reflector, this law does not hold.

11.9 EXPLANATION OF POLARIZATION BY REFLECTION

We may now explain the action of the reflecting surface in producing plane polarized light. The light vectors in the unpolarized light may be resolved at the point of incidence *B* into two components, one perpendicular and the other parallel to the plane of incidence. At the polarizing angle, since the reflected and the refracted rays are just 90° apart, at the point of incidence *B* in Figure 11.9, the vibrations which are in the plane of incidence become parallel to the direction of the reflected ray *BC*, and so they can generate only longitudinal waves along *BC*.

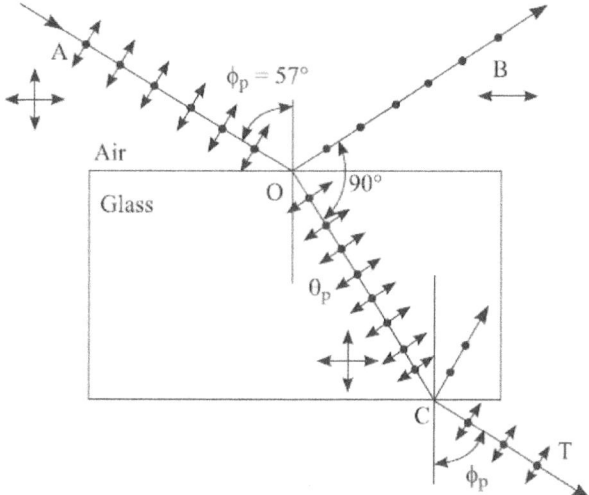

FIGURE 11.9. Illustrates Brewster's law at the polarizing angle.

Since the waves of light are transverse and not longitudinal, we conclude that the wave with vibrations in the plane of incidence is not at all reflected at a polarizing angle but totally (100%) transmitted as a refracted beam. The other wave, in which vibrations are perpendicular to the plane of incidence, is partly reflected (15%) and partly transmitted (85%). The refracted beam is, therefore, strong but only partially polarized, as it always contains both kinds of vibrations, while the reflected beam is weak but completely plane polarized, as it contains only vibrations which are executed perpendicular to the plane of incidence.

We may now explain the variation of the intensity of twice the reflected beam at the polarizing angle owing to the rotation of the glass plate N_2 in the experiment. The plane polarized reflected beam *BC* contains only those vibrations which are perpendicular to the plane of the page, which for the lower plate N_1 is the plane of incidence. When N_1 and N_2 are crossed, the vibrations in *BC* which are perpendicular to the plane of incidence for N_1 will be in the plane of incidence of the ray incident on N_2, and as explained earlier, they are totally transmitted within N_2, thereby making the reflected intensity as zero. But when N_1 and N_2 are parallel, the vibrations in BC are also perpendicular to the plane of incidence of N_2, and so partly transmitted within N_2 and partly reflected along CD. The intensity of the reflected beam is maximum. For the intermediate orientation of N_2, the intensity of twice the reflected beam varies as the square of the cosine of the angle between the two planes of reflections, and so the intensity is intermediate of maximum $(\cos\theta = 1)$ and minimum $(\cos\theta = 0)$ values.

11.10 POLARIZATION BY REFRACTION

Brewster's law applies not only for reflection in a rarer medium but also for reflection in a denser medium. In other words, it can be easily shown that if ordinary light is incident at a polarizing angle on the upper surface of a plane parallel glass plate, the refracted ray will be incident also at a polarizing angle at the lower surface. This remarkable fact is used in producing polarization by refraction through a pile of a large number of glass plates as shown in Figure 11.10.

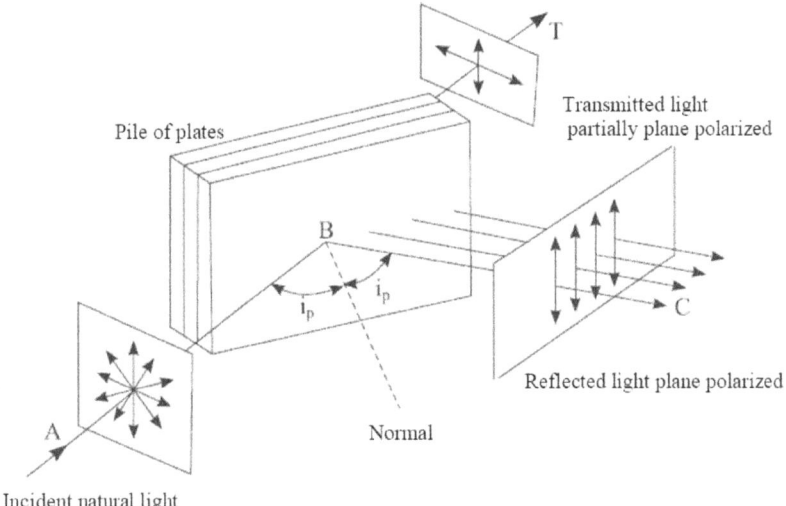

FIGURE 11.10. Polarization of light by a pile of glass plates.

At each reflection, the reflected ray contains 15% of vibrations which are perpendicular to the plane of incidence, while the refracted ray contains vibrations which are parallel (100%) and perpendicular (85%) to the plane of incidence, assuming the μ of glass as 1.5. The larger the number of plates in the pile, the more nearly plane polarized is the transmitted beam, and so in the limit we expect that vibrations which are perpendicular to the plane of incidence will be completely quenched from the transmitted beam, and the vibrations of light in it will be only parallel to the plane of incidence.

The pile of plates when used as a polarizer is mounted in a tube in such a way that the unpolarized light, along the axis of the tube, is incident at the polarizing angle i_p on the pile. The second pile of plates, placed parallel to the first as shown in Figure 11.11, functions as the analyzer.

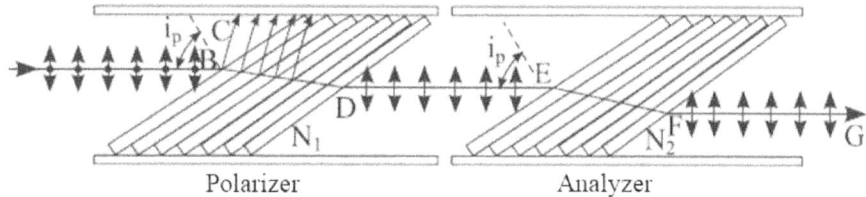

Polarizer   Analyzer

FIGURE 11.11. Pile of glass plates as polarizer and analyzer.

11.11 LAW OF MALUS

In 1809, Etienne Louis Malus experimentally discovered a relation which indicates that the intensity of light transmitted by the analyzer depends upon the inclination of its plane of transmission with that of the polarizer. To derive the relation let us suppose that the angle between the two planes of transmission is θ at any instant. The light vector $AP = a$ in the plane polarized light emerging from the polarizer may be resolved into two components,

$$AE = a\cos\theta \quad \text{and} \quad AO = a\sin\theta \tag{11.3}$$

respectively, along and perpendicular to the plane of transmission of the analyzer as shown in Figure 11.12. The perpendicular component is eliminated in the analyzer while the parallel component is freely transmitted through it. Therefore, the intensity I of light that emerges from the analyzer is given by

$$I = a^2 \cos^2\theta = I_0 \cos^2\theta \tag{11.4}$$

where I_0 is the intensity of the plane polarized light incident on the analyzer. This relation, called Malus's law, states that the transmitted intensity varies as the square of the cosine of the angle between the planes of transmission

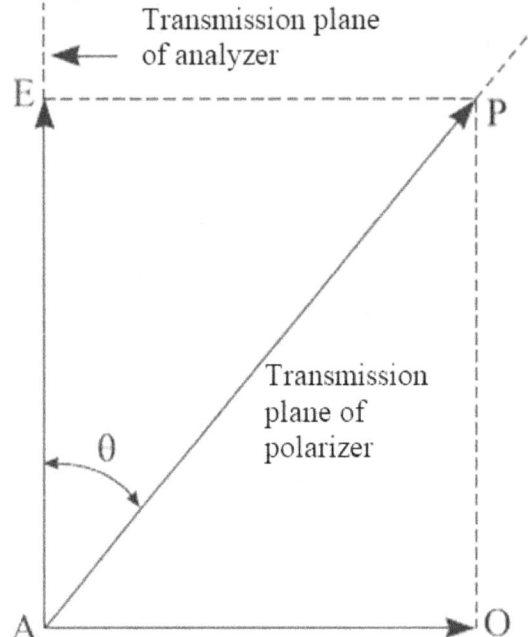

FIGURE 11.12. Resolution of the amplitude of plane-polarized light.

of the analyzer and the polarizer. It may be emphasized that this law holds only when light incident on the analyzer is completely plane polarized, which is true of the combination of glass plates sketched in Figure 11.8, a combination of polaroids, or Nicol's prism, but it is only partially true for the pile of plates. For Equation 11.4 to hold true, it is further assumed that there is no loss of light due to absorption in the analyzer. It should be pointed out that even in the case of absorption, the transmitted intensity depends on $\cos^2 \theta$, and the only effect is the change in the constant in Equation (11.4). The transmitted light is a maximum when $\theta = 0°$ and is zero (if the polarization is complete) when $\theta = 90°$ or when the polarizer and analyzer are crossed.

11.12 DOUBLE REFRACTION

In the year 1669, a Dutch philosopher, Erasmus Bartholinus, discovered that when a ray of ordinary light is incident on a crystal, called Iceland spar or calcite, it splits into two refracted rays, one of which always obeys ordinary laws of refraction and the other, in general, which may not obey them. This phenomenon known as double refraction was called by Huygens as the "strange refraction" of calcite. A slab of calcite and a slab of glass lift the printing underneath; but while the glass lifts only one image, calcite lifts two images. Calcite thus has the peculiar property of exhibiting double refraction. Huygens in 1690 studied this phenomenon more closely and found that both the refracted rays were linearly polarized in mutually perpendicular planes. This phenomenon is also exhibited by other crystals like quartz, except those belonging to the cubic system which are optically anistropic, that is, their optical properties are the same in all directions like glass.

We shall chiefly study this phenomenon in Iceland spar, in which double refraction is very marked and easy to study experimentally.

11.13 GEOMETRY OF CALCITE CRYSTAL

Iceland spar as shown in Figure 11.13, which chemically is hydrated calcium carbonate, $CaCO_3$, was found in large quantities in times gone by in Iceland. $CaCO_3$ crystalizes in a large, colorless, clear hexagonal prism with a blunt pyramid at each end.

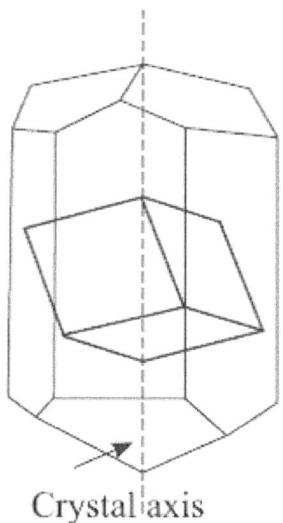

Crystal axis

FIGURE 11.13. Iceland spar crystal.

When struck, it easily cleaves obliquely in three definite planes forming a rhombohedral body whose six rhombic sides make an angle of 45°23' with the lines joining the vertices of the pyramids, the crystallographic axis of the crystal. A spar rhombohedran of this kind is bounded by six faces, each of which is a parallelogram whose angles are $\alpha = 101°55'$ and $\beta = 78°5'$. At the two opposite corners A and H, as shown in Figure 11.14, there are three obtuse angles α, but at the other six corners there are two acute angles, each equal to β, and one obtuse angle α. The length of the edges is not important, with those in the figure being meant to be of equal length.

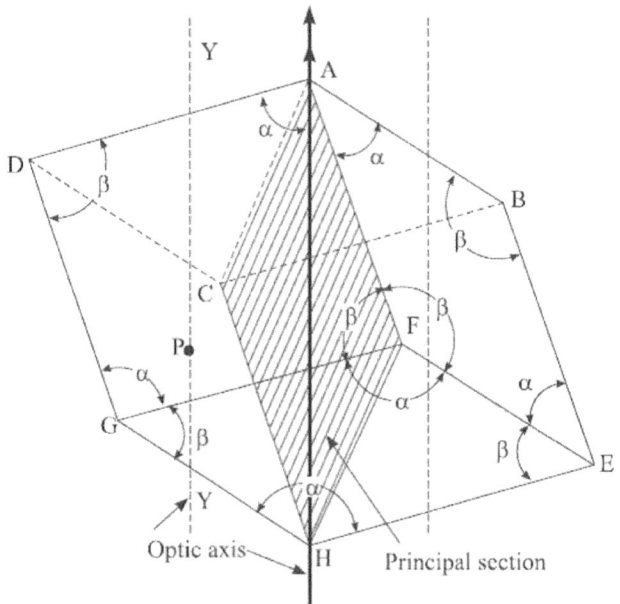

FIGURE 11.14. Rhombohedran of calcite.

11.14 ORDINARY AND EXTRAORDINARY RAYS

The phenomenon of double refraction can be easily demonstrated by allowing a narrow beam of unpolarized light to be incident normally on calcite crystal. The incident beam (in general) splits up into two refracted beams. Double refraction does not occur at all at the second face, and since the two opposite faces of crystal are parallel, the two refracted beams emerge parallel to the incident beam, but they are relatively displaced by a distance which is proportional to the thickness of the crystal as shown in Figure 11.15.

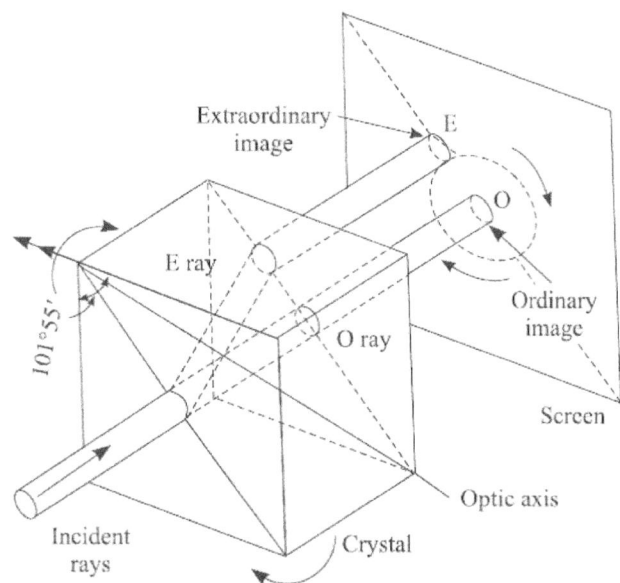

FIGURE 11.15. A narrow beam of natural light can be split into two beams by a doubly refracting crystal.

On a screen, two images, O and E, of the pinhole are thus obtained. One image O lies in the direction of the incident beam. Therefore, it must have been formed in accordance with the ordinary laws of refraction. The other image E is found to be separated from the O image despite perpendicular incidence, in contrast to the ordinary phenomenon. The corresponding refracted beam, therefore, does not, in general, obey the ordinary laws of refraction. On rotating the crystal about the incident beam as an axis, it is observed that the O image remains stationary while the E image revolves in a circular path with its center at the O image, but the line joining the two images is always parallel to the shorter diagonal of the end face of the crystal.

Of the two refracted rays, the one which always obeys ordinary laws of refraction is called the *ordinary ray* or the O ray while the other, which behaves in quite an extraordinary manner, is called the *extraordinary ray* or the E ray. The ratio sin i/sin r physically represents the ratio of the velocity of light within a vacuum to that in the refracting medium. It is therefore obvious that the velocity of the O ray is the same in all directions within the crystal. In the case of an extraordinary ray, since the ratio $\left(\dfrac{\sin i}{\sin r}\right)$ varies with the angle of incidence, its velocity is different in different directions within the crystal. The O ray is always in the plane of incidence, but in general this is not true of the E ray.

11.15 CRYSTALLOGRAPHIC AXIS AND OPTIC AXIS

It will be recalled that two diagonally opposite solid angles of calcite are formed by the junction of three obtuse angles of three faces. A line into the crystal at one of these corners and equally inclined to the three faces is called a crystallographic axis. For a crystal having equal edges, as shown in Figure 11.14, its shortest diagonal *AH* is the crystallographic axis.

One important peculiarity of the calcite crystal lies in the fact that there is one and only one direction through it so that when a ray of light is incident along this direction, the *O* ray and *E* ray do not separate and also traverse the crystal with the same velocity along this direction. Hence, this particular direction is called the *optic axis* and is determined by the crystallographic axis. Any direction in the crystal parallel to the crystallographic axis is called an optic axis. It may be emphasized that the optic axis is a direction and not a line. Through any given point *P* within the crystal, only one line, *YY*, can be drawn parallel to the crystallographic axis. This line gives the direction of the optic axis for this point and for all points lying on it.

11.16 PRINCIPAL SECTION OF THE CRYSTAL AND PRINCIPAL PLANE OF THE *O* RAY AND *E* RAY

Any plane which contains the optic axis and is perpendicular to two opposite faces is called a principal section. This cuts the surfaces of the crystal in a parallelogram with angles of 109° and 71°, as illustrated in Figure 11.16, which represents the principal section through the blunt edges of the crystal. All sections parallel to the principal section depicted in Figure 11.16 are principal sections for points lying on it.

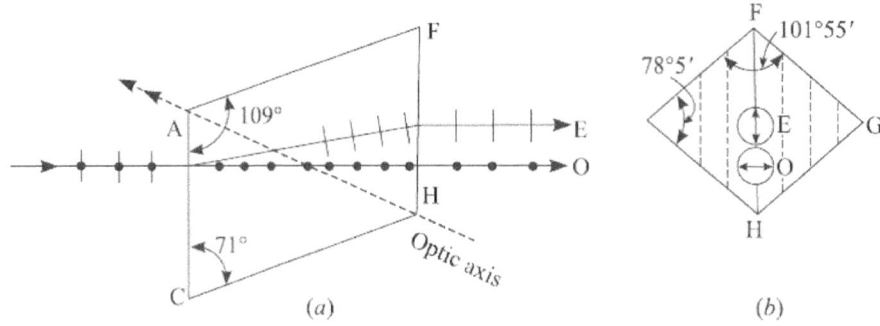

FIGURE 11.16. Principal section in a calcite crystal.

Through every point three principal sections can be drawn corresponding to three pairs of opposite crystal surfaces. An end on view of any principal section is a straight line (shown dotted) in the crystal surface parallel to its shorter diagonal *FH*, which is the end on view of the principal section through the blunt edges.

The principal plane of the ordinary ray is defined as a plane in the crystal drawn so as to contain the optic axis and the ordinary ray. The principal plane of the extraordinary ray is defined as a plane in the crystal drawn through the optic axis and extraordinary ray. The principal planes of the two refracted rays, in general, do not coincide. They do coincide, however, under special circumstances when the plane of incidence is the principal section of the crystal. In a case like this, as illustrated in Figure 11.16, the plane of incidence and the principal planes are all coincident. The line joining the *O* image and the *E* image of the same point is in the principal section. This accounts for the rotation of the *E* image in a circle around the *O* image as the crystal is rotated about the ordinary ray as an axis.

11.17 POLARIZATION BY DOUBLE REFRACTION

When the *O* ray and the *E* ray from the calcite crystal, in the experimental arrangement sketched in Figure 11.15, are examined by a polaroid or tourmaline, it is found that as the analyzer is rotated in its own plane, the intensity of one of the two images, say, the *O* image, gradually increases while that of the *E* image gradually diminishes. It is possible to get a position of tourmaline in which the *O* image acquires maximum intensity while the *E* image completely disappears. In this setting of tourmaline, it will be observed that its longer axis (crystallographic axis), which determines the plane of transmission in it, is exactly parallel to the longer diagonal of the end face of the crystal as shown in Figure 11.17 (a). A rotation of tourmaline from this setting to a position obtained on a 90° rotation will render the intensity of the *E* image maximum while the *O* image will completely vanish. In this setting of tourmaline, it will be observed, its longer axis is parallel to the shorter diagonal of the crystal face as shown in Figure 11.17 (b), which determines the trace of the principal section in the end on view. We therefore conclude that the rays by which the ordinary and extraordinary images are seen must be polarized. Since the *E* ray is intercepted by the tourmaline when its plane of transmission is perpendicular to the principal section of crystal while the *O* ray is intercepted on a 90° rotation, it follows that the vibrations in the *E* ray and the *O* ray are executed respectively along and at right angels to the principal section; that is, the O ray is polarized in the principal section while the E ray is polarized perpendicularly to the principal section.

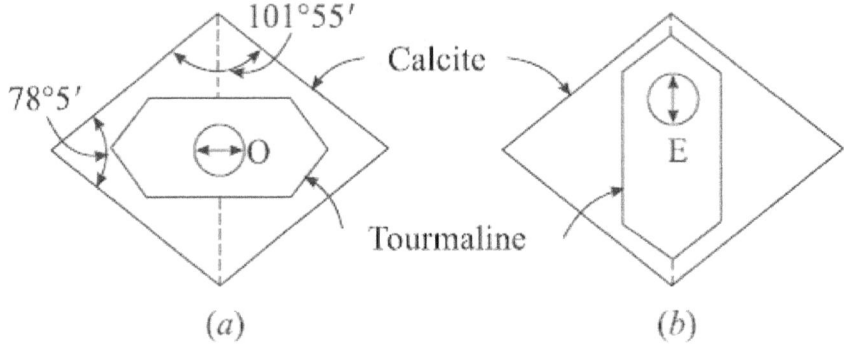

FIGURE 11.17. Polarization by double refraction.

In general, when the plane of incidence is not parallel to the principal section, the vibrations in the wavefront of the E ray are executed in the principal plane of the E ray while those in the wavefront of the O ray are executed at right angles to the principal plane of the O ray. Figure 11.17 fully illustrates this case.

11.18 PARALLEL AND CROSSED NICOLS

Nicol's prism can be used both as a polarizer and an analyzer. When two Nicols are mounted coaxially as shown in Figure 11.18, then the first Nicol N_1, which produces plane polarized light, is called the *polarizer*. The second Nicol N_2, which analyzes the incoming light, is called the *analyzer*. When the principal sections of the two Nicols are parallel, then the vibrations in the E ray, which are in the principal section of the polarizer, are also in the principal section of the analyzer. Consequently, the E ray from N_1 is freely transmitted by N_2 just as it was freely transmitted by the first as shown in Figure 11.18 (a). In this setting of the combination—technically known as parallel Nicols—the intensity of the field of view is maximum.

When the analyzer is rotated about its axis through 90° from this position, the principal sections of the Nicols are mutually perpendicular as shown in Figure 11.19.

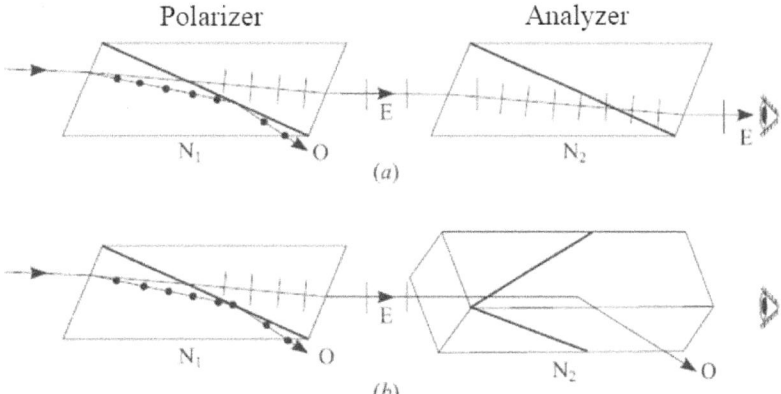

FIGURE 11.18. (a) Parallel Nicols (b) Crossed Nicols.

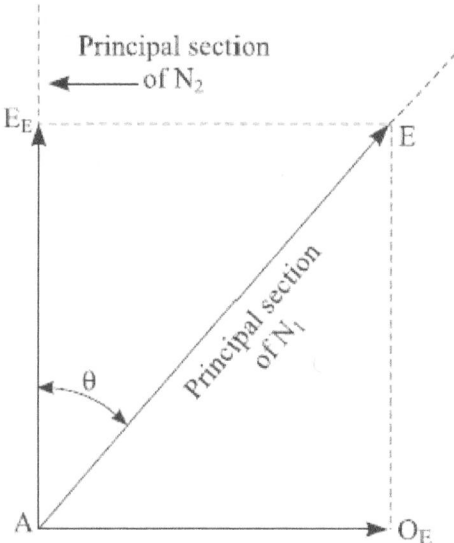

FIGURE 11.19. Analyzer is rotated about its axis through 90° from this position.

Therefore, the vibrations in the E ray, which are in the principal section of N_1, become perpendicular to the principal section of N_2. The E ray from the polarizer, consequently, enters the analyzer as an O ray and, therefore, is totally reflected from the balsam, exactly in the same way as the O vibrations were totally reflected in the polarizer. Thus, no light is transmitted by this setting of the combination, which is technically known as *crossed Nicols*.

After a rotation of 180° from the first position of the analyzer, the principal sections are again parallel, and E vibrations are again transmitted by the analyzer. After a rotation of 270°, Nicols are again crossed, and no light is transmitted by the combination.

In the intermediate position between the crossed and parallel settings of the combination, some light is transmitted through the analyzer. The E vibrations from the polarizer are resolved, just on entering the analyzer, into two component vibrations—the ordinary component $a \sin \theta$, which is perpendicular to the principal section of the analyzer, is totally reflected while the extraordinary component $a \cos \theta$, which is in the principal section, is freely transmitted by the analyzer. Thus, for intermediate positions, the intensity of light is proportional to the square of the cosine of the angle θ between the principal section of the analyzer and the polarizer.

11.19 POLARIZATION BY SELECTIVE ABSORPTION

The production of plane polarized light by selective absorption is exhibited by a certain class of doubly refracting crystals, which not only produce two internal beams polarized at right angles to each other but also absorb one of the polarized components much more strongly than the other. Hence, if crystal is cut of the proper thickness, one of the components is completely extinguished by absorption while the other is transmitted in an appreciable amount as shown in Figure 11.20.

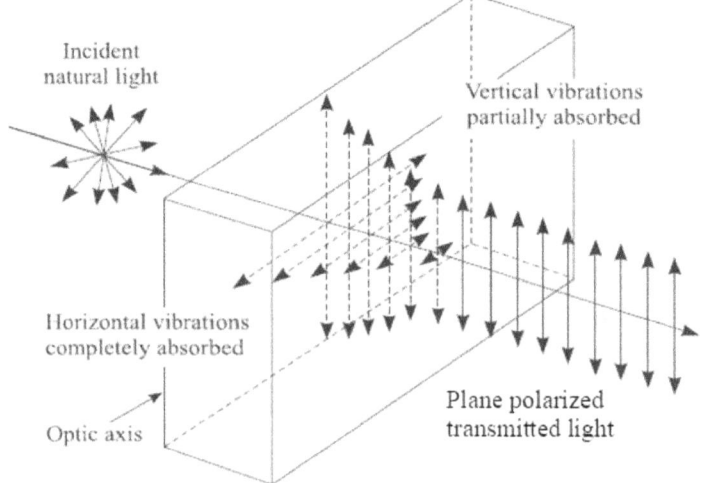

FIGURE 11.20. Plane polarized light transmitted by a dichroic crystal.

This phenomenon is known by the name dichroism, and crystals exhibiting this property are said to be dichroic. The best known dichroic mineral crystal is tourmaline, and Figure 11.20 illustrates its action. A beam of ordinary light incident normally on a thin crystal of tourmaline, cut with its faces parallel to the optic axis, is broken up into O and E rays, traversing the crystal in the same direction but with vibrations respectively in and perpendicular to the plane of incidence. The O beam is completely absorbed in the crystal, while the E beam, only slightly absorbed, is transmitted, as far as the green region of the spectrum is concerned. The emergent polarized light can be analyzed by the second tourmaline as shown in Figure 11.21.

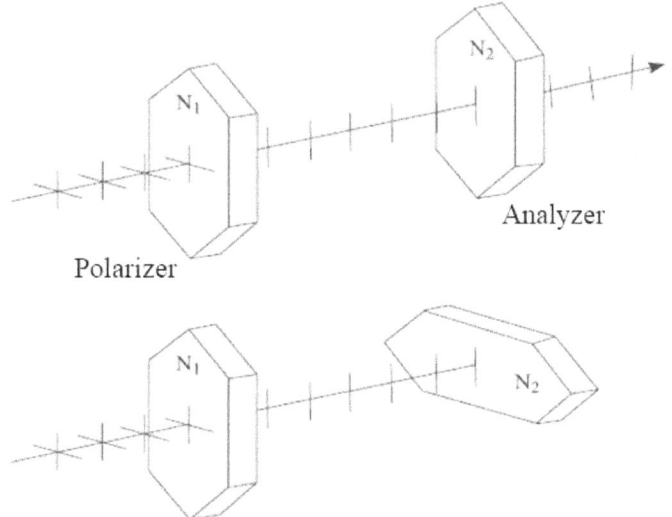

FIGURE 11.21. Tourmaline as polarizer and analyzer.

The use of tourmaline as polarizer is limited, as the plane polarized transmitted light is colored due to unequal absorption of E rays of various wavelengths.

11.20 POLAROID

In 1852, the English physician W. H. Herapath succeeded in producing small crystals of idosulphate of quinine, called *herapathite* after its discoverer, which exhibit strong dichroism—they absorb completely one component of polarization while transmitting the other with little loss. These crystals are ultra-microscopic, pulverizing easily when subjected to relatively small stress, and so they are quite useless as such. In 1932, E. H. Land developed a process which arranges herapathite crystals side by side, oriented with their optic axis all parallel, so that they function as a single crystal of large dimensions. This is achieved by preparing a paste of crystals in nitrocellulose, which is then squeezed out through a fine slit. Obviously, only those crystals pass whose axes are parallel to the length of the slit, thereby producing a fine sheet of millions of tiny crystals, with their optic axes parallel to one another. This is then mounted between two thin sheets of glass, forming what is now called *Polaroid*.

Another variety called *H* Polaroid is formed by stretching Polyvinyl alcohol films so as to orient the complex molecules in the direction of stress, which thereby become doubly refracting and when impregnated with iodine exhibit dichroism. This is entirely colorless, transmitting 33% more light than the herapathite polaroid, while light is 99.99% polarized. Land and Rogers later discovered that when the stretched polyvinyl alcohol film is heated with a dehydrating catalyst, for example, HCl, it slightly darkens but exhibits strong dichroism. This is called *K polaroid* and is extensively employed in automobile headlights and screens.

Polaroids find wide applications in everyday life. The most common use of polaroid is in sunglasses, where it plays the role of the analyzer. We have seen that when unpolarized light is reflected, vibrations perpendicular to plane of incidence are predominantly reflected. Accordingly, when sunlight is reflected from horizontal surfaces such as wet pavement, the plane of incidence is vertical, and in the reflected light horizontal vibrations predominate the vertical vibrations. The direction of the transmission of polaroid is, therefore, kept vertical to cut off completely the horizontal reflected vibrations, while objects are seen by diffusely reflected light which is, however, unpolarized.

Polaroids are fitted in car headlights and windshields, with their planes of transmission inclined at 45° to vertical. The driver can see the light of his own headlight while that from oncoming automobiles is cut off. However, the road and other objects are visible by light scattered by them, which is unpolarized, and a part of it can pass through the windshield polaroid.

Polaroids are used in windows of trains, airplanes, and so on in which there is a fixed outer and rotatable inner polaroid disc for controlling the amount of light entering it from outside.

Polaroids are employed to view stereoscopic motion pictures in which the three-dimensional effect is obtained by first taking two pictures of the same view from slightly different angles—one showing the scene exactly as viewed by the right eye and the other as viewed by the left eye. These pictures are projected side by side on the screen, but light in the two pictures is polarized at right angles to each other. Pictures are now viewed by polarized glasses so that each eye views only the picture which had been taken from its own angle. Polaroid has now largely replaced the Nicol prism for producing and analyzing plane polarized light in the laboratory.

11.21 POLARIZATION BY SCATTERING

The sky is blue while sunset and sunrise are red. Skylight is partially plane polarized. Maximum polarization is observed in a direction perpendicular to the incident light. This can be easily verified by looking at part of the sky overhead through a Nicol prism or a polaroid. All three effects are due to the scattering of sunlight by obstacles such as dust particles, free water molecules, and even by oxygen molecules present in the atmosphere. The process of scattering is simply the absorption and reradiation of energy.

The process of the scattering of light can be easily explained on the basis of electromagnetic theory according to which, in a light wave, periodic electric and magnetic vibrations are executed in mutually perpendicular planes. Furthermore, according to Wiener's experiment, the electric vector is responsible for all optical phenomena. Referring to Figure 11.22, suppose the natural sunlight (unpolarized) is incident on one of the molecules of earth's atmosphere, situated at O just over the observer.

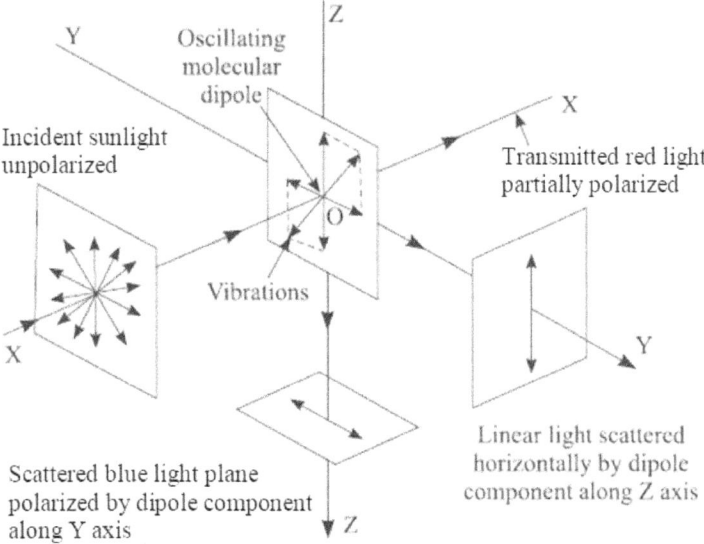

FIGURE 11.22. Polarization by scattering of light by an oscillating molecular dipole.

Every molecule consists of an aggregate of electrically charged particles of opposite signs. Owing to the electric field in the incident light beam, positive charges experience a force in the direction of the field, while on negative charges force is in the opposite direction. Since charges are not rigidly bound in a molecule, a relative displacement of the charges occurs first in one direction and is then periodically reversed, owing to the periodical reversal of the electric field in the light wave. In Figure 11.22, these directions necessarily lie in the YZ plane. An oscillating electric dipole, according to electromagnetic theory, itself radiates electromagnetic waves which form the scattered light. We may here also employ the equivalent components E_Y and E_Z for the entire E vector in the incident beam. The result is that the light incident on the molecule at O produces the equivalent two oscillating molecules along the Y axis and the Z axis, and the frequency of oscillation is the same as that of the incident light, since the oscillations of electric charges in a molecule are (nearly) in unison with the oscillating electric field. The observer looking along the Z axis receives light only from the Y component of incident vibration. This light is, therefore, plane polarized with vibrations parallel to the Y axis. In fact, the light scattered in any direction in a plane transverse to the direction of propagation of the incident light is plane polarized. In all other directions, scattered light will be only partially polarized.

11.22 BLUE COLOR OF SKY

To explain the blue color of sky, we observe that the vibration of electric charges in a molecule under the influence of the electric field in the incident light is a forced vibration. Its amplitude is greater and as a consequence the intensity of scattered light is greater and the frequency of the incident light is closer to the natural frequency of the oscillation of electric charges which, however, is the same as that of a wavelength in the ultraviolet region of the spectrum. The frequencies of light waves in the visible region of the spectrum are less than the natural frequency of a molecule, but the frequency of blue light is very close to it. Therefore, the blue color in sunlight is scattered more than the red by the molecules in the atmosphere.

The exact theory of the scattering process has been worked out by Lord Rayleigh, who showed that the percentage of light scattered is inversely proportional to the 4th power of the wavelength when the size of the particles is of the order of magnitude of the wavelength of a light wave. This is almost 16 times greater for the violet end of the spectrum as compared with the scattering of red light. Consequently, the blue contained in the sunlight is scattered to a much greater extent and more intensely than the red.

In reality, looking at the sky overhead, the blue light due to scattering of incident sunlight from molecules in the neighborhood of O will be only partially plane polarized, with maximum vibrations perpendicular to the plane XOZ. This is due to the fact that the incident light waves are scattered several times by the molecules of the atmosphere before reaching the observer, thus accounting for the blue color of unclouded sky in all directions. Blue color is more brilliant when the atmosphere is free from dust particles and water molecules, so scattering from molecules of gases is to a considerable extent responsible for the blue color of the sky.

If the earth had no atmosphere, we could receive no skylight at the earth's surface. The sky would be perfectly black even in day. This may be proved by flying at high altitudes, where there is less atmosphere above the observer.

The sun and the neighboring sky appear red at sunrise and at sunset. We observe them directly, referring to Figure 11.23, by placing our eye at E.

The rays coming from the sun have to traverse a greater distance approximately horizontally through the earth's atmosphere before reaching the observer. A large proportion of blue is scattered to the side and lost by fine dust and smoke particles near the earth's surface. White light, when blue is removed, is yellow or red in hue. When such light is incident on the cloud near sunset or sunrise, the light reflected from the cloud has the yellow or red hue. Red light is only partially plane polarized.

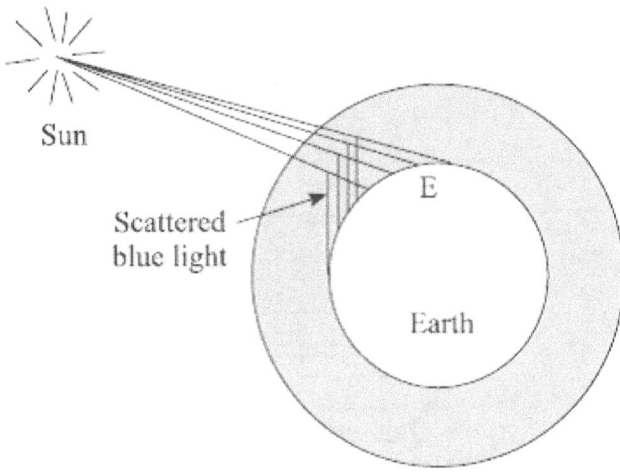

FIGURE 11.23. Blue color of sky.

11.23 SUPERPOSITION OF TWO PLANE POLARIZED WAVES VIBRATING IN TWO MUTUALLY PERPENDICULAR PLANES

Let a beam of plane polarized monochromatic light from a polarizer be incident normally on a calcite plate, cut with faces parallel to the optic axis but oriented in such a way that the linear vibrations make an arbitrary angle θ with the optic axis. Figure 11.24 illustrates a plane polarized light incident normally on a crystal plate cut parallel to the optic axis.

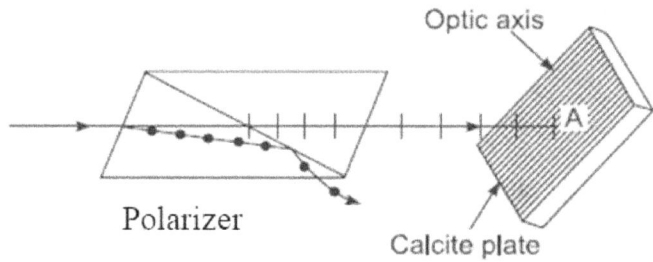

FIGURE 11.24. Plane polarized light incident normally on a crystal plate cut parallel to optic axis.

The amplitude a of the incident wave is resolved by the doubly refracting calcite, as shown in Figure 11.25, into two components: (i) $AE = a\cos\theta$ along the optic axis forming the extraordinary vibrations, and (ii) $AO = a\sin\theta$ perpendicular to optic axis forming the ordinary vibrations.

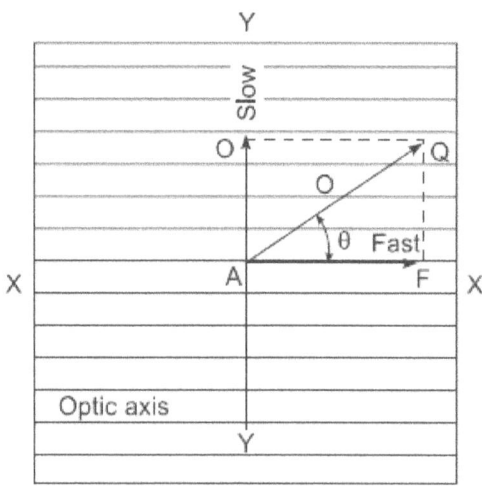

FIGURE 11.25. A beam of plane polarized monochromatic light from a polarizer that is incident normally on a calcite plate.

As they enter the plate, the incident wave and its two components have the same phase. Therefore, they may be expressed by the following equations.

$$S = a \sin\omega t; x = a\cos\theta \sin\omega t; y = a\sin\theta \sin\omega t \tag{11.5}$$

From the wave surface diagram for the normal incidence of plane polarized light, it follows that the O and E vibrations traverse the crystal plate without separation, that is, they traverse the same path but, in calcite, the E wave travels faster than the O wave $(\mu_o > \mu_E)$. Therefore, the wavelength of the E wave $\left(\lambda_E = \dfrac{\lambda}{\mu_E}\right)$ in the crystal is greater than that of the O wave $\left(\lambda_O = \dfrac{\lambda}{\mu_O}\right)$ where λ is the wavelength in a vacuum. Due to this inequality in their wavelengths, the O wave is retarded behind the E wave. The two waves, therefore, emerge from the crystal with a certain relative phase difference, say δ, for a given wavelength, depending upon the thickness of the plate and the magnitudes of μ_O and μ_E.

Therefore, as the two waves leave the plate, they may be represented by the following equations:

$$E \text{ Wave} : x = a \cos\theta \sin(\omega t + \delta) \tag{11.6}$$

$$O \text{ Wave} : y = a \sin\theta \sin\omega t \tag{11.7}$$

The resultant vibration may be readily found by eliminating t between the previous two equations. For this purpose, we proceed as follows:

Equation (11.6) may be written in the form

$$\frac{x}{a\cos\theta} = \sin\omega t \cos\theta + \sqrt{1 - \sin^2\omega t} \sin\delta \tag{11.8}$$

Substituting the value of $(\sin\omega t)$ from Equation (11.7) in the previous equation and rearranging, we get

$$\frac{x}{a\cos\theta} - \frac{y}{a\sin\theta}\cos\delta = \sqrt{1 - \frac{y}{a^2 \sin^2\theta}}\sin\delta \tag{11.9}$$

Squaring and simplifying, we get

$$\frac{x^2}{a^2 \cos^2\theta} - \frac{2xy}{a^2 \cos\theta \sin\theta}\cos\delta + \frac{y^2}{a^2 \sin^2\theta} = \sin^2\delta \tag{11.10}$$

This is an equation of the central conic, and since we know from the condition of the problem that x and y are never infinite, the conic is an ellipse. Equations (11.6) and (11.7) show that x varies from $+a\cos\theta$ to $-a\cos\theta$ and that y varies from $+a\sin\theta$ to $-a\sin\theta$. Hence, the ellipse represented by these equations or Equation (11.10) is inscribed in a rectangle with sides of length $2a\cos\theta$ and $2a\sin\theta$ and whose diagonal represents the rectilinear vibrations of the incident wave. Thus, during the passage of the wave, the vibrating particle describes, in general, an ellipse about its position of rest in the plane perpendicular to the direction of propagation. The exact nature of the resultant motion and therefore of the light emerging from calcite, however, depends upon the value of δ. We consider the following special cases:

1. **Resultant Vibration Rectilinear**—Emergent Light Plane Polarized.

   a. If $\delta = 2n\pi$ where $n = 0, 1, 2, 3$, etc. Equation (11.10) reduces to

   $$\frac{x^2}{a^2\cos^2\theta} - \frac{2xy}{a^2\cos\theta\sin\theta} + \frac{y^2}{a^2\sin^2\theta} = 0 \qquad (11.11)$$

   which easily simplifies to the form

   $$y = (\tan\theta)x \qquad (11.12)$$

   Thus, when the two mutually perpendicular superposed plane polarized waves are in phase, the ellipse degenerates into a straight segment, inclined at an angle θ to the axis of x, coinciding with the diagonal of the rectangle that lies in the first and third quadrants as shown Figure 11.26.

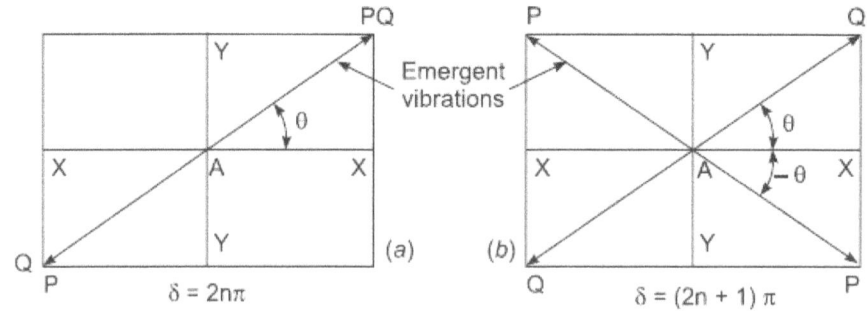

FIGURE 11.26. Emergent light plane polarized corresponding to $\delta = 2n\pi$ and $\delta = (2n+1)\pi$.

We express this fact by saying that the light emerging from the crystal is plane polarized, vibrating in the same plane as that of the plane polarized light as shown for QAQ in Figure 11.26 (a) incident on the plate.

b. If $\delta = (2n + 1)\pi$ where $n = 0, 1, 2, 3$, etc. Equation (11.10) easily reduces to the form

$$y = -(\tan\theta)x = \tan(-\theta)x \qquad (11.13)$$

The light emerging from the crystal plate is again linearly polarized, but now the vibration PAP is parallel to the diagonal of the rectangle that lies in the second and fourth quadrants as shown in Figure 11.26 (b). It is, therefore, inclined at an angle 2θ with the plane of vibration as shown for QAQ in Figure 11.26 (b) in the plane polarized light incident on the calcite plate.

2. ***Resultant Motion Elliptical*** —Emergent Light Elliptically Polarized. If δ is an odd multiple of $\left(\dfrac{\pi}{2}\right)$, that is, $\delta = (2n+1)\left(\dfrac{\pi}{2}\right)$, the product term in Equation (11.10) vanishes and the equation reduces to

$$\frac{x^2}{a^2\cos^2\theta} + \frac{y^2}{a^2\sin^2\theta} = 1 \qquad (11.14)$$

This is an equation for an ellipse, and its principal axes $2a\cos\theta$ and $2a\sin\theta$ are parallel and perpendicular to the optic axis of the crystal plate as shown for XAX in Figure 11.27 (a). The light emerging from the crystal plate is, therefore, elliptically polarized; the tip of the resulting light vector continuously sweeps out an ellipse in a plane perpendicular to the direction of propagation. At any one time the direction of the resultant varies from point to point along the direction of propagation. It has a space-periodicity λ.

If $\delta = \dfrac{\pi}{2}, \dfrac{5\pi}{2}, \dfrac{9\pi}{2}$, and so forth, the components of the elliptic motion along the major and minor axes are obtained by putting this value in Equations (11.6) and (11.7).

$$x = a\cos\theta\cos\omega t \text{ and } y = a\sin\theta\sin\omega t \qquad (11.15)$$

The ellipse is described counterclockwise with respect to an observer toward whom the wave travels, and the light is described as left-handed elliptically polarized light, as shown in Figure 11.27 (a). This statement can be easily verified by plotting the previous equation on graph paper. For example, we proceed as follows:

(i) When $\omega t = 0$, $x = a\cos\theta$ and $y = 0$

(ii) When $\omega t = \dfrac{\pi}{4}$, $x = a\cos\dfrac{\theta}{\sqrt{2}}$ and $y = a\sin\dfrac{\theta}{\sqrt{2}}$

(iii) When $\omega t = \dfrac{\pi}{2}$, $x = 0$ and $y = a\sin\theta$

But, if $\delta = \dfrac{3\pi}{2}, \dfrac{7\pi}{2}, \dfrac{11\pi}{2}$, and so forth, the components of elliptical motion along the major and minor axis are

$$x = -a\cos\theta\cos\omega t \text{ and } y = a\sin\theta\sin\omega t \qquad (11.16)$$

the ellipse, however, is now described in the clockwise direction, and the light is described as right-handed elliptically polarized light, as shown in Figure 11.27 (b).

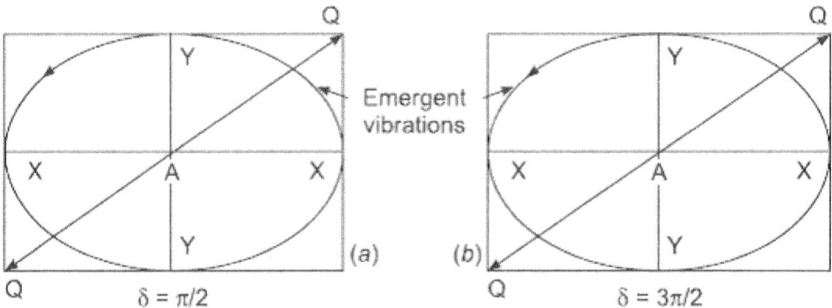

FIGURE 11.27. Emergent light elliptically polarized with principal axes along and perpendicular to optic axis.

For every value of δ, other than $n\pi$ and $\dfrac{(2n+1)\pi}{2}$, the principal axes of the ellipse are inclined to the x and y axis.

3. ***Resultant Motion Circular*** — Emergent Light Circularly Polarized. In the special case when $\theta = 45°$ in addition to $\delta = \dfrac{(2n+1)\pi}{2}$, the ellipse reduces to a circle,

$$x^2 + y^2 = \dfrac{a^2}{2} \qquad (11.17)$$

and the emergent light is said to be circularly polarized. In this case, the optical vector representing the optical disturbance at a given point in space rotates with uniform angular speed without change of magnitude.

When $\delta = \dfrac{\pi}{2}, \dfrac{5\pi}{2}, \dfrac{9\pi}{2}$, and so on, the circle is described counterclockwise with respect to an observer toward whom the wave travels as shown in Figure 11.28 (a). But if $\delta = \dfrac{3\pi}{2}, \dfrac{7\pi}{2}, \dfrac{11\pi}{2}$, and so on, the circle is described in the clockwise direction as shown in Figure 11.28 (b).

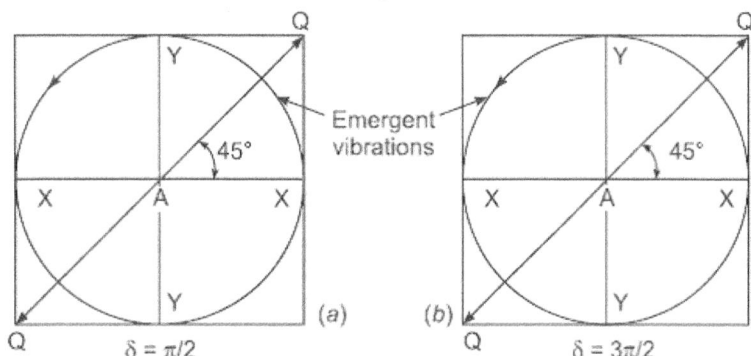

FIGURE 11.28. Emergent light circularly polarized when $\theta = 45°$ in addition to $\delta = \dfrac{(2n+1)\pi}{2}$.

It is now easy to discuss the change in the polarization of the plane polarized beam as it penetrates deeper and deeper normally to the optic axis within the crystal plate, cut with faces parallel to the optic axis. This is illustrated in Figure 11.29 for a special case when the plane of vibration of the incident linear light is inclined at 45° to the principal plane of the crystal. At the point of incidence ($t = 0$), the components E and O of the incident vibration are in phase ($\delta = 0$), hence equivalent to the original linear vibration. In the second figure as shown Figure 11.29, the thickness traversed by the incident beam is such that a phase difference of $\left(\delta = \dfrac{\pi}{4} \right)$ is introduced between the two components.

Hence, at this point of the crystal, the two components combine to form an elliptically polarized light. In the third figure, the phase difference advances to $\left(\dfrac{\pi}{2} \right)$ and since $\theta = 45°$, at this point of the crystal the light is circularly polarized. Thus, as the beam penetrates deeper and deeper, δ gradually increases until in the fifth figure $\delta = \pi$ is reached, and again we get plane polarized light, but the plane of vibration is coincident with the diagonal of the square that lies in the second and fourth quadrants. As the beam penetrates further, the resultant vibrations go through the same cycle of figures, but the figures are now described in the clockwise direction with respect to an observer toward whom the light travels, and ultimately when $\delta = 2\pi$, the resultant vibration is linear in the same plane as that of the incident beam. Beyond this thickness, the figures go through the same cycle and again the first figure is repeated when $\delta = 4\pi$ radians and so on.

In concluding our discussion of the superposition of two plane polarized coherent light waves vibrating in two mutually perpendicular planes, it should be emphasized that they never produce destructive interference, whatever the phase difference.

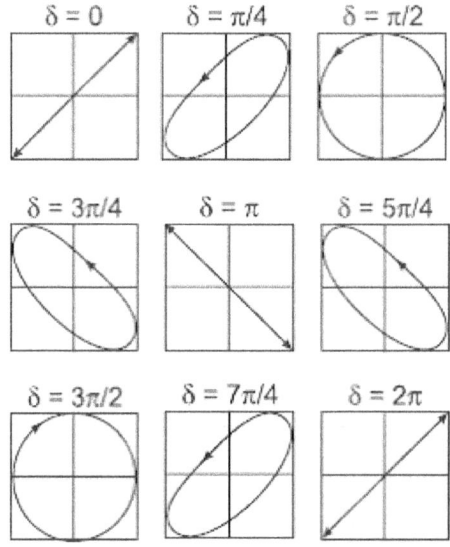

FIGURE 11.29. States of polarization corresponding to different values of the phase difference from 0 to 2π and $\theta = 45°$.

11.24 RETARDATION PLATES

A plate cut from a doubly refracting crystal by sections parallel to the optic axis and employed to introduce a given phase difference between the ordinary and extraordinary waves in transmission normally through it is called a *retardation plate*. This phase difference may be deduced as follows:

Let the thickness of the plate be t in the direction of propagation, let μ_o be the index for the O ray, and let μ_E be the index of the plate for the E ray. Within the plate, the optical path for the E ray is simply $\mu_E t$ and that for the O ray is $\mu_o t$. The path difference is therefore

$$\Delta = \left(\mu_o - \mu_E\right)t, \ \mu_o > \mu_E \tag{11.18}$$

The corresponding phase difference between the two waves is, therefore, expressed by

$$\delta = \frac{2\pi}{\lambda}\left(\mu_o - \mu_E\right)t \tag{11.19}$$

Half-wave plate—There are two most useful retardation plates are called (a) a *half-wave plate* and (b) a *quarter-wave plate*. The former is of such a thickness that in traveling through the plate, a relative phase difference of π comes in between the O and E waves, and accordingly one wave drops behind the other by just one-half a wavelength. This action is illustrated in more detail in Figure 11.30. In a schematic cutaway view of the half-wave plate, shown in part (c) of Figure 11.30, two and a half E waves and three full O waves are marked, thereby indicating that the O wave is just one-half a wavelength behind the E wave on emergence. The two waves on emergence are shown in parts (d) and (e) of Figure 11.30.

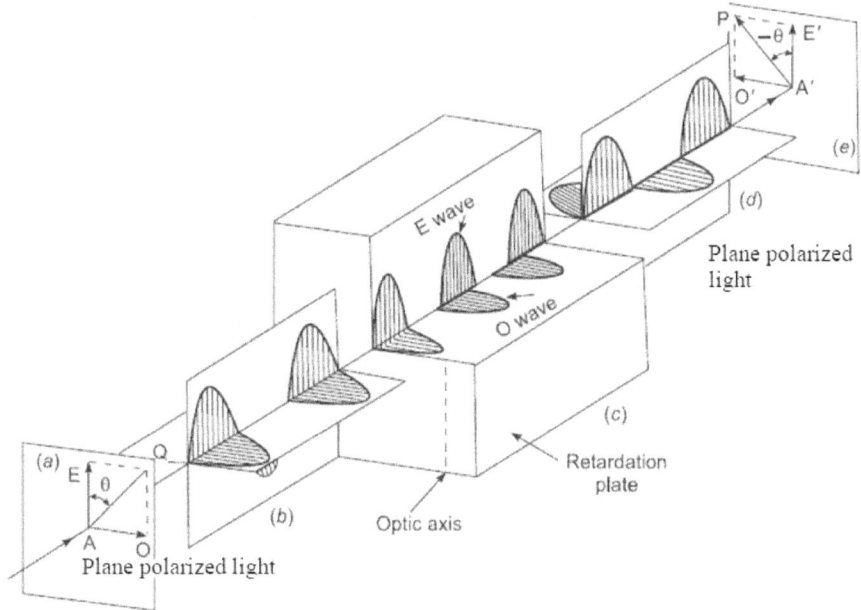

FIGURE 11.30. A half-wave plate rotates the direction of polarization of linear light by 2θ when light is incident on the plate at an angle θ with the optic axis.

Evidently, they combine to form a linearly polarized light with vibration direction inclined at 2θ to that in the incident wave, where θ is the angle between the plane of vibration of the incident wave and the principal section of the plate. Owing to this property of rotating the plane of vibration of plane polarized light by 2θ, this plate is employed in Laurent's polarimeter to divide the field of view into two halves presented side by side.

This thickness of the half-wave plate may be computed by substituting $\delta = \pi$ in Equation (11.19), and thus, we get

Calcite plate: $$ t = \frac{\lambda}{2(\mu_o - \mu_E)} \tag{11.20}$$

In particular, a plate whose thickness satisfies the equation $t(\mu_o - \mu_E) = (2n+1)\dfrac{\lambda}{2}$ behaves like a *half-wave plate*.

Quarter-wave plate—It is of such a thickness that a relative phase difference of $\dfrac{\pi}{2}$ comes in between the O and E waves in transmission through its full length. The physical conception of this is that one wave lags behind the other by just one quarter of a wavelength on emergence. The quarter-wave plate is obviously only half as thick as the half-wave plate. For a given wavelength, thickness of this plate may be computed from the relation

Calcite plate: $$ t = \frac{\lambda}{4(\mu_o - \mu_E)} \tag{11.21}$$

In particular, a plate whose thickness satisfies the equation $t(\mu_o - \mu_E) = (4n+1)\dfrac{\lambda}{4}$ behaves like a quarter-wave plate.

The ordinary and extraordinary waves on emergence from this plate are shown in Figures 11.31 and 11.32, while the action within the plate may be imagined by considering the retardation plate in Figure 11.30, only half as thick as that actually shown in Figure 11.30. The O wave, therefore, lags behind by just one-quarter of a wavelength. The nature of the resultant wave is circular as shown in Figure 11.32 for a special case, that is, $\theta = 45°$; for other angles, it is an ellipse.

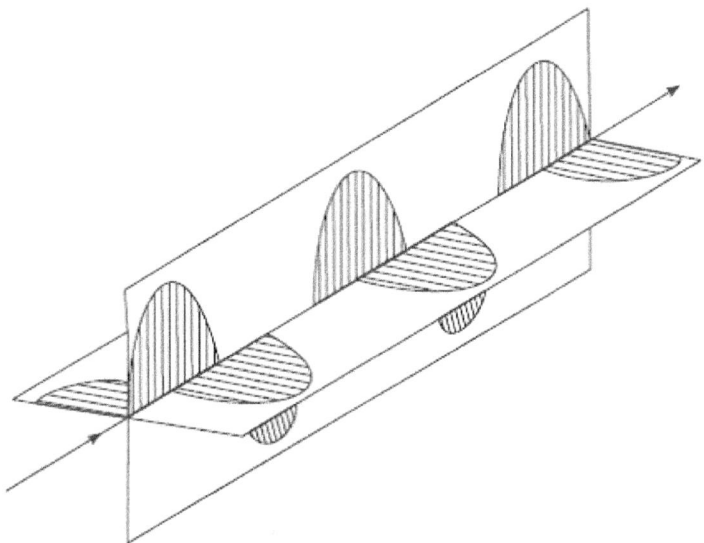

FIGURE 11.31. Emergent waves from $\frac{\lambda}{4}$ plate.

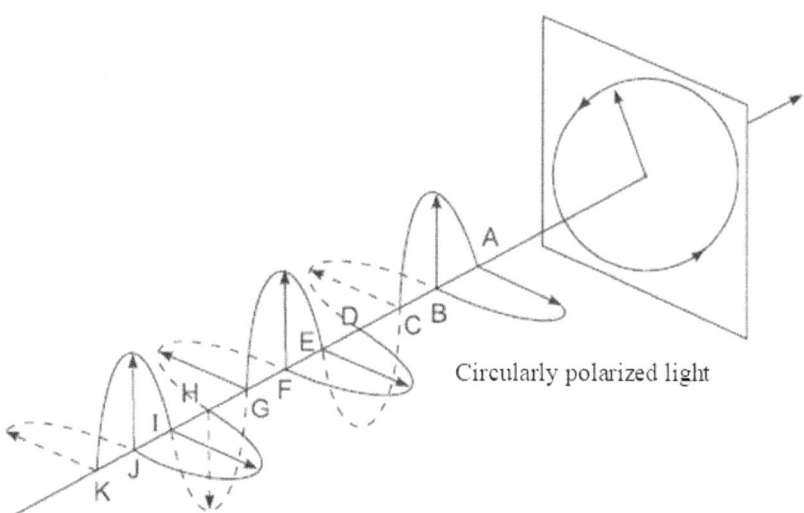

Circularly polarized light

FIGURE 11.32. A quarter-wave plate converts linear light to circular light, when linear light is incident on the plate at an angle of 45° with the optic axis.

The quarter-wave plate is commonly employed for the production and detection of circularly and elliptically polarized light.

A quarter-wave plate for sodium light is not a quarter-wave plate for light of any other wavelength, and the same holds for a half-wave plate.

Quarter- and half-wave plates are often made of thin sheets of split mica or of quartz cut parallel to the optic axis. In quartz the *O* ray travels faster than the *E* ray, hence $\mu_o < \mu_E$. The thickness of the quartz half-wave plate is, therefore, computed from the relation

$$2t\left(\mu_E - \mu_o\right) = \lambda \tag{11.22}$$

and that of the quarter-wave plate from the relation

$$4t\left(\mu_E - \mu_o\right) = \lambda \tag{11.23}$$

Mica is a negative biaxial crystal; that is, it has two optic axes. Nevertheless, there are some forms of mica for which the angle between the two axes is small. Mica has the advantage of having a natural cleavage plane, and so a thin sheet of any desired thickness with optically flat faces can be easily cut from a thick sheet. Quartz has no natural cleavage planes and has to be cut and the faces polished to optical flatness.

11.25 PRODUCTION OF CIRCULARLY POLARIZED LIGHT

We are now in a position to describe the experimental technique employed for the production of circularly polarized light. It would be recalled that a circular motion results when two mutually perpendicular coherent linear vibrations of equal amplitude and period but differing in phase by $\dfrac{\pi}{2}$ are compounded together. For example, consider two mutually perpendicular coherent linear vibrations,

$$x = \left(\frac{a}{\sqrt{2}}\right)\sin\left(\omega t + \frac{\pi}{2}\right) = \left(\frac{a}{\sqrt{2}}\right)\cos\omega t \tag{11.24}$$

$$y = \left(\frac{a}{\sqrt{2}}\right)\sin\omega t \tag{11.25}$$

which, on squaring and adding yield

$$x^2 + y^2 = \frac{a^2}{2} \tag{11.26}$$

an equation to a circle.

Two such mutually perpendicular vibrations can be very easily obtained by allowing a beam of plane polarized monochromatic light to be incident normally on a quarter-wave plate, the vibration direction in the incident light being inclined at 45° to the direction of optic axis of the plate. The incident vibration is resolved into two equal components—the E and O vibration—respectively along and perpendicular to the optic axis. Initially both are in the same phase, but on emergence, the O wave drops behind the E wave by one quarter of a wavelength (if the $\frac{\lambda}{4}$ plate is of calcite), which is equivalent to a phase difference of $\frac{\pi}{2}$ between the O and E waves. Since the amplitudes of the E and O waves are equal, on emergence from the quarter-wave plate they compound into a circular wave; that is, the emergent light is circularly polarized.

The physical picture of the formation of a circularly polarized wave can be very easily seen by referring to Figure 11.32, in which the E and O waves are shown on emergence from the plate for the sake of explaining the rotation of the resultant light vector representing the optical disturbance at a given point. At points $A, C, E, G, I,$ and K, the magnitude of the horizontal component is a positive or negative maximum, while that of the vertical component is a zero. The resultant light vector at these points is therefore shown by a horizontal arrow, and its magnitude is equal to the maximum of the O wave. At points $B, D, F, H,$ and J, the magnitude of the horizontal component is zero, and hence the resultant light vector is vertical, equal in magnitude to the maximum of the E wave, and hence equal in magnitude to the maximum of the O wave ($\theta = 45°$). At a point midway between the points A and B, the magnitude of each component is 0.707 times the maximum of each wave; that is, it is equal to $\frac{0.707a}{\sqrt{2}}$, where a is the amplitude of the incident vibration. Obviously, the resultant light vector at this point is inclined at 45° to the horizontal and vertical planes and its magnitude is $\frac{a}{\sqrt{2}}$, the maximum of the E and O waves. Hence, it is equal to the resultant at the points A and B. In a similar way, it is easy to see that the resultant light vector at a point midway between B and C is equal in magnitude to that at $A, B,$ and C, but inclined at 45° to the vertical plane on the opposite side of it. It is now obvious that the magnitude of the resultant light vector is the same at all points, and it rotates continuously around the direction of propagation, completing one revolution in the interval in which the wave advances one wavelength. In a plane perpendicular to the direction of propagation, through any point in this direction, the light vector rotates with uniform angular velocity as the wave advances, the tip of the vector describing a circular path. The light is therefore said to be circularly polarized.

11.26 PRODUCTION OF ELLIPTICALLY POLARIZED LIGHT

It will be recalled that the elliptically polarized light arises when two mutually perpendicular plane polarized coherent light waves of unequal amplitudes but differing in phase by a value other than $n\pi$ are compounded together. We are, however, more interested in the special case arising when $\delta = \dfrac{\pi}{2}$.

For example, consider coherent linear vibrations,

$$x = a\cos\theta \sin\left(\omega t + \frac{\pi}{2}\right) \text{ or } \frac{x}{a\cos\theta} = \cos\omega t \tag{11.25}$$

$$y = a\sin\theta \sin\omega t \text{ or } \frac{y}{a\sin\theta} = \sin\omega t \tag{11.26}$$

which on squaring and adding yield

$$\frac{x^2}{a^2\cos^2\theta} + \frac{y^2}{a^2\sin^2\theta} = 1 \tag{11.27}$$

an equation to ellipse.

Elliptically polarized light can be, therefore, conveniently produced by passing a beam of plane polarized monochromatic light through a quarter-wave plate, its optic axis being inclined at about $\theta = 30°$ to the plane of vibration of the incident light.

The physical picture of the formation of an elliptically polarized wave can be obtained by referring to Figure 11.33, in which the E and O waves of unequal amplitudes are shown on emergence from a quarter-wave plate.

When the light vectors at every point in the E and O waves are added up it will be found, as with circular light, that while the resultant light vector rotates around the direction of the propagation of the ray, its magnitude varies periodically. If we imagine a plane perpendicular to the ray, then as the wave advances, the resultant light vector in this plane rotates, the tip of the vector describing an ellipse, its principal axes being along and perpendicular to the principal section of the plate. Thus, a quarter-wave plate converts plane

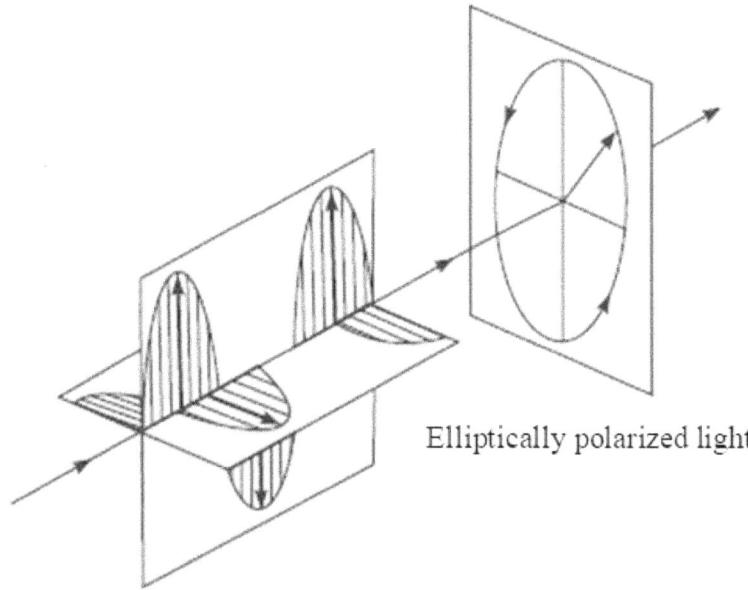

Elliptically polarized light

FIGURE 11.33. A quarter-wave plate converts linear light into elliptic light, when linear light is incident on the plate at an angle (excluding 0°, 45°, and 90°) with the optic axis.

polarized light to elliptically polarized light when the vibration direction in the incident light is inclined at an angle other than 0°, 45°, and 90° with the optic axis. The principal axes of the ellipse are parallel and perpendicular to the optic axis of the $\frac{\lambda}{4}$ – plate.

11.27 ANALYSIS OF LIGHT

Let us suppose that as a result of some experiment we have a beam of light, and it is desired to ascertain its nature from the following seven possibilities.

    (i) Unpolarized light
   (ii) Plane polarized light
  (iii) Elliptically polarized light
  (iv) Circularly polarized light
   (v) Mixture of unpolarized and plane polarized light
  (vi) Mixture of unpolarized and elliptically polarized light
 (vii) Mixture of unpolarized and circularly polarized light

Before giving a systematic qualitative investigation to ascertain the nature of a given beam, we shall first give the characteristics exhibited by every one of the previously listed kinds when examined by the analyzer, if one knows beforehand the nature of light.

Unpolarized Light

The unpolarized light on entering the Nicol prism is decomposed into ordinary and extraordinary beams of equal intensity. The E vibration is freely transmitted and the O vibration is totally reflected at the balsam surface, whatever may be the orientation of Nicol, which is, however, rotated about the direction of the incoming beam as an axis. Thus, whatever may be the orientation of Nicol, the transmitted intensity is the same.

Plane Polarized Light

The plane polarized light when examined by the Nicol can be completely extinguished for one setting of the analyzer. In this setting, the direction of vibration of the incident light is perpendicular to the principal section of the analyzer—that is to the shorter diagonal of its end face. For each complete rotation of the analyzer, there are two positions (180° from each other) at which the incoming beam is completely extinguished and two positions (90° from the former) at which the transmitted intensity is a maximum.

Elliptically Polarized Light

The elliptically polarized wave may be regarded as the resultant of two mutually orthogonal plane polarized waves. The amplitudes and the phase difference, however, depend on the choice of these planes of vibration. In the special case when the phase difference is $\frac{\pi}{2}$, the major and minor axes of the ellipse are along the planes of vibration of the component waves as shown in Figure 11.34.

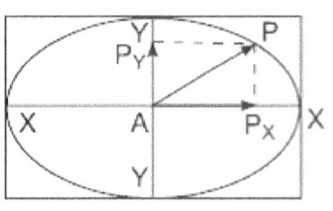 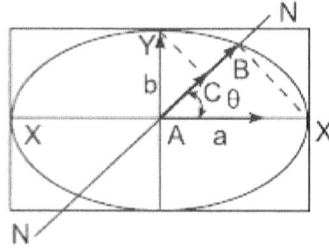

FIGURE 11.34. Analysis of elliptically polarized light.

Conversely, a given elliptically polarized wave can be resolved into two linearly polarized waves of unequal amplitudes, vibrating in planes parallel to the major and minor axes of the ellipse and differing in phase by $\dfrac{\pi}{2}$. The amplitude of one wave is equal to the semi-minor axis b, and that of the other equal to the semi-major axis a of the elliptic vibration. Let the principal section NN of the analyzing Nicol make an angle θ with the major axis of the ellipse as shown in Figure 11.34. The resolved components along NN of the amplitudes of component waves, namely, $AB = a \cos\theta$ and $AC = b \sin\theta$, are freely transmitted as extraordinary vibrations, but they differ in phase by $\dfrac{\pi}{2}$. The transmitted intensity by Nicol is, therefore, given by

$$I\alpha R^2 = a^2 \cos^2\theta + b^2 \sin^2\theta + 2ab\cos\theta\sin\theta\cos\frac{\pi}{2}$$

$$= \left(a^2 - b^2\right)\cos^2\theta + b^2 \tag{11.28}$$

when $\theta = 0$, $I_{max} = a^2$ and when $\theta = \dfrac{\pi}{2}$, $I_{min} = b^2$

Thus, for each complete rotation of the analyzer about an axis parallel to the incident beam, there are two positions of NN (180° from each other) at which the transmitted intensity is a minimum and two positions (at 90° from the former) at which the transmitted intensity is maximum; the beam is never completely extinguished as with plane polarized light.

Circularly Polarized Light

Circular polarization is a special case of elliptic polarization, that is, when $a = b$, the ellipse degenerates into a circle. Therefore, when circularly polarized light is examined by a Nicol, the transmitted intensity is

$$I = a^2 \tag{11.29}$$

which is the same whatever may be the orientation of the analyzer. In this respect the circularly polarized light resembles the unpolarized light.

It should be remarked that the circularly polarized wave is equivalent to two mutually perpendicular plane polarized waves of equal amplitude and differing in phase by $\dfrac{\pi}{2}$. Because of symmetry of the circular wave about the direction of propagation, all mutually perpendicular directions of resolution are equivalent.

Mixture of Unpolarized and Plane Polarized Light

We have explained above that when unpolarized light is examined by Nicol, the transmitted intensity is the same for all orientations of the analyzer, but linear light is extinguished for two settings (180° from each other) of the analyzer. Consequently, when a mixture of these lights is examined by the analyzer, the transmitted intensity will never become zero, but it will periodically fluctuate between a maximum and a minimum. In this respect this mixture resembles elliptically polarized light.

Mixture of Unpolarized and Circularly Polarized Light

It will be recalled that unpolarized and circularly polarized light when examined separately by the analyzing Nicol exhibit no variation in intensity. Obviously, when the mixture of these lights is examined by the Nicol, the transmitted intensity is the same in all orientations of the analyzer.

Mixture of Unpolarized and Elliptically Polarized Light

For one complete rotation of the analyzer, there are two settings at which the transmitted intensity is a maximum and two settings at which the transmitted intensity is a minimum. In this respect this mixture resembles elliptically polarized light and also a mixture of unpolarized and plane polarized light.

Systematic Analysis of Light

We are now in a position to give a scheme for the systematic analysis of the nature of a given light. The first step is to examine the oncoming beam of light by the Nicol prism. Any of the following three phenomena will be observed:

a. Light can be completely extinguished for one setting of the analyzer. This indicates that light is plane polarized.

b. The transmitted intensity is the same for all settings of the analyzer. This indicates that light is either (i) unpolarized or (ii) circularly polarized or (iii) a mixture of circularly polarized and unpolarized light. To distinguish between these lights, we make use of a quarter-wave plate in a manner given later.

c. The transmitted intensity varies as the Nicol is rotated, and for each complete rotation of the analyzer we get two positions (180° from each other) at which the transmitted intensity is a maximum and two positions at which the transmitted intensity is a minimum. This indicates that the light is either (i) elliptically polarized or (ii) a mixture of unpolarized and plane polarized light or (iii) a mixture of unpolarized and elliptically polarized light. To distinguish between these lights, we employ a quarter-wave plate in the manner described later.

Distinction between Unpolarized, Circularly Polarized, and a Mixture of Unpolarized and Circularly Polarized Light

To accomplish this identification a quarter-wave plate is inserted in the path of an oncoming beam of light, and the light emerging from the plate is examined by the Nicol prism, rotating it about the direction of propagation of the beam. Any of the following three phenomena will be observed:

(i) The light emerging from the $\frac{\lambda}{4}$-plate can be completely extinguished for two settings (180° from each other) of the analyzer in its one complete rotation. This indicates that the original light is circularly polarized. The reason for this conclusion is as follows: Circularly polarized light, just on entering the $\frac{\lambda}{4}$-plate, can be resolved into two components of equal amplitudes—one along and other perpendicular to the optic axis—differing in phase by $\frac{\pi}{2}$ radians. In transmission through the $\frac{\lambda}{4}$-plate, a further phase difference of $\frac{\pi}{2}$ is introduced so that on emergence the resultant phase difference between the phases of two components becomes $\frac{\pi}{2} + \frac{\pi}{2} = \pi$ if the original light circle is described in the counterclockwise direction as shown in Figure 11.35 (a), and if the circle is described in the clockwise direction with respect to an observer toward whom the wave travels, the resultant phase difference will be $\frac{\pi}{2} - \frac{\pi}{2} = 0$. Consequently, in either case they compound into rectilinear vibrations *PAP*, that is, the emergent light is plane polarized. This may be completely extinguished by the Nicol prism when its principal section *NN* is at 45° with the optic axis of the quarter-wave plate as shown in Figure 11.35 (a).

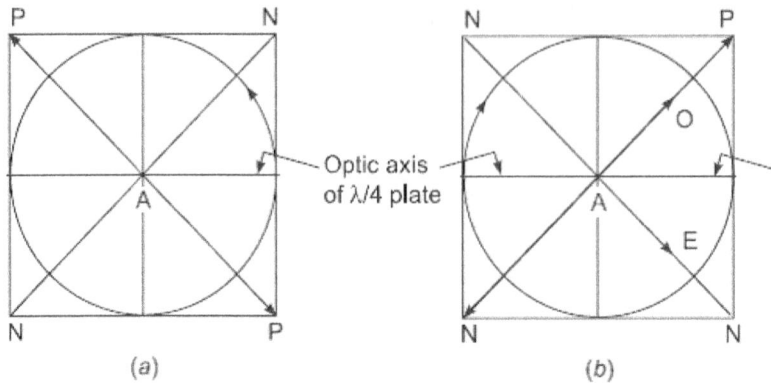

FIGURE 11.35. (a) Analysis of circularly polarized light (b) Analysis of a mixture of circularly and unpolarized light.

(ii) The light emerging from the quarter-wave plate exhibits no variation in intensity as the analyzing Nicol is rotated about its axis. This
indicates that the original light is unpolarized. The reason for this
conclusion is as follows: unpolarized light is theoretically equivalent
to two mutually perpendicular linear lights having no constant phase
relation to each other. That is, incoherent O and E disturbances are
propagated through the crystal. Hence, the $\frac{\lambda}{4}$-plate cannot introduce any constant phase difference between the two components
of unpolarized light. No matter what phase difference is introduced
between the two components of unpolarized light, if the components
recombine, the resultant will always be unpolarized light. One can, in
fact, be said to have unpolarized light at all points through the plate.
Hence, we conclude that unpolarized light is not at all changed in
transmission through the quarter-wave plate. Hence, we observe no
variation in the intensity of emergent light when examined by the
analyzer.

(iii) The emergent light exhibits two maxima and two minima in one
complete rotation of the analyzer. This indicates that the original
beam is a mixture of unpolarized and circularly polarized light. The
reason for this conclusion is as follows: the circularly polarized part of
the original beam on transmission through the $\frac{\lambda}{4}$-plate, as explained
earlier, is converted into plane polarized light *PAP* as shown in
Figure 11.35 (b), while the unpolarized part is transmitted without

any modification. Thus, the light emerging from the $\frac{\lambda}{4}$-plate is a mixture of unpolarized and plane polarized light. Obviously, when the principal section NN of the Nicol prism is at right angles to PAP, shown in Figure 11.35 (b), the transmitted intensity is a minimum. For, while the linear vibration PAP is completely extinguished, the unpolarized part gives rise to E vibrations as shown for AE in Figure 11.35 (b) in transmission through the analyzer. Hence, we observe the variation in the intensity as stated earlier.

Distinction between Elliptically Polarized Light, Mixture of Elliptically Polarized and Unpolarized Light, and Mixture of Plane Polarized and Unpolarized Light

We insert a quarter-wave plate between the analyzer and the incoming beam of light. The quarter-wave plate is rotated gradually in steps of $1°$ or so, and for each setting of the plate the analyzer is given one rotation. Then any of the following three phenomena will be observed:

a. It is possible to get one setting of the $\frac{\lambda}{4}$-plate with respect to the original light so that the emergent light can be completely extinguished for one setting of the analyzer. The principal section of the analyzer, in other words, the shorter diagonal of the end face of Nicol, is inclined to the optic axis of the plate. This indicates that the original beam is elliptically polarized, and the axes of the ellipse are respectively parallel and perpendicular to the optic axis when the emergent light is extinguished. The reason of this conclusion is as follows:

Suppose the optical vector in the elliptically polarized wave rotates in the counterclockwise direction with respect to an observer toward whom the wave travels, and that the optic axis (say the x-axis) of the plate is parallel to the major axis of the ellipse as shown in Figure 11.36 (a). The incident wave, just on entering the plate, is resolved into two linearly polarized waves, vibrating in the x (major axis) and y (minor axis) directions, that is, along and perpendicular to the optic axis of the plate, differing in phase by $\frac{\pi}{2}$ radians. The equations of component vibrations are respectively,

$$E \text{ Wave} : x = a \cos\omega\, t = a \sin\left(\omega\, t + \frac{\pi}{2}\right) \tag{11.30}$$

$$O \text{ Wave} : y = b \sin\omega\, t \tag{11.31}$$

There is a relative phase advancement of $\dfrac{\pi}{2}$ of the first wave with respect to the second. If we assume that $\mu_E < \mu_o$, that is, $V_E > V_o$, the quarter waveplate produces an additional relative phase advancement of $\dfrac{\pi}{2}$ of the first wave with respect to the second. Therefore, as the waves emerge from the plate, they have a relative phase difference of $\dfrac{\pi}{2} + \dfrac{\pi}{2} = \pi$, and so they combine into a plane polarized wave. Therefore, the light emerging from the plate can be completely extinguished when the principal section of the analyzer is perpendicular to the plane of vibration of the emergent light.

If in the original beam, the ellipse is described in the clockwise direction, then the resolved plane polarized vibrations along the major and minor axes are

$$E \text{ Wave}: x = -a \cos\omega\, t = a\sin\left(\omega\, t - \frac{\pi}{2}\right) \tag{11.32}$$

$$O \text{ Wave}: y = b\sin\omega\, t \tag{11.33}$$

The first wave lags behind the second by a phase angle $\dfrac{\pi}{2}$. If we assume as before that $V_E > V_o$, the $\dfrac{\lambda}{4}$-plate produces a relative phase advance of $\dfrac{\pi}{2}$ of the first wave with respect to the second. Therefore, the two waves emerging from the plate have a phase difference of $\dfrac{1}{2\pi} - \dfrac{1}{2\pi} = 0$. Consequently, they combine into a plane polarized wave, the plane of vibration being parallel to the diagonal of the rectangle that lies in the first and third quadrants, and so the emergent light can be completely extinguished for the setting of the analyzer marked in Figure 11.36 (b).

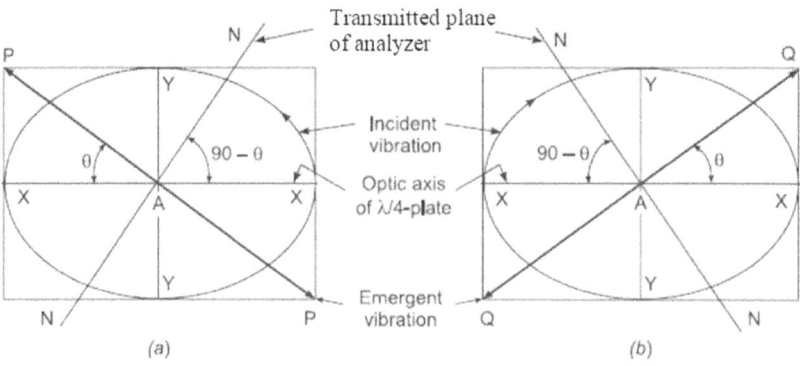

FIGURE 11.36. Detection of elliptically polarized light.

b. For one complete rotation of the analyzer, there are two settings (180° from each other) at which the transmitted intensity is minimum (not equal to zero), and in this setting its principal section (transmission plane) is inclined to the optic axis of the $\frac{\lambda}{4}$-plate as shown in Figure 11.36 (a). This indicates that the original light is a mixture of unpolarized light and elliptically polarized light. The reason of this conclusion is as follows:

When the optic axis of the $\frac{\lambda}{4}$-plate is parallel to the major or minor axis of the elliptically polarized part of the mixture, then on transmission through the plate, as explained earlier, the elliptic vibration degenerates into a linear vibration. The plane of vibration as *PAP* in Figure 11.37 (a) is inclined to the optic axis of the plate. On the other hand, the unpolarized part of the mixture is transmitted without any modification. Obviously, when the principal section of the analyzer is at right angles to the plane of vibration of plane polarized part of the mixture emerging from $\frac{\lambda}{4}$-plate and therefore inclined to the optic axis of the $\frac{\lambda}{4}$-plate, the transmitted intensity is a minimum. For, while the plane polarized vibration *PAP* is completely extinguished but the unpolarized part gives rise to plane polarized light as for *E* vibrations, *AE* in Figure 11.37 (a) in transmission through the analyzer.

c. The principal section of the analyzer in the setting for a minimum of transmitted intensity (not equal to zero) is either parallel to or perpendicular as shown in Figure 11.37 (b) to the optic axis of the quarter-wave plate. This indicates that the original light is a mixture of unpolarized light and plane polarized light. The reason for this conclusion is as follows: The plane polarized part labeled as *QAQ* in Figure 11.37 (b) of the mixture is converted into an elliptically polarized wave on emergence from the $\frac{\lambda}{4}$-plate while the unpolarized part is transmitted without any modification. The principal axes of the ellipse are respectively parallel and perpendicular to the optic axis of the $\frac{\lambda}{4}$-plate. It will be recalled that when the elliptically polarized light is separately examined by the Nicol prism, the transmitted intensity is a minimum when the principal section is parallel to the minor axis of the ellipse, while the unpolarized light exhibits no variation in intensity. It is now obvious that the mixture emerging from the $\frac{\lambda}{4}$-plate will exhibit the variation in the intensity in the manner stated earlier.

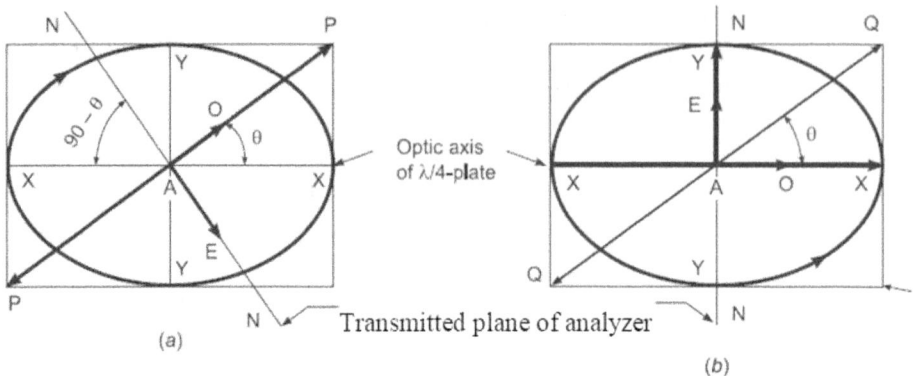

FIGURE 11.37. (a) Detection of mixture of unpolarized light and elliptically polarized light
(b) Detection of mixture of unpolarized light and plane polarized light.

The analysis given previously may be summarized as follows:

Detection:

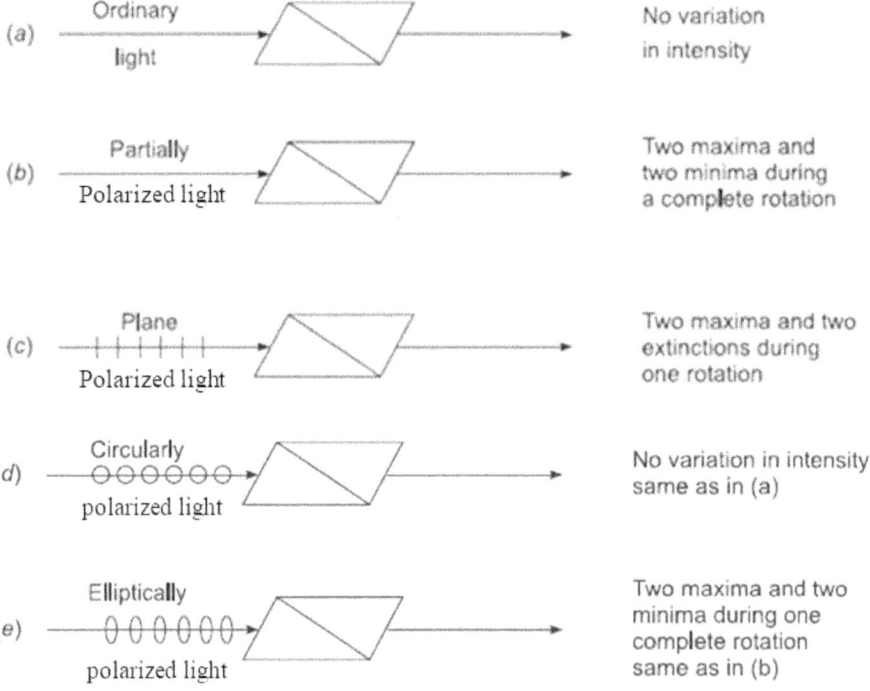

FIGURE 11.38. Detection analysis.

Distinction:

(i) Between circularly polarized and unpolarized light.

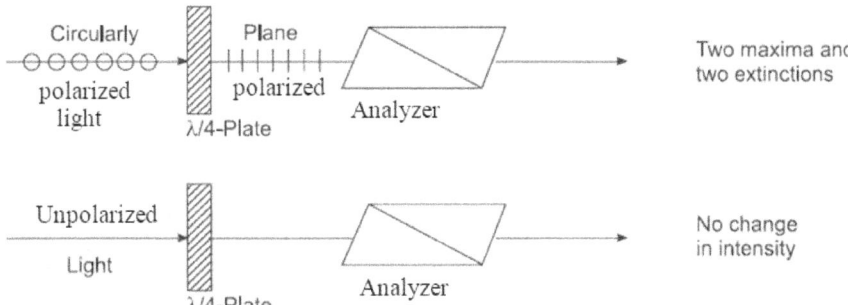

FIGURE 11.39. Distinction analysis between circularly polarized and unpolarized light.

(ii) Between elliptically polarized and partially polarized light.

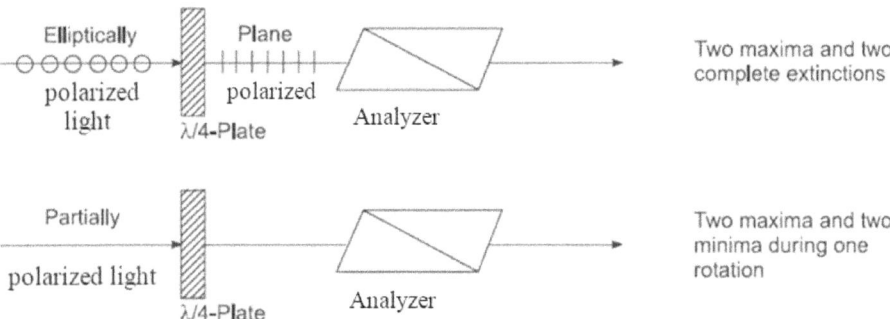

FIGURE 11.40. Distinction analysis between elliptically polarized and partially polarized light.

11.28 BABINET COMPENSATOR

In the production and analysis of elliptically polarized light, the use of a given quarter-wave plate is limited only to a narrow range of wavelengths for which the path difference between the E and O rays on transmission through it is $\dfrac{(4m+1)\lambda}{4}$, where $m = 0, 1, 2, 3$, and so on. The Babinet compensator is an apparatus which has no such limitation of wavelength when in use, and it has been successfully employed by Jamin for accurate measurement of constants of elliptic vibration in elliptically polarized light. In reality, it measures only the phase difference between the two components transmitted by any aniso-tropic plate.

The Babinet compensator consists of two quartz wedges of equal acute angles. The optic axis in one wedge is parallel to while the other perpendicular to the longer edge of their free rectangular faces. The optic axes are mutually perpendicular when the wedges are placed in contact with each other as shown in Figure 11.41 so as to form a thin plate of the rectangular cross-section ABCD. The wedge angles are much exaggerated in Figure 11.41 for the sake of convenience in drawing the figure. As modified by Jamin, the instrument has fixed cross-wires in front of the upper wedge and a microme-ter screw to displace the lower wedge relative to the upper fixed wedge along their plane of contact, thus forming a plane parallel plate whose thickness can be varied.

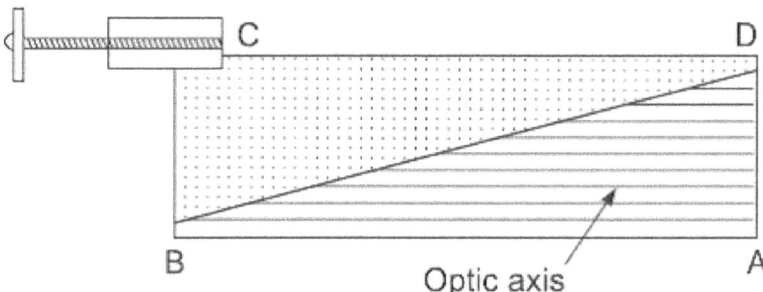

FIGURE 11.41. Babinet-Jamin compensator.

11.28.1 Calibration of Micrometer Screw of Babinet Compensator

It is first essential to calibrate the micrometer screw in terms of phase dif-ference or path difference between the E and O rays originating from the monochromatic light which is to be employed in the experiment. The analyzer and polarizer are crossed, and the compensator is placed between them, ori-ented so that a parallel beam of plane polarized monochromatic light incident normally on the compensator has its plane of vibration at 45° to the optic axes of the wedges. Just on entering the first wedge, the incident vibration is resolved into two components, namely, the E vibration along the optic axis and the O vibration perpendicular to it. The O ray travels faster than the E ray since quartz is a positive uniaxial crystal. The acute angles of the wedges are so small that we can neglect the separation of the E and O rays. Since the optic axis of the second wedge is perpendicular to that of the first, the O and E rays in the first wedge are transmitted as E and O rays respectively in the second wedge, that is, the two rays exchange their character and so change their velocities on passing from one wedge into the other. Now let n_E and n_o be the refractive indices of quartz for the E and O rays respectively. Let us consider a point on the compensator where the upper wedge is of thickness h_1 and the lower wedge of thickness h_2. The optical path difference introduced between the E and O rays in the transmission through the first wedge at this point is $h_1(n_E - n_o)$ and that introduced by the second wedge is $-h_2(n_E - n_o)$. The negative sign arises because the rays that are E rays and O rays in the first wedge become, respectively, O rays and E rays in the second wedge. Hence, the total optical path difference is

$$\Delta = (h_1 - h_2)(n_E - n_o) \tag{11.34}$$

and the resultant phase difference is

$$\delta = \left(\frac{2\pi}{\lambda}\right)(h_1 - h_2)(n_E - n_o) \tag{11.35}$$

Due to the increasing thickness of one component wedge and the decreasing thickness of the other, Δ and δ increase linearly on one side of the center $(h_1 = h_2)$ and decrease on the other side; that is, the retardation is of opposite signs on the two sides of the zero point.

The compensator has constant thickness along lines perpendicular to the longer edge of the wedge. For the central ray $h_1 = h_2$ and $\delta = 0$; that is, in effect the plate is of zero optical thickness at the center. The emergent light at the central line is plane polarized with its plane of vibration parallel to that of the incident light—in effect the original vibration is transmitted. This is also true for the emergent light at equidistant parallel lines, on either side of the central ray, where $\delta = \pm 2\pi, \pm 4\pi, \pm 6\pi, \ldots = \pm 2m\pi$. Therefore, the light is extinguished as if no birefringence medium were present between the crossed analyzer and polarizer. At intermediate parallel lines where $\delta = \pm(2m+1)\pi$, the emergent light is plane polarized, but its plane of vibration is inclined at $\frac{\pi}{2}$ to that of the incident light; that is, the emergent vibrations are along the principal section of the analyzer, and so they are freely transmitted through it. At other parallel lines the emergent light is either circularly polarized or elliptically polarized.

The polarization of emergent light varies along the length of the compensator. When the compensator is viewed through the analyzer (crossed) by the help of a low-power microscope focused on the fixed cross-wires, a series of equally spaced alternate dark and bright bands, perpendicular to the long edge, are seen in the field of view. If the plane of vibration of the incident light is not exactly at 45° to the optic axes, the emergent vibration at points where $\delta = \pm 2m\pi$ will not be exactly perpendicular to the principal section of the analyzing nicol. The fringes will be dark but not black, and the contrast in the fringes will be reduced.

When the lower wedge is displaced with respect to the other by rotating the micrometer screw, the position of zero optical thickness moves across the wedge. Therefore, the fringes move laterally across the field of view. Thus, any desired dark fringe can be brought under the wire. The movable wedge is adjusted until one of these dark bands is on the cross-wires. The movement $2c$ of the quartz wedge as read on the micrometer screw or its angular rotation ϕ necessary to produce a shift of one fringe spacing is that which corresponds to a change of phase difference by 2π or a change of path difference of one wavelength. The screw is thus calibrated in terms of phase difference or path difference. Now, the central difference of path can be altered by a known amount by displacing the wedge by the known amount.

In order to locate the central black band corresponding to $\delta = 0$, it is only necessary to illuminate the compensator with plane polarized white light. Only the central fringe at $h_1 = h_2$ is dark, and on either side of it the Newtonian colors of the thin plate appear.

11.28.2 Measurement of the Birefringence ($n_E - n_o$)

The experimental arrangement is the same as sketched in Figure 11.42, and monochromatic light is employed.

By moving the lower wedge, a dark fringe is brought under the cross-wires. A doubly refracting plate cut from the given crystal, with its faces parallel to the optic axis, is introduced between the compensator and one of the nicols. The plate is oriented so that its optic axis is inclined at 45° with the principal sections of the nicols. Thus, its optic axis is parallel and perpendicular to the optic axis of the wedges. An extra phase difference

$$\delta' = \left(\frac{2\pi}{\lambda}\right)(n_E - n_o)h \qquad (11.36)$$

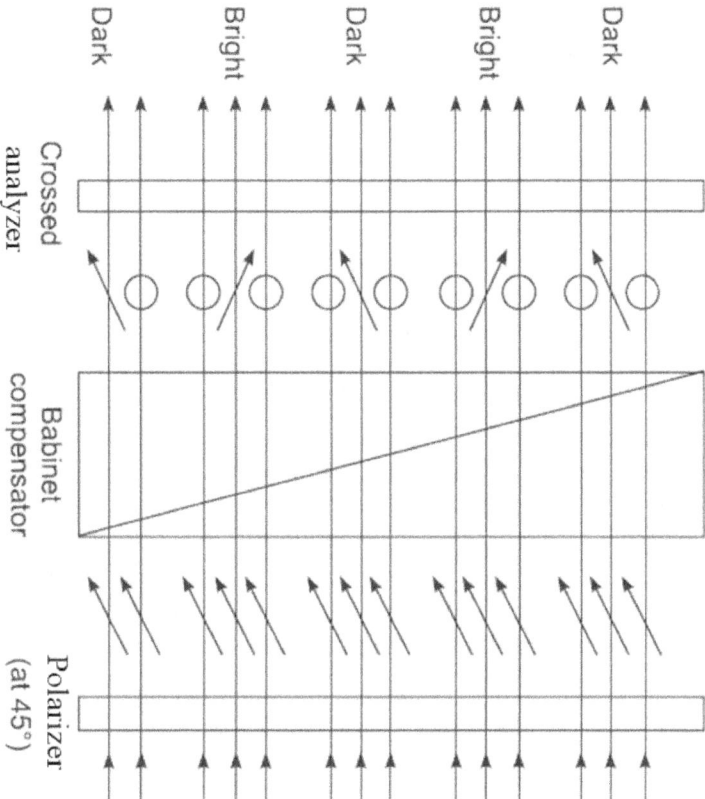

FIGURE 11.42. Calibration experiment.

11.28.3 Determination of Constants of Elliptic Vibration in Elliptically Polarized Light

We shall now describe the use of the Babinet compensator to determine the constants of elliptic vibration.

Phase Difference

When an elliptically polarized monochromatic light is incident normally on the compensator, we may consider the elliptic vibration as resolved at the surface into two components, namely

$$x = A\sin(\omega t + \alpha); \quad y = B\sin(\omega t + \beta) \tag{11.37}$$

along and perpendicular to the optic axis of the first quartz wedge. The phase difference between the two components is $(\alpha - \beta)$. Transmission through the compensator changes this phase difference by

$$\delta = \left(\frac{2\pi}{\lambda}\right)(h_1 - h_2)(n_E - n_o) \qquad (11.38)$$

The total phase difference $\alpha - \beta + \delta$ will be integral multiples of π along a system of equally spaced lines. The light emerging along these lines will be plane polarized. Therefore, when the compensator is observed through a properly oriented Nicol, bright and dark bands will be seen in the field of view, and the central band again corresponds to $\alpha - \beta + \delta = 0$.

To begin with we employ a plane polarized light, and the wedge is moved to bring the central band at the cross-wires. The phase difference between the E and O rays at this point is zero. On substituting the elliptically polarized light, the central band shifts to a point where the initial phase difference $\alpha - \beta$ between the components of elliptic vibration is exactly neutralized by the phase difference δ introduced by the compensator. The lower wedge is now displaced by turning the micrometer screw until the central dark band is again under the cross-wires. If the displacement is x, we have

$$\frac{\alpha - \beta}{\pi} = \frac{x}{c} \text{ or } \alpha - \beta = \frac{\pi x}{c} \qquad (11.39)$$

where $2c$ is the distance through which the wedge must be moved to introduce a phase difference of 2π at the cross-wires. The value of $2c$ has been previously determined in the calibration experiment.

Position and Ratio of Axes

The components of elliptic vibrations along the major and minor axes differ in phase by $\frac{\pi}{2}$. With this knowledge we proceed to determine the position and the ratio of the axes as follows:

The analyzer and polarizer are crossed and adjusted with their principal sections at $45°$ to the optic axes of the quartz wedges. A dark band is brought at the cross-wires by moving the lower wedge. In this setting, the phase difference between the ordinary ray and the extraordinary ray, emerging under the cross-wires, is $2m\pi$. The micrometer screw is turned through $\frac{\phi}{4}$ turn to introduce a phase difference of $-\frac{\pi}{2}$ at the cross-wires so that the total phase difference becomes $2m\pi - \frac{1}{4\pi}$ and, therefore, there results a displacement of fringes in the field of view. The polarizing nicol is removed, and the elliptically polarized light under examination is allowed to fall normally on the compensator, which is now rotated in its own plane about the line of vision until the same dark band again moves on to the cross-wires. This will happen when the phase difference at the cross-wires again becomes $2m\pi$. The analyzer must be adjusted to make this dark band as black as possible, that is, to achieve the extinction perfectly at the cross-wires. The situation is now shown in Figure 11.43.

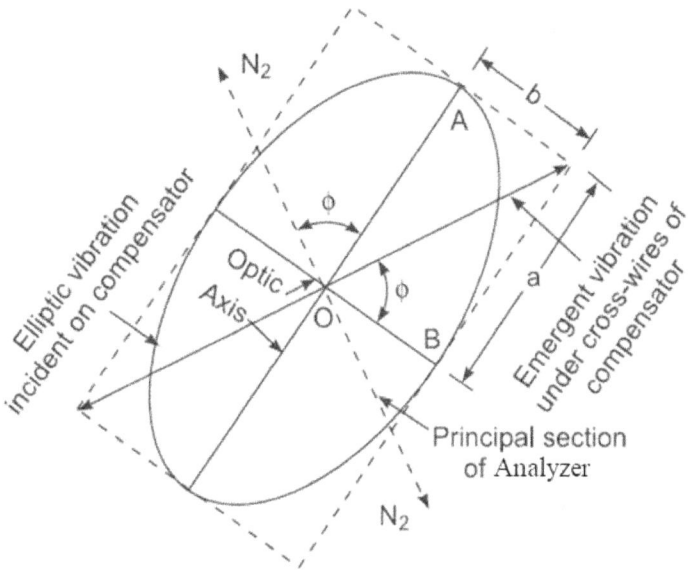

FIGURE 11.43. Analyzer of elliptically polarized light.

The optic axes of the two quartz wedges now give the directions of the axes of the elliptic vibration. The reason of this conclusion is as follows: when the major and minor axes of the elliptic vibration are along the optic axes of the quartz wedges, the components of the elliptic vibration along the optic axes differ in phase by $\dfrac{\pi}{2}$ all over the surface of the compensator. In traversing the compensator, a further phase difference $2m\pi - \dfrac{1}{2\pi}$ is introduced between these components at cross-wires so that the resultant phase difference becomes $2m\pi$. Thus, the light emerging under the cross-wires is plane polarized. Hence, the same dark fringe should appear under the cross-wires.

Suppose the analyzing Nicol is in the setting of complete extinction of the dark fringe under the cross-wires. Its principal section is perpendicular to the direction of the resultant of the (emerging) two components of the elliptic vibration whose phase difference has been compensated by the compensator at the cross-wires. The tangent of the angle which the principal section by the analyzer makes with the optic axis of the quartz wedge gives the ratio of the axis of the elliptic vibration, which is

$$\tan \phi = \frac{a}{b}$$

11.29 DETERMINATION OF DIRECTIONS AND RATIO OF THE MAJOR AND MINOR AXES OF ELLIPTICALLY POLARIZED LIGHT

The elliptically polarized light is examined by a Nicol prism, which is rotated until the transmitted intensity is a maximum. In this setting of the analyzer, its principal section is parallel to the major axis of the ellipse. Accordingly, the shorter diagonal of the end face of the Nicol is parallel to the major axis of the ellipse. Obviously, its minor axis is parallel to the longer diagonal of the end face. This setting of the analyzer is left undisturbed, and a quarter-wave plate is placed in front of it in such a way that the optic axis is parallel to the shorter diagonal of the end face of Nicol and therefore parallel to the major axis of the ellipse. Suppose, for example, that the optical vector of the elliptically polarized light rotates in the anticlockwise direction. Then, the light emerging from the quarter-wave plate is a plane polarized wave. The plane of vibration of this wave is parallel to the diagonal of the rectangle that lies in the second and fourth quadrants as in Figure 11.36 (a). The shorter diagonal of the end face of Nicol (which determines the transmission axis of the analyzer) is initially parallel to the optic axis of the quarter-wave plate. The transmitted light is extinguished by rotating the analyzer in the anticlockwise direction through an angle equal to $(90 - \theta)$, where θ is the angle which the plane of vibration of the plane polarized emergent wave makes with the optic axis of the plate. If the rotation of Nicol is β as measured by the help of a graduated circle and vernier, then

$$\beta = 90 - \theta \text{ or } \theta = 90 - \beta$$

The tangent of the angle θ so determined gives the ratio of the minor axis to the major axis of the ellipse.

$$\tan \theta = \frac{b}{a}$$

11.30 CHARACTER OF NATURAL LIGHT

We have explained that vibrations in unpolarized light can be resolved into transverse linear components in any two arbitrary mutually perpendicular directions. The average amplitudes of two components are equal, and these components have no definite mutual phase relationship. So, they must be considered to be incoherent—the phase difference between them is constantly changing in quite random fashion. Therefore, at any given instant, these components correspond to a particular elliptically polarized light. Due to incoherent components of unpolarized light, we can regard it as elliptically polarized light, in which the shape of the ellipse as well as positions of its axes change rapidly and irregularly with time, circularly polarized and plane polarized light being included as special cases.

Example 11.1

A beam of linearly polarized light is changed into circularly polarized light by passing it through a slice of crystal 0.003 cm thick. Calculate the difference in the refractive index of the two rays in the crystal assuming this to be the minimum thickness that will produce the effect and that the wavelength is 6×10^{-5} cm.

Solution:

It will be recalled that when the least phase difference between the O and E waves on emergence from the doubly refracting plate is $\dfrac{\pi}{2}$ and the incident linear vibrations are inclined at $45°$ to the optic axis, then the O and E waves recombine to form circularly polarized light. Obviously, the crystal plate is a quarter-wave plate, and its thickness is computed from Equation (11.21), which when solved for $\left(\mu_o - \mu_E \right)$

$$\mu_o - \mu_E = \frac{\lambda}{4\pi} = \frac{6 \times 10^{-5}}{4 \times 0.003} = 0.005.$$

Example 11.2

Plane polarized light is incident on a piece of quartz cut parallel to the axis. Find the least thickness for which the ordinary and extraordinary rays combine to form plane polarized light given that

Solution:

$$\mu_o = 1.5442, \ \mu_E = 1.5533, \ \lambda = 5 \times 10^{-5} \text{ cm}.$$

It will be recalled that if the least phase difference between the ordinary and extraordinary waves in transmission through the quartz plate is λ, they combine to form plane polarized light. Obviously, the quartz plate is a half-wave plate, and its thickness is computed from Equation (11.22).

$$t = \frac{\lambda}{2(\mu_E - \mu_o)} = \frac{5 \times 10^{-5}}{2(1.5533 - 1.5442)} = 0.00274 \text{ cm.}$$

Example 11.3

Quartz has a refractive index of 1.554 for the ordinary ray and 1.553 for the extraordinary ray when measured with sodium light. What thickness of quartz between a crossed polarizer and analyzer will produce annulment of the light, the quartz being cut parallel to the optic axis?

Solution:

It will be recalled that no light is transmitted when the principal section of the analyzer is perpendicular to that of the polarizer. If the quartz is of such a thickness that the phase difference between the O and E waves is 2π in transmission through the plate, then on emergence they combine to form linearly polarized light, the plane of vibration coinciding with that of the incident linearly polarized light. Obviously, this light will be again extinguished by the analyzer. To compute the thickness of the quartz plate, we substitute $\delta = 2\pi$ in the equation

$$\delta = \left(\frac{2\pi}{\lambda}\right)(\mu_E - \mu_o)t \text{ and solve it for } t.$$

$$t = \frac{\lambda}{\mu_E - \mu_o} = \frac{5896 \times 10^{-8}}{1.553 - 1.544} = 0.0065 \text{ cm.}$$

Example 11.4

A quartz plate is a half-wave plate for light whose wavelength is λ. Assuming that the variations in the indices of refraction with wavelength can be neglected, how would this behave with respect to light of wavelength $\lambda_1 = 2\lambda$?

Solution:

The thickness of the half-wave plate is computed from Equation (11.20)

$$t = \frac{\lambda}{2(\mu_o - \mu_E)}$$

which can be easily rewritten as

$$t = \frac{2\lambda}{4(\mu_o - \mu_E)} = \frac{\lambda_1}{4(\mu_o - \mu_E)}$$

which is identical to Equation (11.21). The given plate, therefore, behaves like a quarter-wave plate for light of wavelength λ_1.

11.31 EXERCISES

1. (a) What is meant by the term "polarization of light"?
    (b) Comment on the statement "polarization requires that vibrations are transverse."

2. (a) What is Brewster's law? Show that at Brewster's angle, the reflected ray is perpendicular to the refracted ray.
    (b) Explain how you would obtain and detect a beam of plane polarized light by reflection. How does this prove the transverse character of light vibrations?

3. Describe the construction and use of a Nicol's prism and explain how it produces plane polarized light. Would a similar prism prepared from quartz serve a similar purpose?

4. Describe the polarization of light by scattering. Explain:
    (a) Blue of the sky.
    (b) Red color of sunset.

5. Describe different methods for producing plane polarized light.

6. A ray of light is incident on the surface of a glass plate of refractive index 1.5 at the polarizing angle. Calculate the angle of refraction of the ray.

7. When the angle of incidence on a certain material is 60°, the reflected light is almost completely polarized. Find the refractive index of the material.

8. Discuss the phenomenon of superposition of two rectangular S.H. wave Vibrations of the same period and show how it may be used for the production of linearly, circularly, and elliptically polarized lights.

9. (a) What is meant by plane polarized, circularly polarized, and elliptically polarized lights?
    (b) Show that plane and circularly polarized lights are special cases of elliptically polarized light.

10. Describe, giving relevant theory, how linearly, circularly, and elliptically polarized lights are produced and detected.

11. How will you find whether a beam of light is plane polarized, circularly polarized, or elliptically polarized?

12. What is circularly polarized light? How is it produced? How can you distinguish between circularly polarized and unpolarized lights?

13. What is elliptically polarized light? How is it produced? How will you distinguish between elliptically polarized light and a mixture of plane and unpolarized lights?

14. (a) Explain the construction and use of a quarter-wave plate and a half-wave plate.
    (b) If a quarter-wave plate and a half-wave plate are given to you, how would you proceed to distinguish them from each other?

15. Give the construction of the Babinet's compensator and explain how you would use it to analyze elliptically polarized light.

16. Calculate the thickness of a quarter-wave plate of quartz for sodium light ($\lambda = 5893$ A.U.). The refractive indices of quartz for the E and O waves are 1.5533 and 1.5442 respectively.

17. Calculate the thickness of a doubly refracting crystal required to introduce a path difference of $\dfrac{\lambda}{2}$ between the ordinary and extraordinary rays when $\lambda - 6000$ A.U., $\mu_o = 1.55$, and $\mu_o = 1.54$.

18. Calculate the thickness of a calcite plate which would convert plane polarized light into circularly polarized light, given $\mu_o = 1.658$, $\mu_e = 1.486$, and $\lambda = 5890$ A.U.

19. (a) Plane polarized light falls normally on a quarter-wave plate. Explain what the nature of the emergent light will be if the plane of polarization of the incident light makes the following angles with the principal plane of the quarter wave plate:

    $0°, 30°, 45°, 90°$

    (b) Elliptically polarized light falls normally on a quarter-wave plate. Explain the nature of the emergent light if the major axis of the ellipse makes the following angles with the principal plane of the quarter-wave plate:

    $0°, 30°, 90°$

20. Show that the resultant of two coherent beams of elliptically polarized light is in general another beam of elliptically polarized light.

21. How may a quarter-wave plate by made to produce a
(a) right-handed and (b) left-handed circularly polarized light.
How may a Nicol prism and a quarter-wave plate be used to distinguish between right-handed and left-handed circularly polarized lights?

22. How may a doubly refracting plate be made to change (a) a right-handed elliptically polarized beam into a left-handed beam (b) a right-handed circularly polarized light into left-handed beam?

23. (a) How will you produce right-handed circularly polarized light?
(b) How will you distinguish between elliptically polarized light and partially polarized light?
(c) Two nicols are oriented with their principal planes making an angle of 30°. What percentage of incident polarized light will pass through the system?

24. A parallel beam of plane polarized light of wavelength 5890 A.U. (in a vacuum) is incident on a quartz crystal. Find the wavelengths of the ordinary and extraordinary waves in the crystal ($\mu_o = 1.5418$, $\mu_E = 1.5508$).

LASERS

Chapter Outline

12.0 INTRODUCTION

As we know light is an electromagnetic wave consisting of electric ($\vec{E}$) and magnetic ($\vec{B}$) fields oscillating at right angles to each other and to the direction of propagation of the wave traveling at a velocity of 3×10^8 m/sec. Vectors $\vec{E}$ and $\vec{B}$ are of equal significance to the wave. However, photochemical, photoelectrical, and several other effects are mainly due to the electric component. The actual light wave may be treated as a collection of plane monochromatic polarized waves of various frequencies, direction of propagation, and polarization planes. Light is defined as a stream of tiny parcels (quantum) of energy, called "photons" while explaining quantum phenomena. Photons have zero rest mass and zero electric charge. Photons cannot be pasteurized, and they can only be defined by knowing their energy (E) and direction of motion or, in other words, their momentum $\vec{p}$. The direction of $\vec{p}$ indicates the direction of travel of the photon and the modulus of $\vec{p}$ gives the energy of the photon divided by the velocity of light in vacuum, that is

$$p = \frac{\varepsilon}{c} \tag{12.1}$$

In addition, polarization of the photon should also be specified. The photon energy

$$\varepsilon = h\upsilon \tag{12.2}$$

in terms of its frequency υ where h is Planck's constant having value 6.626×10^{-34} Js. Putting Equation (12.2) in (12.1) we get

$$p = \frac{h\upsilon}{c} = \frac{h}{\lambda} \tag{12.3}$$

where λ is the wavelength of light. Equations (12.2) and (12.3) reflect the dual nature of radiation, that is, energy ε and momentum $\vec{p}$ of a quantum, and frequency υ and wavelength λ of a wave. Both the descriptions of light are complementary to each other and are connected mathematically by Planck's constant h.

The more photons are in a given state, the higher the probability that new photons will occupy this particular state. The boson nature of the photon statistics is of extreme importance for optical phenomena.

12.1 QUANTUM TRANSITIONS IN ABSORPTION AND EMISSION OF LIGHT

The energy of an atom or molecule can take on only definite (discrete) values E_0, E_1, E_2, ..., E_n. These are the energy levels of the atom (molecule). Any species cannot remain in an excited state for an indefinitely long time. Sooner or later, it inevitably comes down to a less excited state. This time is called the "'lifetime" of the energy level. The transition of an atom or molecule from one energy level to another occurs in a jump and is called the *quantum transition*. Quantum transitions may be induced by various causes. In particular, they can occur when atoms interact with optical radiation.

Absorption of Light

Consider two energy levels of an atom. Let the energy of the lower level be E_1 and that of the upper level E_2. Assume that the atom is in the lower level, and a photon of energy $(E_2 - E_1)$ is incident on the atom. The atom can absorb this photon and rise from level E_1 to E_2, thus making a transition by absorption of a quantum of light.

Stimulated Emission of Light

When the atom lies in the upper energy level, then the same incident photon may play the role of a trigger (stimulus) and induce the transition from E_2 to E_1, and the atom falls into the lower level. The transition causes an emission of a photon. Both the stimulating and stimulated photons have the same energy $(E_2 - E_1)$.

Moreover, both of them have identical direction of their momenta and identical polarization. In other words, the secondary photon finds itself in the same state as the primary photon. This result is a sequel of the bosonic behavior of photons—they tend to accumulate in the same state. This is the phenomenon of *stimulated emission*. Stimulated emission is illustrated in Figure 12.1.

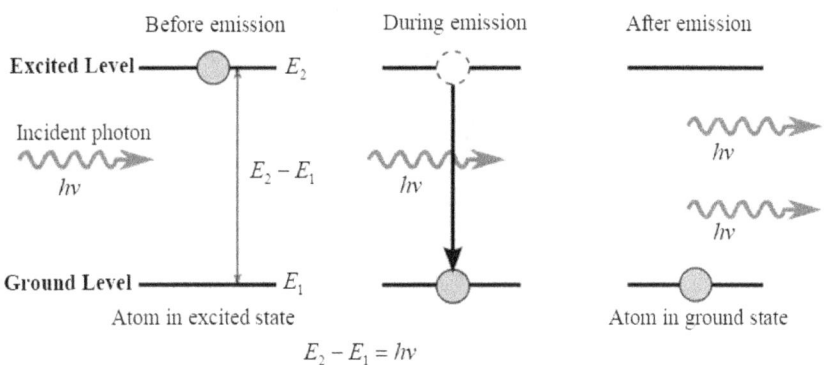

FIGURE 12.1. Stimulated emission.

The more the primary photons incident, the higher the probability that the atom lying in level E_2 will be forced to undergo a transition to E_1. Thus, there is a certain similarity between the stimulated emission and absorption processes, namely, the probabilities of both processes are proportional to the number of primary photons.

Thus, if the atom is in ground state E_1, a photon of energy $(E_2 - E_1)$ leads to the absorption process, whereas if the atom is in the higher level E_2, the same photon, with the same probability as in the absorption process, stimulates the atom to fall from E_2 to E_1 with the emission of another photon.

If there are many atoms in level E_2 of the material, an e.m. wave of the energy $(E_2 - E_1)$ is capable of stimulating $E_2 \rightarrow E_1$, transitions in many atoms; that is, the primary photon may initiate an avalanche of secondary photons. All of them will be emitted in the same state as the stimulating primary photon.

Spontaneous Emission

An atom lying in level E_2 will tend to decay to level E_1 spontaneously, that is, without any stimulus, say, in the form of a photon applied to it from outside. The photon emitted in the spontaneous $E_2 \rightarrow E_1$ emission has the energy $(E_2 - E_1)$, while its other characteristics—momentum direction, polarization—are arbitrary.

The unpredictable release of photon energy by an atom is called *spontaneous emission*. Spontaneous emission is illustrated in Figure 12.2.

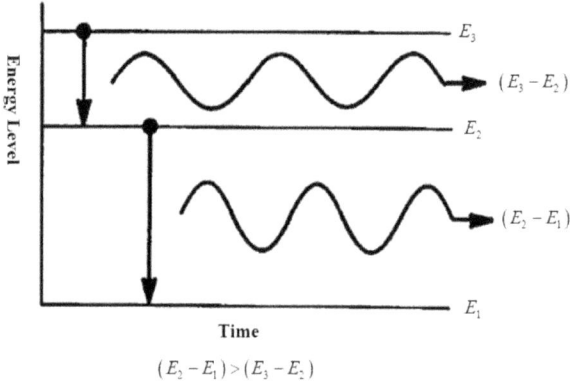

FIGURE 12.2. Spontaneous emission.

The probability of the spontaneous emission of a photon is determined only by the properties of the transition. Thus, there are two types of emission of light, namely, stimulated and spontaneous emission. The former may be viewed as a controllable process, as it is stimulated by the primary photon, which not only induces the transition but also governs the characteristics of the new emitted photon. The second process proceeds spontaneously and is random in nature. The instant of transition and direction of the emitted photon are both random. Strictly speaking, an element of chance is present in the process of stimulated emission too, as the primary photon may or may not initiate the transition. Therefore, the probability of transition processes are the quantities which are considered for writing an equation related to transitions.

Generation of Coherent Light Waves

Any light wave is characterized by a certain degree of coherence, however small it may be. Coherent waves can be obtained by division of amplitude or division of wavefront as seen earlier.

The pertinent questions are how can sufficiently coherent waves be generated and how can the degree of coherence of a wave be increased. The answer lies in the necessity of the photons to occupy only selected states. It means that the photon population is a controlled characteristic.

Such control is possible owing to the "boson nature" of photons, that is, their tendency to populate predominantly just those states that already have a sufficiently high population density.

Competition between Absorption and Stimulated Emission

Earlier we have seen that two different and competitive processes of absorption and stimulated emission are possible when photons with energy $(E_2 - E_1)$ interact with active centers. An active medium is a solid, liquid, or gas containing atoms, ions, or molecules which are capable of decaying from their higher energy states in a radiative manner, that is, by emitting electromagnetic waves. These atoms (ions or molecules) are sometimes called "active species," or "active centers." The active medium is the heart of any laser. The greater the number of active centers in the appropriate initial state, the higher the probability of each of these two processes. If the number of active centers on level 1 is greater than that on level 2, absorption becomes a more probable process. If on the contrary there are more active centers on level 2, stimulated emission becomes more probable.

Consequently, population inversion between levels 1 and 2 is a necessary condition for stimulated emission to exceed absorption.

Each photon emitted by inevitable spontaneous transitions $2 \rightarrow 1$ may be absorbed or may stimulate the emission of new photons. The emission of new photons predominates in a medium with an inverted population of levels 1 and 2. Thus, spontaneously emitted photons trigger numerous stimulated emissions. "Spontaneous" photons are independently emitted and therefore have arbitrary direction of their momenta and therefore produce corresponding stimulate emissions. Directional selectivity can be achieved by preparing the active medium in the shape of long rod (in the case of a solid medium) or a long tube (in the case of a gaseous medium) with a comparatively small cross-section. Spontaneous photons with momentum parallel to the rod's axis are favored, since they can interact with a higher number of active centers and therefore stimulate an intensive avalanche of stimulated photons. On the other hand, those spontaneous photons with momenta at an angle to the rod's axis will soon leave the active medium. The path within the active medium traveled by favored photons may be increased still further by placing partially reflecting mirrors at the end faces of the rods. The mirrors reflect the radiation back and therefore enhance the effect of stimulated emission. From here originates the idea of an optical resonator, an essential element of a laser. The selectivity of photon energies can also be facilitated by appropriate optics.

12.2 THE LASER

A Light Amplification by Stimulated Emission of Radiation (laser) is a device that emits light through a process of optical amplification based on the stimulated emission of electromagnetic radiation. A laser is triggered by initiating its pumping system. This system provides excitation of active centers, and an inverted population of lasing level builds up. The optical resonator (together with some additional elements, namely, the frequency selective element) provides selectivity of direction, polarization, energy, and so forth. As a result, a highly coherent radiation, called laser radiation, appears along the axis xx'.

The active medium and the additional elements are located inside the optical resonator. Due to the resonator, the emitted radiation propagates along the axis xx'. Note that a laser can emit radiation both in a single direction and in two opposite directions along the resonator axis.

(i.) Active Medium and Methods of Excitation

The following active media are used in lasers:

a. gases and mixture of gases (gas lasers).

b. crystal and glasses doped by special ions (solid-state lasers).

c. liquids, e.g., dyes (dye-laser).

d. semiconductors (semiconductor laser).

Normally, the active medium of a gas laser is a mixture of several gases; atoms or molecules of one of them are active centers, while other gaseous components serve to produce population inversion on the lasing levels of the active centers. One possible mixture, for example, is helium and neon. Neon atoms are active centers. Helium helps in excitation and thus population inversion in Ne. The mixture is placed in a gas discharge tube at low pressure: neon at a pressure of about 10 Pa and helium at about 100 Pa.

The active medium of a solid-state laser is normally a rod with circular cross-section doped with special ions that play the role of active centers. A classical example of a solid lasing medium is a ruby rod, 3 to 20 mm in diameter and 5 to 30 cm long. Ruby is crystalline alumina (Al_2O_3) doped with chromium ions (from 0.05 to 0.5%). Note that it is this impurity (dopant) that gives ruby its typical color (from pink to deep red).

Systems of pumping vary and to a large extent depend on the type of active medium. Excitation in gas lasers is realized in the simplest manner by means of electric discharge in the active medium. In this case the energy of excitation is transferred to active centers as a result of collisions with particles in the gas discharge plasma.

Excitation in solid-state lasers is carried out by irradiating the lasing rod with light from a sufficiently powerful light source such as an optical flash lamp. In this case active centers are excited owing to the absorption of photons emitted by the pump lamp. Thus, lasers with optical pumping can be considered as converters of optical radiation, converting, for instance, the incoherent radiation of a pulsed lamp into a highly coherent laser emission.

(ii.) Optical Resonator

An optical resonator is realized as a system of mirrors. Figure 12.3 shows (a) a linear resonator, (b) a confocal resonator, and (c) a resonator consisting of a plane mirror and a concave mirror. Mirrors may be coated with dielectric or metal layers. At least one of the mirrors must be partially transparent with respect to the emitted radiation. Resonator mirrors of gas lasers are usually mounted at both ends of the gas discharge tube and are not linked to it in a rigid manner. Mirrors in solid-state lasers are typically formed on specially prepared end faces of the active medium rod. An optical resonator defines the direction in which radiation is emitted.

FIGURE 12.3. (a) a linear resonator, (b) a confocal resonator, and (c) a resonator consisting of a plane mirror and a concave mirror.

When radiation inside a real laser is considered, it is necessary to take into account not only stimulated emission and photon absorption processes causing radiation losses such as scattering, diffraction, and so on. Laser oscillation is sustained only if stimulated emission by active centers compensates not only for the absorption by these centers but for all other losses as well.

The planes of the end faces of gas-laser discharge tubes are usually tilted at an angle so that the normal to the end face and the optical axis are at the Brewster angle corresponding to the refractive index of the material of which the end face window is made, as shown in Figure 12.4.

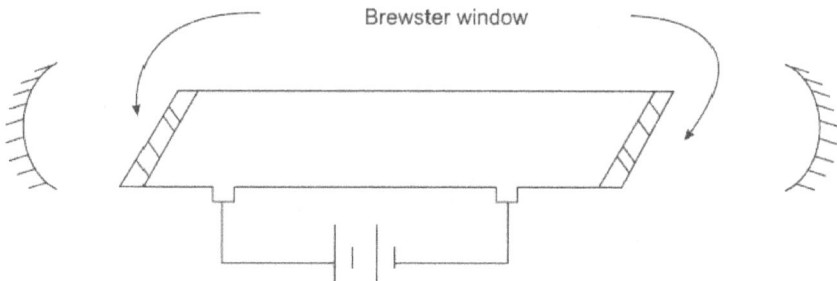

Brewster window

FIGURE 12.4. A typical He–Ne–Laser with external mirrors. The ends of the discharge tube are fitted with Brewster windows.

The end face of the discharge tube is oriented at the Brewster angle in order to realize the conditions of polarization selectivity of emission. We should recall that when an unpolarized light wave is incident on the plane-parallel window (end face) of the tube.

The light polarized in the plane perpendicular to the plane of incidence is reflected, while that with parallel polarization is transmitted. Thus, the former is lost, and the latter is repeatedly reflected by the resonator mirrors and therefore passes repeatedly through the active medium and gets amplified. Thus, we obtain plane-polarized laser emissions.

(iii) Basic Modes of Laser Oscillation

The three basic modes of laser oscillation are continuous oscillation, pulsed mode of free oscillation, and pulsed mode with controlled losses (e.g., Q – switching). The continuous mode is typical for gas lasers, and pulsed mode is mostly employed in solid-state lasers. However, if necessary, any type of laser can be made to operate in any of these modes of oscillation.

(iv) The Wonderful Laser Beam

Imagine a helium-neon laser operating in a dark room. The rich red color of the beam is a wonderful sight in the semidarkness of the room. There is no divergence (widening), and intensity is practically constant. One can place a number of reflecting mirrors in its way and make the beam trace a zig-zag path in the room, and the look is magnificent. If the beam diameter is magnified using a lens and then it falls on a screen such as a sheet of paper, an unusual light spot is observed which "speckles," and dark and bright spots appear and vanish. Such a behavior of the laser beam is exclusively due to its high degree of coherence. Light and dark specks appear because of the interference of coherent beams reflected to the observer's eyes from different points of the spot. Slight unconscious motions of the observer's head change the angle at the parts of the spot which are seen and modify the conditions of interference, changing the bright spots into dark ones and vice versa.

Characteristics of a Laser

The significant feature of a laser is the enormous difference between the character of its light and the light from other sources such as the sun, a flame, or an incandescent lamp. The characteristic features of lasers are (a) directionality, (b) high intensity, (c) monochromaticity, and (d) high degree of coherence.

Directionality

A directional beam from any conventional light source can be obtained with the help of an aperture. Lasers emit in one direction only. The directionality is expressed as "full angle beam divergence." It is twice the angle that the outer edge of the beam makes with the axis of the beam. The outer edge of a beam is defined as a point at which the strength of the beam has dropped to $\left(\dfrac{1}{e}\right)$ times its value at the center. For a typical laser, the beam divergence is less than 1 milli-radian.

Intensity

From an ordinary incandescent lamp, light spreads more or less uniformly in all directions. The laser gives a narrow light beam of very small diameter, 1 mm or less. Thus, energy is concentrated in a small region. This concentration of energy, both spatially and spectrally, accounts for the high intensity of lasers. Even a 1 mW laser can damage the eyes if one looks directly into it. The emission of photons/sec in lasers is around $10^{16} - 10^{28}$, whereas in thermal sources it is $\sim 10^{12}$.

Monochromaticity

The light emitted by a laser is far more monochromatic than that by any conventional monochromatic source. However, no light source including lasers is perfectly monochromatic. Only better and better approximation to the ideal can be sought. Figure 12.5 shows an ideal monochromatic source of light with a group of photons in exactly one frequency, while Figure 12.6 shows the linewidth of the light emitted from a laser and a conventional light source.

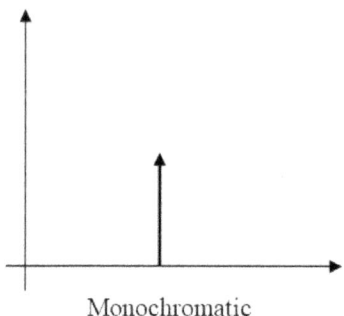

Monochromatic

FIGURE 12.5. Ideal monochromatic source of light with a group of photons in exactly one frequency.

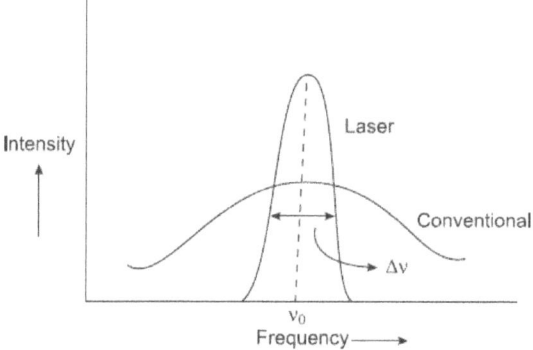

FIGURE 12.6. Linewidth of the light emitted from a laser and a conventional light source.

The degree of non-monochromaticity of a wave may be defined as relative bandwidth $\left(\dfrac{\Delta v}{v_0}\right)$, where v_0 is the central frequency of emission and Δv is full width at half maximum (FWHM) of the laser line describing the spread of frequencies about the mean value.

Coherence

Laser radiation has a high degree of coherence, and therefore laser radiation can be focused to a very small spot creating high power density, namely 10^{13} watts in a spot of 1 μm diameter. This facilitates cutting of metals, microwelding, microdrilling through diamond crystals, and so on. The coherence time τ of laser radiation may be as high as 10^{-3} sec. Therefore, the coherence length τc may be as long as 10^5 m up to 100 km. This value is seven orders of magnitude higher than the coherence length of ordinary light sources. Interference patterns in experiments such as the Fresnel biprism, Michelson interferometer, Newton's ring, and thin film could be observed in the pre-laser era, because the path length difference was very small—in the order of a millimeter at most. Figure 12.7 shows an ideal monochromatic source of light with a group of photons in the same relative phase.

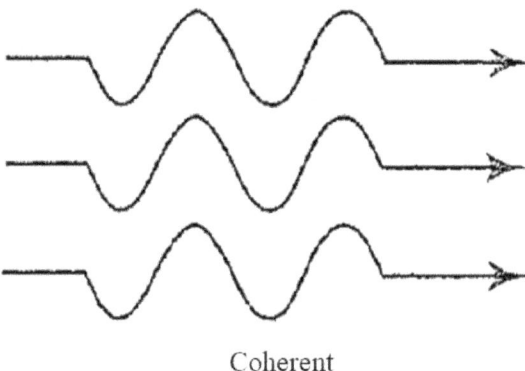

Coherent

FIGURE 12.7. Ideal monochromatic source of light with a group of photons in the same relative phase.

The Einstein Coefficients

We consider two levels of an atomic system as shown in Figure 12.8, and let N_1 and N_2 be the number of atoms per unit volume present in the energy levels E_1 and E_2 respectively.

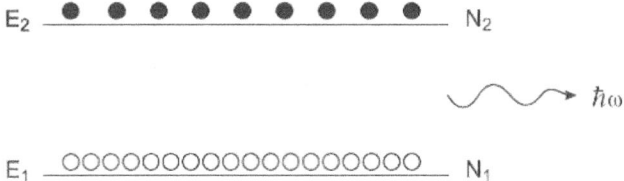

FIGURE 12.8. Two states of an atom with energies E_1 and E_2; their corresponding population densities are N_1 and N_2 respectively. At thermal equilibrium, $N_2 < N_1$ and $\dfrac{N_2}{N_1} = e^{-\frac{(E_2 - E_1)}{k_B T}}$.

If radiation at a frequency corresponding to the energy difference $(E_2 - E_1)$ falls on the atomic system, it can interact in three distinct ways:

1. An atom in the lower energy level E_1 can absorb the incident radiation and be excited to E_2. This excitation process requires the presence of radiation. The rate at which absorption takes place from level 1 to level 2 will be proportional to the number of atoms present in level E_1 and also the energy density of the radiation at the frequency $\omega = \dfrac{(E_2 - E_1)}{\hbar}$. Thus, if $u(\omega)$ represents the radiation energy per unit volume between ω and $\omega + d\omega$, then we may write the number of atoms undergoing absorptions per unit time per unit volume from level 1 to level 2 as

$$\Gamma_{12} = B_{12} u(\omega) N_1 \tag{12.4}$$

where B_{12} is a constant of proportionality and depends on the energy levels E_1 and E_2. Notice here that $u(\omega)$ has the units of energy density per frequency interval.

2. For the reverse process, namely the deexcitation of the atom from E_2 to E_1, Einstein postulated that an atom can make a transition from E_2 to E_1 through two distinct processes, namely stimulated emission and spontaneous emission. In the case of stimulated emission, the radiation which is incident on the atom stimulates it to emit radiation, and the rate of transition to the lower energy level is proportional to the energy density of radiation at the frequency ω. Thus, the number of stimulated emissions per unit time per unit volume will be

$$\Gamma_{12} = B_{21} u(\omega) N_2 \tag{12.5}$$

where B_{21} is the coefficient of proportionality and depends on the energy levels.

3. An atom which is in the upper energy level E_2 can also make a spontaneous emission; this rate will be proportional to N_2 only, and thus we have for the number of atoms making spontaneous emissions per unit time per unit volume

$$U_{12} = A_{21} N_2 \tag{12.6}$$

At thermal equilibrium between the atomic system and the radiation field, the number of upward transitions must be equal to the number of downward transitions. Hence, at thermal equilibrium

$$N_1 B_{12} u(\omega) = N_2 A_{21} + N_2 B_{21} u(\omega)$$

or

$$u(\omega) = \frac{A_{21}}{\left(\dfrac{N_1}{N_2}\right) B_{12} - B_{21}} \tag{12.7}$$

Using Boltzmann's law, the ratio of the equilibrium populations of levels 1 and 2 at temperature T is

$$\frac{N_2}{N_1} = e^{-\frac{(E_2 - E_1)}{k_B T}} = e^{\frac{\hbar\omega}{k_B T}} \tag{12.8}$$

Where $k_B = 1.38 \times 10^{-23}$ J/K is the Boltzmann's constant. Hence

$$u(\omega) = \frac{A_{21}}{B_{12} e^{\frac{\hbar\omega}{k_B T}} - B_{21}} \tag{12.9}$$

Now according to Planck's law, the radiation energy density per unit frequency interval is given by

$$u(\omega) = \frac{\hbar\omega^3 n_0^3}{\pi^2 c^3} \frac{1}{e^{\frac{\hbar\omega}{k_B T}} - 1} \tag{12.10}$$

where c is the velocity of light in free space and n_0 is the refractive index of the medium.
Comparing Equations (12.9) and (12.10), we obtain

$$B_{12} = B_{21} = B \tag{12.11}$$

and

$$\frac{A_{21}}{B_{21}} = \frac{\hbar\omega^3 n_0^3}{\pi^2 c^3} \tag{12.12}$$

Thus, the stimulated emission rate per atom is the same as the absorption rate per atom, and the ratio of spontaneous to stimulated emission coefficients is given by Equation (12.12). The coefficients A and B are referred to as the Einstein A and B coefficients.

At thermal equilibrium, the ratio of the number of spontaneous to stimulated emissions is given by

$$R = \frac{A_{21}N_2}{B_{21}N_2 u(\omega)} = e^{\frac{\hbar\omega}{k_B T}} - 1 \tag{12.13}$$

Thus, at thermal equilibrium at a temperature T, for frequencies $\left(\omega \gg \dfrac{k_B T}{\hbar}\right)$, the number of spontaneous emissions far exceeds the number of stimulated emissions.

Example 12.1

Let us consider an optical source at $T = 1000\ K$. Determine the ratio of the number of spontaneous to stimulated emissions.

Solution:

At this temperature,

$$\frac{k_B T}{\hbar} = \frac{1.38 \times 10^{-23}\ (J/K) \times 10^3\ (K)}{1.054 \times 10^{-34}\ (Js)} \approx 1.3 \times 10^{14}\ \text{Hz}$$

Thus for $\omega \gg 1.3 \times 10^{14}$ Hz, the radiation would be mostly due to spontaneous emissions.

For $\lambda \simeq 5000 A^{\circ}$, $\omega \approx 3.8 \times 10^{15}$ Hz and $R \approx e^{29.2} \approx 5.0 \times 10^{12}$.

Thus, at optical frequencies the emission is predominantly due to spontaneous transitions, and hence the light from usual light sources is incoherent.

12.3 TYPES OF LASERS

Different types of lasers can be classified into solid, liquid, gas, and semiconductor lasers, depending on the laser medium. The solid laser was the first one to be designed, that is, the ruby laser. Next came the He-Ne laser, a gas laser. Then came the diode laser, dye laser, chemical lasers, excimer lasers, gas dynamic lasers, free electron lasers, and so on, which have importance owing to their characteristics in terms of wavelengths, method of excitation, tunability, power levels, cost, and so forth. The frequency of a laser can also be "up-converted" or "down-converted" using standard techniques.

The following tables summarize different types of lasers and their characteristic wavelengths.

1. Solid Lasers:

Name of the Laser	Active Medium	Operating Wavelength
Ruby	$Al_2O_3 : Cr^{3+}$	694.3 nm
Calcium fluoride Uranium laser	$CaF_2 : u^{3+}$	2.49 μm
Nd–YAG laser	$Y_3Al_5O_{12} : Nd^{3+}$	1.064 μm
Nd–glass laser	$BeAl_2O_4 : Cr^{3+}$	701 to 818 nm
Titanium Sapphire laser	$Al_2O_3 : Ti$	660 to 1180 nm
Color Center lasers	KF	1.06 μm
	KCl	1.06 μm
	KCl : Tl	1.06 μm
	KCl : Li	0.514 μm
	KCl : Na	0.514 μm
	LiF	0.647 μm
	NaCl	1.06 μm
Fiber glass laser	Glass fiber: Er	1.53 μm
Dye laser (tunable)	Dicyanomethylene	610–705 nm
	Coumarin 102	460–515 nm
	Coumarin 6	506–558 nm
	Coumarin 30	485–535 nm
	Polyphenyl –2	363–410 nm
	Stilbene 1	391–435 nm
	Fluoroscein	500–550 nm
	Carbazines	600–700 nm

2. Gas lasers:

Name of the Laser	Active Medium	Operating Wavelength
He-Ne laser	He–Ne	0.633 μm (most popular)
		1.15 μm and 3.39 μm
Argon ion laser	Ar	1.15 μm and 3.39 μm
		About 25 different wavelengths in visible region between 408.9 to 686.1 nm

(continued)

(continued)

Name of the Laser	Active Medium	Operating Wavelength
Krypton in laser	Kr	406.7 to 676.4 nm
He-Cd laser	He-Cd	441.6, 352, 354 nm
Copper vapor laser	Cu	510.5 and 578.2 nm
Lead vapor laser	Pb	722.9 nm
Gold vapor laser	Au	627.8 nm
CO_2 laser	$CO_2 + N_2 + He$	10.6 and 9.5 μm
Excimer laser	F_2	153 nm
	Ar – F	193 mm
	Kr – F	248 mm
	Xe – Cl	308 mm
	Xe – F	353 mm
Nitrogen laser	N_2	337 nm
Far infrared laser	Water vapor Hydrogen cyanide and others	30 μm to 1.8 mm

3. Chemical lasers:

Name of the Laser	Active Medium	Operating Wavelength
Hydrogen chloride laser	$H + Cl_2 \rightarrow HCl^\circ$ + Cl + heat	3.77 μm
Hydrogen fluoride laser	$H + F_2 \rightarrow HF^\circ + F$	2.7–2.9 μm

4. X-ray lasers:

Name of the Laser	Active Medium	Operating Wavelength
X-ray lasers	Pumping laser and a solid target (like Se)	Discrete wavelengths from 3.56 to 46.9 nm

5. Free electron laser:

Name of the Laser	Active Medium	Operating Wavelength
Free electron laser	Free electrons moving at relativistic speed and undergoing acceleration and deceleration	Tunable continuously from X-rays to sub-millimeter range

12.4 THE HELIUM-NEON LASER

The first continuously operated gas laser was the He-Ne laser. The main problem in a gas laser is how the atoms can be selectively excited to proper levels in quantities sufficient to achieve the required population inversion. The primary mechanism for excitation used in gas lasers is by electron impact. Suppose atoms of a kind A are exited in a discharge tube, by this method, to a metastable state. Such a state can be populated appreciably at moderate electron densities. If the discharge tube also contains atoms of another kind B, whose excited state lies very close to that of the metastable state of A, a resonant energy transfer may take place and the atoms B will be lifted from their ground state to the excited state. If the rate of transfer of excitation is larger than the rate of radiative decay of the excited state, the population of the excited state of the atom B will steadily increase, and the state will be so populated that inversion may exist between it and another level.

This process was used in the He-Ne laser, the first continuously operated gas laser. The energy level schemes of He and Ne are shown in Figure 12.9.

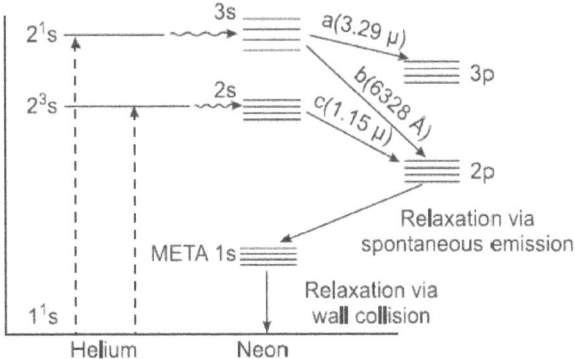

FIGURE 12.9. Energy levels of the He-Ne system.

He atoms are found to be much more readily excited by electron impact than *Ne* atoms. The atoms can be excited by d.c. or a.c. power supply. The 2^1S and 2^3S levels of *He* have a relatively long lifetime, that is, they are metastable levels. Laser action, however, occurs between energy levels of *Ne*. The role of the *He* atoms is to assist in the pumping process.

The metastable levels 2^1S at ~ 20.5 eV and 2^3S at ~ 19.81 eV of helium coincide in energy with sets of excited levels 3*s* and 2*s* respectively of neon. The notation used here to represent *Ne* levels is Paschen's notation. The 2*s* group comprises four levels denoted by $2s_2, 2s_3, 2s_4$, and $2s_5$. They may also be represented using the Russell-Saunders terminology as $1P_1, 3P_0, 3P_1, 3P_2$ respectively. However, $L - S$ coupling is not suitable for the description of the *Ne* system, as many transitions which are forbidden by the selection rules for $L - S$ coupling are actually observed.

When a helium atom in the metastable state collides with a neon atom in the ground state, an exchange of energy takes place, and a neon atom is lifted to the 2*s* or 3*s* level, and the helium atom drops back to the ground state. This provides a selective population mechanism which supplies continuously Ne atoms to 2*s* and 3*s* levels, increasing their population. The possible transitions allowed by selection rules are those to the ground state and to the 2*p* states. The decay time of *s* levels ($\tau_s = 100$ nsec) is an order of magnitude longer than the decay time of the *p* states ($\tau_p = 10$ nsec). Population inversion, therefore, can be produced between *s* and *p* states, thus satisfying the condition for operation as a four-level laser system.

Since 2*s* states are also radiatively connected to the ground state, they would decay quickly to the ground state. Is this not a disadvantage as far as the population inversion between *s* and *p* levels is concerned? It is indeed so at very low pressures, when the probability of decay to the ground state vastly exceeds that to the 2*p* levels. However, at a high pressure of the order of a mm Hg, transitions to the ground state undergo complete resonance trapping; that is, every time a photon is emitted, instead of escaping from the gas, it is absorbed by another atom in the ground state, which thereby ends up in the excited state 2*s*, and since the population of the ground state is large, the process is continuous. Thus, the high population in the 2*s* levels is maintained. The lifetime of *s* levels, therefore, is determined primarily by the radiative decay to the 2*p* levels.

Figure 12.9 shows the transitions on which laser oscillations can be expected. Of the various transitions from the component levels of the 3*s* and 2*s* groups to the levels of 3*p* and 2*p* groups, the following are more prominent:

(a) $3s_2 \rightarrow 3p_4$ $\lambda = 3.39$ μm

(b) $3s_2 \rightarrow 2p_4$ $\lambda = 0.633$ μm

(c) $2s_2 \rightarrow 2p_4$ $\lambda = 1.15$ μm

For effective working of the laser, certain conditions will have to be satisfied which put severe restrictions on the values of some operational parameters. For instance, one has to ensure that (i) $2p$ levels are not excited by inelastic collisions between electrons and the atoms in the metastable level $1s$; and (ii) the atoms in the $2s$ levels are not de-excited to the ground state. Both these processes, namely

$$e + Ne\ (1s) \rightarrow Ne\ (2p) + e \qquad (12.14)$$

and

$$e + Ne\ (2s) \rightarrow Ne + e \qquad (12.15)$$

become prominent at high current densities and upset the population inversion between $2s$ and $2p$ levels. One cannot ignore the possibility of these processes occurring, because 1s levels have no allowed downward transition and may also be radiation trapped on the transition $2p \rightarrow 1s$. Consequently, they tend to build up a sizable population on the $2p$ level. The current density, therefore, must be adjusted to its optimum value. The population of the $1s$ level also must be kept at a minimum value, and this necessitates the adjustment of the total and partial pressures of the gas mixture and also the size of the tube. The Ne pressure is usually kept much below that of He ($P_{Ne} \sim 0.1$ Torr, $P_{He} \sim 1$ Torr) and the diameter of the tube is small (~2 mm) so that 1s atoms may collide with the wall and depopulate the level. There was enough evidence to show that the measures, as previously, maximize the performance of lasers. Thus, when the discharge was switched off, the $2s$ levels were found to decay with the time variation identical to that of 2^3S states, and the amplification was found to be higher during the afterglow than during the current pulse. This is obviously because during the afterglow, no electrons are available for the excitation of 1s atoms to $2p$ levels.

The continuous wave laser oscillations in a discharge tube containing Ne and He at pressures 0.1 Torr and 1 Torr respectively were observed. The chamber consisted of a long quartz tube of length 80 cm and inside diameter 1.5 cm, as shown in Figure 12.10.

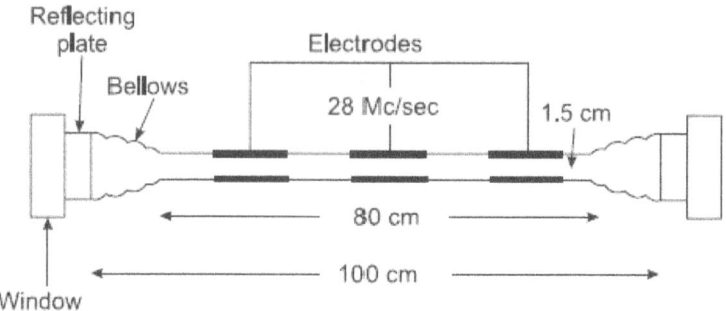

FIGURE 12.10. Schematic diagram of the first He-Ne laser.

At each end of the tube, there was a metal chamber containing high reflection plates. Flexible bellows were used in the end chambers to allow external mechanical tuning of the Fabry-Perot plates. This enabled them to align the reflectors for parallelism to within 6 seconds of arc. Two optically flat windows were provided at the ends of the system, which allowed the laser beam to be transmitted without any distortion. The end plates were separated by a distance of 100 cm. The discharge was excited by means of external electrodes using a 28 Mc/sec generator. The input power was around 50 watts. The flat plates were of fused silica and were flat to a hundredth of a wavelength. The high reflectance was achieved by means of 13-layer evaporated dielectric films. The reflectance was 98.9% in the wavelength range 11000 Å–12000 Å.

Javan et al. found that five different infrared wavelengths, namely 1.118, 1.153, 1.160, 1.199, and 1.207 μm, corresponding to the different $2s - 2p$ transitions, could be made to oscillate. The strongest oscillation was found to occur at 1.153 μm corresponding to $2s_2 \rightarrow 2p_4$ transition, with an output power of 15 mW. The beam divergence was about one minute of arc, which is close to the theoretical diffraction limited value. A screen containing nine slits with spacing of 0.125 cm and each with slit width 0.005 cm was placed across the full 1.125 cm aperture of the beam. The resulting diffraction pattern indicated that there was a very small phase variation over the aperture, which means that the beam consisted of perfectly spatially coherent light.

The difficulty of aligning the F.P. interferometer with a spacing of 100 cm is not small. Hence, an alternative system is now adopted in which spherical mirrors are employed in a confocal or near confocal configuration, as shown in Figure 12.11.

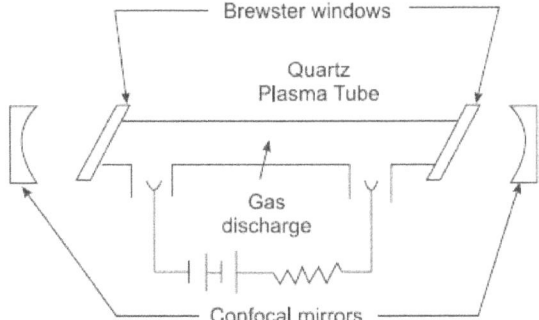

FIGURE 12.11. Confocal mirror system with Brewster angle window.

In a confocal resonator, the radius of curvature of each mirror is equal to the cavity length. The diffraction losses of such a configuration are low, and the alignment is much less critical. The divergence, however, is large, but can be reduced by placing a circular aperture in front of the exit mirror. The mirrors are placed external to the discharge tube. With such an arrangement, it is necessary to minimize unwanted reflections from the windows at the end of the discharge tube. This was found to be possible by the use of Brewster angle windows, as shown in Figure 12.11.

The Brewster angle is given by

$$\tan \theta_B = \eta \tag{12.16}$$

where η is the refractive index. An unpolarized wave incident on a plate can be considered as a resultant of two superposed plane polarized waves, one of which is polarized in the plane drawn through the normal to the window and the tube axis (plane of oscillation) and the other normal to this plane. The wave polarized normal to the "plane of oscillations" is completely reflected by the window plate and, hence, is shut out of the picture; whereas the wave polarized in the plane of oscillation is transmitted in the same direction and is repeatedly reflected by the mirrors without causing any losses. An additional advantage of this arrangement is that one obtains a plane polarized laser emission. The spectral purity of the output of such a laser is remarkably high. For brief periods, the oscillation frequency, which is around 3×10^{14} c/sec, does not vary by more than 1 c/sec.

It can be seen from Figure 12.9 that depending on certain factors, laser oscillation can also be achieved on the transitions of the types "a" and "b," and with the rapid development in the field, laser action was discovered at 3.39 μm (a) 6328 Å (b). The 6328 Å He-Ne laser is one of the most popular and most widely used lasers. The commercial model of this type of laser requires 5 to 10 W of excitation power and produces 0.5 to 50 mW of cw laser output.

The He-Ne laser is made to oscillate on a particular transition by making the multilayer dielectric mirrors in such a way that they have a maximum reflectivity at the desired wavelength. Another interesting way of obtaining laser oscillations at a particular wavelength is to use a dispersion prism, as shown in Figure 12.12.

FIGURE 12.12. Wavelength selection.

A prism placed inside the resonator provides wavelength selectivity. The light waves of different frequencies emitted by the active medium are spatially separated by the prism. If the right-hand mirror of the resonator is perpendicular to the propagation direction of, say, a λ_1 wave, then λ_2 is incident on the mirror obliquely and is not returned into the active medium after reflection. One can change the wavelength by rotating the mirror about an axis perpendicular to the plane.

Application of Lasers

Studies of interaction of laser radiation with matter are of extreme scientific significance. Lasers are widely used for fundamental studies in physics, chemistry, and biology and also in engineering sciences.

High power laser radiation can alter, in a reversible manner, the physical behavior of material leading to diverse nonlinear optical phenomena.

Lasers facilitate high concentration of light power within narrow bandwidths and frequency tuning over considerable ranges. Therefore, they are useful light sources in investigations of optical spectra of materials.

Tunable lasers facilitate selective excitation of atoms or molecules or for selective rapture of chemical bonds. This opens up the possibility for initiating the desired chemical reactions, controlling these reactions, and studying their kinetics.

Picosecond pulses available from lasers render possible the study of ultrafast processes, namely biomedicals, photosynthesis, and so forth.

The numerous fields of laser applications may be broadly categorized into two groups. One involves applications where laser beams—as a rule of high power—are exploited to produce a targeted effect on material such as welding, cutting, and so on. The other group involves data transmission and processing, measurements, and quality control.

12.5 EFFECTS OF STRONG LASER RADIATION ON MATERIALS

Using optical lenses, a laser beam can be focused into a light spot of 10–100 µm diameter on the surface of a material, thereby leading to a very high irradiance. If 1 kW output of a CO_2 laser is focused in a spot of 100 µm diameter, the resultant irradiance will be 10^7 W/cm^2. Around 10^5 W/m^2 the material starts melting. At still higher levels of irradiance, it starts boiling and an intense evaporation occurs. As the beam irradiance grows further, vapor is ionized by the light to produce high temperature plasma. The plasma may be absorptive and prevent the beam from entering the material. Accordingly, power densities realized in material working systems are decided.

For welding applications relatively low peak power but longer duration pulses (10^{-3} to 10^{-2}sec) are suitable. For perforation and hole drilling where intense evaporation is essential, intense and short pulses ($\sim 10^{-4}$ sec) are required. The wavelength of the laser is also important, as it governs the portion of light absorbed by the surface of the material. However, a beam of shorter wavelength may be focused into a smaller light spot.

Laser Welding

The welding is contactless, thereby precluding any possibility of impurity in the weldment. Unlike electron-beam welding running in a vacuum, laser welding is performed in the atmosphere. It offers a possibility to weld in inaccessible regions too. Laser welding is capable of a fast and accurate local melting at a given point or along a line. The heat affected area is very small, and this is very useful where welding is to be made in the vicinity of heat-sensitive components, namely in micro-electronics. Lasers used are Nd: YAG, Nd: glass, ruby, and CO_2. Laser welding lends itself well to automatic processes in automotive production lines, for joining titanium and aluminum sheets in ship building, for trunk pipeline construction, and so forth. Welding of nonmetallic solids is also an interesting application of lasers; for example, glasses are welded by lasers of 100 W output while quartz needs 300 W.

Laser heat treatment of surfaces enhances the strength of the treated area, for example, in the automotive industry: cylinder blocks, valves, gears, and so on. Most commonly used is a 1 kW CO_2 laser operating in continuous wave mode.

Laser Cutting

Lasers can be used to cut paper, cloth, plywood, glass, asbestos sheets, ceramics, sheet metals, and so forth in two as well as three dimensions, according to a complicated profile with easy automation of the process and high production rates. For example, 50 mm thick board may be cut with a 200 W CO_2 laser, width of the cut being 0.7 mm; sheets of plywood require a 8 kW CO_2 laser; glass of 10 mm thickness requires 20 kW lasers. Metals can be cut with 100 to 500 W power, provided the beam-heated material is blown with a jet of oxygen so that gas-laser cutting results.

Drilling and Perforating Holes

Hole perforation by lasers relies on intense evaporation of material heated by powerful light pulses of 10^{-4} to 10^{-3} sec duration and power density ~ 10^7 W/cm^2 at the material surface. Nd :YAG lasers are suited for metallic targets only. CO_2 lasers are equally suited to metallic and non-metallic (plastic, ceramic, glass) targets. Very large aspect ratios (length/diameter) are possible; for example, lasers can perforate very small diameter (0.2 mm) holes to large values of depths. Further, the advantage is the possibility of drilling holes in close vicinity with each other and near the item edge.

Lasers in Medicine

The focused laser beam proves to be a unique scalpel, capable of bloodless surgery, since the beam not only cuts but also "welds" blood vessels being cut. Such surgery is sterile because it is contact-less. It is also painless as it is very fast. It finds use in ophthalmology in treating the detachment of retinas, cataracts, and glaucoma.

Isotope Separation

Several applications in the medicine and research industries require substances enriched in a particular isotope, namely deuterium against hydrogen.

Laser separation of isotopes exploits the fact that various isotopes of an element exhibit different absorption bands in their spectra. These absorption bands are fairly narrow and lie close to each other in the spectrum. To excite one isotope without disturbing the other, one has to irradiate their mixture by a source of narrow bandwidth centered on the wavelength of the required isotope. It is desirable that this source be tunable so that the radiation may be tuned to the desired wavelength. This facility is offered by tunable lasers. Thus, the excited atoms of the required isotope are raised to the upper level, while the other isotope remains in the ground state. The mixture of isotopes is then irradiated by another radiation, which is absorbed by the excited atoms only to ionize them. As a result, the desired isotope is obtained in the form of ions easily separable by applying a dc electric field. This is essentially the idea of one of the laser separation techniques, called *two-step photo-ionization*.

Optical Communication

Besides using optical waveguides (optical fibers), laser communication in open space can be established between satellites, satellites and aircraft, and satellites and stations on the ground. Unguided communications can exist also between stations on the earth's surface and underwater.

Ranging and Measurement

Similar to conventional microwave radar, the optical radar using lasers, also called "rangefinder" by militaries, has been designed to detect distant objects and retrieve information about these objects from the radar signal reflected from these objects. High carrier frequency, possibility of extremely narrow directionality of the radiation, operation in the range of nanosecond and picosecond pulses—all these features predetermine a number of advantages of optical radar systems over ordinary radars. The high resolution of the method makes possible the determination of the size and shape of the object and its orientation.

For velocity measurements, laser rangefinders measure Doppler shift in the signal frequency due to the velocity of the ranging object collinear with the laser beam. The higher the carrier frequency, the greater the detected Doppler shift and the accuracy of velocity evaluation.

The disadvantage is that the laser wavefront is degraded and attenuated in the medium where it propagates, and then an adaptive optics method has to be used to cope with this drawback.

Interferometric distance measurements are performed using a frequency-stabilized He-Ne laser operating in the central longitudinal mode. This technique controls tool motions, automatically compensating for errors due to wear, precisely aligns fixtures in aircraft engine manufacturing, and so forth. Flow velocity measurements are also allowed using a 10 mW He-Ne laser.

The Laser Gyroscope

Such a gyroscope measures angular velocity with a precision of 10^{-3} degree/hour and is capable of tracking very small angular velocities. This precision is comparable with that available using the most sophisticated and expensive conventional gyroscopes. The laser gyros are useful for altitude control of satellites orbiting spacecrafts and aircrafts.

It involves a co-propagating and a counter-propagating laser beam which interfere, and a shift in interference fringes is used to detect the angular rotation.

Laser Monitoring of the Environment

Laser properties of directionality and monochromaticity are put to good use for measurement of the concentration of various atmospheric pollutants. These include oxides of nitrogen, carbon monoxide, sulphur dioxide, and a variety of particulate matter such as dust, smoke, and fly ash. The laser techniques perform these measurements by remotely sensing the composition of the atmosphere with a light beam without the necessity of sample collection or any chemical processing. The results of the measurements are available readily, and there is no distortion of the quantities being measured. Consequently, these techniques yield real-time data and are extremely suitable for sounding time variations of the atmosphere, that is, for environmental monitoring.

12.6 EXERCISES

1. Calculate the energy of a photon that has a wavelength of 500 nm.

2. What is the energy of a photon at the He-Ne output wavelength $\lambda = 546.8$ nm?

3. Define a laser.

4. What is the active medium?

5. What are the active media used in lasers?

6. What is the difference between stimulated and spontaneous emissions?

7. What is the energy of the photon that is spontaneously emitted between the two states $E_2 = 2.40$ eV and $E_1 = 1.0$ eV? What is the wavelength of the photon?

8. What are the four active media used in lasers?

9. Let two energy states, $E_2 = 2.30$ eV and $E_1 = 1.0$ eV, and there are 1×10^{16} electrons/cm^3 in E_1. At a temperature of 1000 K, how many electrons are in state E_2?

10. If the energy difference between the ground state and an excited state in an atom is 4 eV at a temperature of 300 K, determine the fraction of the atoms in the excited state.

11. What are the four different types of lasers?

12. Why are lasers useful light sources in investigations of optical spectra of materials?

CHAPTER 13

FIBER OPTICS AND SENSORS

Chapter Outline

13.0 INTRODUCTION

Optical fiber is the medium to transmit light as a carrier for communicating signals between the two ends of the fiber. Furthermore, optical fiber is the waveguide for light. The light can be guided through thin fibers of glass or plastic. Fiber optic cables transmit data through very small cores at the speed of light. Fiber optic cables provide high bandwidths and low losses, which allow high data-transmission rates over long distances. There are three common types of fiber optic cables: single-mode, multimode, and graded-index.

13.1 ADVANTAGES OF OPTICAL FIBERS

The following are the main advantages of optical fibers:

1. Attenuation in a fiber is markedly lower than that of coaxial cable or twisted pairs and is constant over a very wide range. So, transmission within a wide range of distance is possible without repeaters, etc.

2. Smaller size and lighter weight. Optical fibers are considerably thinner than coaxial cable or bundled twisted-pair cable. So, they occupy much less space.

3. Electromagnetic isolation. Electromagnetic waves generated from electrical disturbances or electrical noises do not interfere with light signals. As a result, the system is not vulnerable to interference, impulse noise, or cross-talk.

4. No physical electrical connection is required between the sender and the receiver.

5. The fiber is much more reliable, because it can better withstand environmental conditions such as pollution and radiation, and salt produces no corrosion. Moreover, it is nominally affected by nuclear radiation. Its life is longer in comparison to copper wire.

6. There is almost no cross-talk in optical fibers, and hence transmission is more secure and private, as it is very difficult to tap into a fiber.

7. Greater bandwidth. Bandwidth of the optical fiber is higher than that of an equivalent wire transmission line.

8. As fibers are very good dielectrics, isolation coating is not required.

9. Data rate is much higher in a fiber, and hence much more information can be carried by each fiber than by equivalent copper cables. For example: at the immensely high frequencies of an optical fiber, data rates of 2 Gbps over tens of kilometers have been demonstrated. Compare this to the practical maximum of hundreds of Mbps over about 1km for coaxial cable and just a few Mbps over 1 km for twisted pairs.

10. The cost per channel is lower than that of an equivalent wire cable system. It is expected that in the near future, the optical fiber communication system will be more economical than any communication system of other types.

11. Due to the non-inductive and non-conductive nature of a fiber, there is no radiation and interference on other circuits and systems.

12. Greater repeater spacing: Fewer repeaters indicate lower cost and fewer sources of error. It has been observed that a fiber transmission system can achieve a data rate of 5 Gbps over a distance of 111 km without repeaters, whereas coaxial and twisted-pair systems generally have repeaters every few kilometers.

13. The raw material is available in plenty.

Fiber as a Guiding Medium

Optical frequencies are extremely large ($\sim 10^{15}$ Hz) as compared to conventional radio waves ($\sim 10^{6}$ Hz) and microwaves ($\sim 10^{10}$ Hz), and a light beam acting as a carrier wave is capable of carrying far more information in comparison to radio waves and microwaves.

With the discovery of the laser, some experiments on the propagation of information-carrying light waves through the open atmosphere were carried out, but it was realized that because of the vagaries of the terrestrial atmosphere, for example, rain, fog, and so forth, in order to have an efficient and dependable communication system, one would require a guiding medium in which the information-carrying light waves could be transmitted. This guiding medium is the optical fiber, which is hair-thin and guides the light beam from one place to another.

In addition to the capability of carrying a huge amount of information, fibers fabricated with recently developed technology are characterized by extremely low losses (~ 0.2 dB/km), and as a consequence, the distance between two consecutive repeaters (used for revamping the attenuated signals) can be as large as 250 km. In a recently developed fiber optic system, it has been possible to send 140 Mbit/s information through a 220 km link of one optical fiber; this is equivalent to about 450,000 voice channels. In comparison, the copper cables used today have repeaters spaced every few kilometers or so.

In addition to the long-distance communication systems, optical fibers are also being extensively used for Local Area Networks (LANs)—networks that wire up telephones, televisions, computers, or robots in offices and cities.

13.2 TYPES OF FIBER

An optical fiber is a dielectric waveguide that operates at optical frequencies. This fiber wave guide is normally cylindrical in form. It confines electromagnetic energy in the form of light to within its surface and guides the light in a direction parallel to its axis. The transmission properties of an optical waveguide are dictated by its structural characteristics, which have a major effect in determining how an optical signal is affected as it propagates along the fiber. The structure basically establishes the information-carrying capacity of the fiber and also influences the response of the waveguide to environmental perturbation.

The propagation of light along a waveguide can be described in terms of a set of guided electromagnetic waves called the modes of the waveguide. These guided modes are referred to as the bound or trapped modes of the waveguide. Each guided mode is a pattern of electric and magnetic field lines that is repeated along the fiber at intervals equal to the wavelength. Only a certain discrete number of modes are capable of propagating along the guide; these modes are those electromagnetic waves that satisfy the homogeneous wave equation in the fiber and the boundary condition at the waveguide surfaces.

A circular solid core of refractive index n_1 is surrounded by a cladding having a refractive index $n_2 < n_1$. In principle, a cladding is not necessary for light to propagate along the core of the fiber, but it serves several purposes. The cladding reduces scattering loss resulting from dielectric discontinuities at the core surface, it adds mechanical strength to the fiber, and it protects the core from absorbing surface contaminants with which it could come in contact.

In low- and medium-loss fibers, the core material is generally glass which is surrounded by either a glass or a plastic cladding. Higher-loss plastic core fibers with plastic claddings are also widely in use. In addition, most fibers are encapsulated in an elastic, abrasion-resistant plastic material. This material adds further strength to the fiber and mechanically isolates or buffers the fibers from small geometrical irregularities, distortions, or roughness of adjacent surfaces. These perturbations could otherwise cause scattering losses induced by random microscopic bends that can arise when the fibers are incorporated into cables or supported by other structures.

Variations in the material composition of the core give rise to the two commonly used fiber types. In the first case, the refractive index of the core is uniform throughout and undergoes an abrupt change (or step) at the cladding boundary. This is called a step-index fiber.

In the second case, the core refractive index is made to vary as a function of the radial distance from the center of the fiber. This type is a graded-index fiber.

Both the step and the graded index fibers can be further divided into single-mode and multimode classes. As the name implies, a single-mode fiber sustains only one mode of propagation, whereas multimode fibers contain many hundreds of modes. Multimode fibers offer several advantages compared to single-mode fibers. The larger core radii of multimode fibers make it easier to launch optical power into the fiber and facilitate the connecting together of similar fibers.

A disadvantage of multimode fibers is that they suffer from intermodal dispersion. When an optical pulse is launched into a fiber, the optical power in the pulse is distributed over all (or most) of the modes of the fiber. Each of the modes that can propagate in a multimode fiber travels at a slightly different velocity. This means that the modes in a given optical pulse arrive at the fiber end at slightly different times, thus causing the pulse to spread out in times as it travels along the fiber. This effect, which is known as intermodal dispersion, can be reduced by using a graded-index profile in the fiber core: this allows graded-index fibers to have much larger bandwidths (data rate transmission capabilities) than step-index fibers.

Refractive Index Profiles

Figure 13.1 illustrates the fiber cross-section and ray paths for step-index fibers, graded-index fibers, and single-mode fibers.

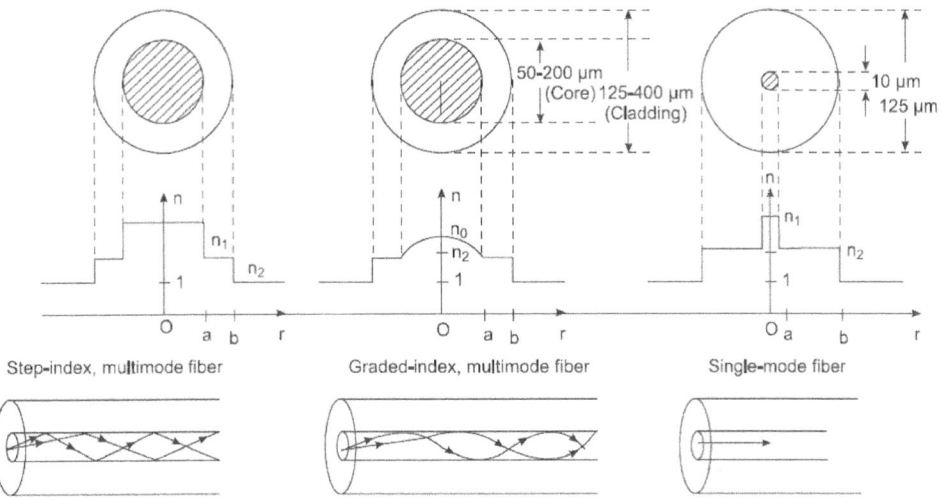

FIGURE 13.1. Fiber cross-section and ray paths.

Total Internal Reflection

A ray of light travels more slowly in an optically dense medium than in one that is less dense, and the refractive index gives a measure of this effect. When a ray is incident on the interface between two dielectrics of differing refractive indices (e.g., glass-air), refraction occurs, as illustrated in Figure 13.2(a).

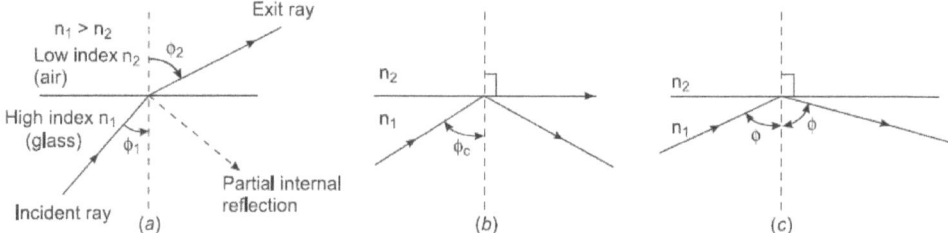

FIGURE 13.2. Internal reflection.

It may be observed that the ray approaching the interface is propagating in a dielectric of refractive index n_1 and is at an angle ϕ_1 to the normal at the surface of the interface. If the dielectric on the other side of the interface has a refractive index n_2 which is less than n_1, then the refraction is such that the ray path in this lower index medium is at an angle ϕ_2 to the normal, where ϕ_2 is greater than ϕ_1. The angles of incidence ϕ_1 and refraction ϕ_2 are related to each other and to the refractive indices of the dielectrics by *Snell's law of refraction*, which states that

$$\left(n_1\sin\phi_1 = n_2\sin\phi_2\right) \rightarrow \frac{\sin\phi_1}{\sin\phi_2} = \frac{n_2}{n_1} \tag{13.1}$$

It may also be observed in Figure 13.2 (a) that a small amount of light is reflected back into the originating dielectric medium (partial internal reflection). As n_1 is greater than n_2, the angle of refraction is always greater than the angle of incidence. Thus, when the angle of refraction is 90° and the refracted ray emerges parallel to the interface between the dielectric, the angle of incidence must be less than 90°. This is the limiting case of refraction, and the angle of incidence is now known as the critical angle ϕ_c as shown in Figure 13.2 (b).

From Equation (13.1) the value of the critical angle is given by

$$\sin\phi_c = \frac{n_2}{n_1} \tag{13.2}$$

At angles of incidence greater than the critical angle, the light is reflected back into the originating dielectric medium (TIR) with high efficiency (around

99.9%). Hence, it may be observed in Figure 13.2 (c) that total internal reflection occurs at the interface between two dielectrics of differing refractive indices when light is incident on the dielectric of lower index from the dielectric of higher index, and the angle of incidence of the ray exceeds the critical value. This is the mechanism by which light at a sufficient shallow angle (less than $90° - \phi_c$) may be considered to propagate down an optical fiber with low loss.

Acceptance Angle

The geometry concerned with launching a light ray into an optical fiber is shown in Figure 13.3, which illustrates a meridional ray A at the critical angle ϕ_c within the fiber at the core-cladding interface.

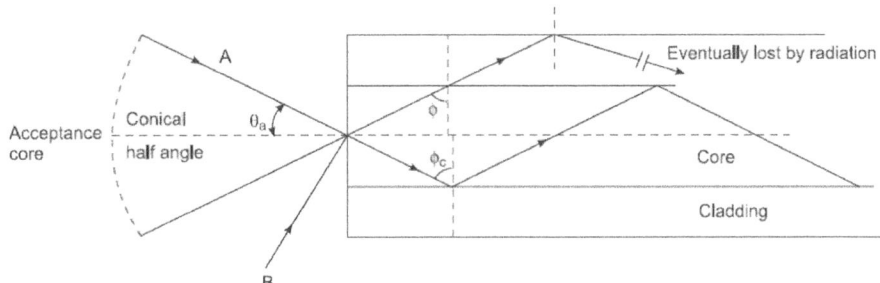

FIGURE 13.3. Acceptance angle.

It may be observed that this ray enters the fiber core at an angle θ_a to the fiber axis and is refracted at the air-core interface before transmission to the core-cladding interface at the critical angle. Hence, any rays which are incident into the fiber core at an angle greater than θ_a will be transmitted to the core-cladding interface at an angle less than ϕ_c, and will not be totally internally reflected. This situation is shown in Fig. 13.3, where the incident ray B at an angle greater than θ_a is refracted into the cladding and eventually lost by radiation. Thus, for rays to be transmitted by total internal reflection within the fiber core, they must be incident on the fiber core within an acceptance core defined by the conical half angle θ_a. Hence, θ_a is the maximum angle to the axis at which light may enter the fiber in order to be propagated, and is often referred to as the acceptance angle for the fiber.

Numerical Aperture

A light ray is incident on the fiber core at an angle θ_1 to the fiber axis, which is less than the acceptance angle for the fiber θ_a. The ray enters the fiber from a medium (air) of refractive index n_0, and the fiber core has a refractive index n_1, which is slightly greater than the cladding refractive index n_2. Assuming the entrance face at the fiber core to be normal to the axis, then considering the refraction at the air-core interface and using *Snell's law*

$$n_0 \sin\theta_1 = n_1 \sin\theta_2 \tag{13.3}$$

considering the right-angled triangle *ABC* indicated in Figure 13.4,

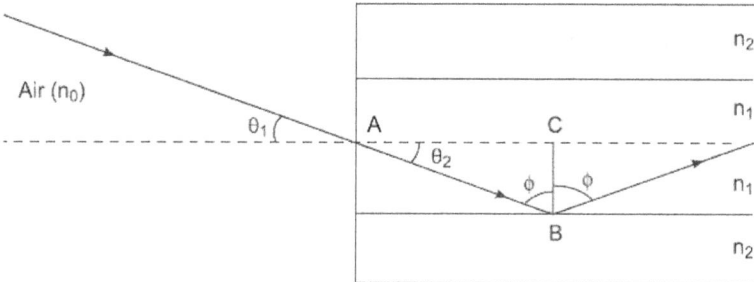

FIGURE 13.4. Snell's law and numerical aperture.

then

$$\phi = \frac{\pi}{2} - \theta_2 \tag{13.4}$$

where ϕ is greater than the critical angle at the core-cladding interface. Hence, Equation (13.3) becomes

$$n_0 \sin\theta_1 = n_1 \sin\left(-\phi + \frac{\pi}{2}\right) = n_1 \cos\phi \tag{13.5}$$

using the trigonometrical relationship $\sin^2\phi + \cos^2\phi = 1$, Equation (13.5) may be written in the form

$$n_0 \sin\theta_1 = n_1 \sqrt{(1 - \sin^2\phi)} \tag{13.6}$$

when the limiting case for total internal reflection is considered, ϕ becomes equal to the critical angle for the core-cladding interface and is given by

$$\sin\theta_c = \frac{n_2}{n_1}$$

Also in this limiting case, θ_1 becomes the acceptance angle for the fiber θ_a. Combining these limiting cases gives

$$n_0 \sin\theta_a = \sqrt{(n_1^2 - n_2^2)} \tag{13.7}$$

Equation (13.7), apart from relating the acceptance angle to the refractive indices, serves as the basis for the definition of the important optical fiber parameter, the numerical aperture (*NA*). Hence, the *NA* is defined as:

$$NA = n_0 \sin\theta_a = \sqrt{\left(n_1^2 - n_2^2\right)} \tag{13.8}$$

Since the *NA* is often used with the fiber in air where n_0 is unity, it is simply equal to $\sin\theta_a$. It may also be noted that incident meridional rays over the range $0 \leq \theta_1 \leq \theta_a$ a will be propagated within the fiber.

The numerical aperture may also be given in terms of the relative refractive index difference Δ between the core and the cladding, which is defined as

$$\Delta = \frac{n_1^2 - n_2^2}{2n_1^2} = \frac{\left(n_1 - n_2\right)\left(n_1 + n_2\right)}{2n_1^2} = \frac{\left(n_1 - n_2\right)2n_1}{2n_1^2} \simeq \frac{n_1 - n_2}{n_1} \text{ for } \Delta \ll 1 \tag{13.9}$$

Hence, combining Equation (13.8) with Equation (13.9), we can write

$$NA = n_1 \sqrt{2\Delta} \tag{13.10}$$

The relationships given in Equation (13.8) and Equation (13.10) for the numerical aperture are a very useful measure of the light-collecting ability of a fiber.

They are independent of the fiber core diameter and will hold for diameters as small as 8 μm. However, for smaller diameters they break down, as the geometric optics approach is invalid. This is because the ray theory model is only a partial description of the character of light.

Example 13.1

A silica optical fiber with a core diameter large enough to be considered by ray theory analysis has a core refractive index of 1.50 and a cladding refractive index of 1.47. Determine (a) the critical angle at the core cladding interface, (b) the NA for the fiber, and (c) the acceptance angle in air for the fiber.

Solution:

a. The critical angle ϕ_c at the core cladding interface is given by

$$\phi_c = \sin^{-1}\left(\frac{n_2}{n_1}\right) = \sin^{-1}\left(\frac{1.47}{1.50}\right) = 78.5°$$

b. The numerical aperture $NA = \sqrt{\left(n_1^2 - n_2^2\right)} = \sqrt{\left(1.50\right)^2 - \left(1.47\right)^2} = 0.30$

c. The acceptance angle in air $\theta_a = \sin^{-1}\left(NA\right) = \sin^{-1}\left(0.30\right) = 17.4°$

Example 13.2

A typical relative refractive index difference for an optical fiber designed for long distance transmission is 1%. Estimate the *NA* and the solid acceptance angle in air for the fiber when the core index is 1.46. Further, calculate the critical angle at the core cladding interface within the fiber. It may be assumed that the concepts of geometric optics hold for the fiber.

Solution:

With $\Delta = 0.01$ the numerical aperture is

$$NA = n_1\sqrt{2\Delta} = 1.46\sqrt{0.02} = 0.21$$

For small angles the solid acceptance angle in air G is given by

$$G = \pi\theta_a^2 = \pi\sin^2\theta_a = \pi(NA)^2 = \pi(0.04) = 0.13 \text{ rad}$$

The relative refractive index difference Δ;

$$\Delta \simeq \frac{(n_1 - n_2)}{n_1} = 1 - \frac{n_2}{n_1}$$

Hence

$$\frac{n_2}{n_1} = 1 - \Delta = 1 - 0.01 = 0.99$$

The critical angle at the core cladding interface is

$$\phi_c = \sin^{-1}\left(\frac{n_2}{n_1}\right) = \sin^{-1}(0.99) = 81.9°$$

13.3 CONCEPT OF MODES

The electromagnetic light field that is guided along an optical fiber can be represented by a superposition of bound or trapped modes. Each of these guided modes is composed of a set of simple electromagnetic field configurations which form a standing wave pattern in the transverse direction, that is, transverse to the waveguide axis.

Maxwell's Equations

Maxwell's equations that give the relationships between the electric and magnetic fields are:

$$\nabla \times \vec{E} = -\frac{\partial\vec{B}}{\partial t} \tag{13.11}$$

$$\nabla \times \vec{H} = \frac{\partial \vec{D}}{\partial t} \tag{13.12}$$

$$\nabla \bullet \vec{D} = 0 \tag{13.13}$$

$$\nabla \bullet \vec{B} = 0 \tag{13.14}$$

where,

$\vec{D} = \varepsilon \vec{E}$ and $\vec{B} = \mu \vec{H}$

ε is the permittivity (or dielectric constant), μ is the permeability of the medium,

$\vec{E}$ is electric field, $\vec{B}$ is magnetic flux density, $\vec{H}$ is magnetic field, and $\vec{D}$ is electric displacement field

Taking the curl of Equation (13.11) and making use of Equation (13.12) yields

$$\nabla \times \left(\nabla \times \vec{E} \right) = -\nabla \times \frac{\partial \vec{B}}{\partial t} = -\frac{\partial}{\partial t} \left(\nabla \times \vec{B} \right) = -\frac{\partial}{\partial t} \left(\nabla \times \mu \vec{H} \right)$$
$$= -\mu \frac{\partial}{\partial t} \left(\nabla \times \vec{H} \right) = -\mu \frac{\partial}{\partial t} \frac{\partial \vec{D}}{\partial t} = -\mu \varepsilon \frac{\partial^2 \vec{E}}{\partial t^2} \tag{13.15}$$

Using the vector identity $\nabla \times \left(\nabla \times \vec{E} \right) = -\nabla \left(\nabla \bullet \vec{E} \right) - \nabla^2 \vec{E} = -\nabla^2$, we get

$$\nabla^2 \vec{E} = \varepsilon \mu \frac{\partial^2 \vec{E}}{\partial t^2} \tag{13.16}$$

Similarly, by taking the curl of Equation (13.12), it can be shown that

$$\nabla^2 \vec{H} = \varepsilon \mu \frac{\partial^2 \vec{H}}{\partial t^2} \tag{13.17}$$

Equations (13.16) and (13.17) are the standard *wave equations*.

Field Components in Optical Waveguides

Now consider electromagnetic waves propagating along a cylindrical fiber shown in Figure 13.5. For this fiber a cylindrical coordinate system (r, ϕ, z) is defined with the z-axis lying along the axis of the waveguide.

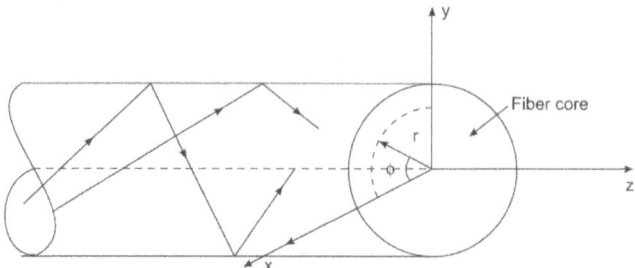

FIGURE 13.5. Electromagnetic waves propagating along a cylindrical fiber.

If the electromagnetic waves are to propagate along the z-axis, they will have a functional dependence of the form

$$\vec{E} = \vec{E}_0(r,\phi)e^{j(\omega t - \beta z)} \tag{13.18}$$

$$\vec{H} = \vec{H}_0(r,\phi)e^{j(\omega t - \beta z)} \tag{13.19}$$

For the curl in cylindrical coordinates refer to

$$\nabla \times \vec{A} = \left(\frac{1}{r}\frac{\partial A_z}{\partial \phi} - \frac{\partial A_\phi}{\partial z}\right)a_r + \left(\frac{\partial A_r}{\partial z} - \frac{\partial A_z}{\partial r}\right)a_\phi + \frac{1}{r}\left(\frac{\partial}{\partial r}(rA_\phi) - \frac{\partial A_r}{\partial \phi}\right)a_z \tag{13.20}$$

which are harmonic in time t and coordinate z. The parameter is the z-component of the propagation vector and will be determined by the boundary conditions on the electromagnetic fields at the core-cladding interface.

When Equations (13.18) and (13.19) are substituted into Maxwell's curl equations, we have, from Equation (13.11),

$$\nabla \times \vec{E} = -\frac{\partial \vec{B}}{\partial t} = -\frac{\partial}{\partial t}\mu\vec{H}_0 e^{j(\omega t - \beta z)} = -j\omega\mu\vec{H}_0 e^{j(\omega t - \beta z)} = -j\omega\mu\vec{H} \tag{13.21}$$

On transforming Maxwell's field equation in the polar coordinate system (r, ϕ, z) we get in the cylindrical system of coordinates, we have

$$\nabla \times \vec{E} = -\frac{\partial \vec{B}}{\partial t} = -j\omega\mu\vec{H} = \begin{vmatrix} \dfrac{1}{r}\hat{r} & \hat{\phi} & \dfrac{1}{r}\hat{z} \\[2mm] \dfrac{\partial}{\partial r} & \dfrac{\partial}{\partial \phi} & \dfrac{\partial}{\partial z} \\[2mm] \vec{E}_r & \vec{E}_\phi & \vec{E}_z \end{vmatrix} \tag{13.22}$$

$$= -\left(\mu\frac{\partial \vec{H}_r}{\partial t}\hat{r} + \mu\frac{\partial \vec{H}_\phi}{\partial t}\hat{\phi} + \mu\frac{\partial \vec{H}_z}{\partial t}\hat{z}\right)$$

Thus,

$$-j\omega\mu\vec{H}_r = \frac{1}{r}\frac{\partial \vec{E}_z}{\partial \phi} + j\beta\vec{E}_\phi \qquad (13.23)$$

$$j\omega\mu\vec{H}_\phi = \frac{\partial \vec{E}_z}{\partial r} + j\beta\vec{E}_r \qquad (13.24)$$

$$-j\omega\mu\vec{H}_z = \frac{1}{r}\frac{\partial}{\partial r}\left(r\vec{E}_\phi\right) - \frac{1}{r}\frac{\partial \vec{E}_r}{\partial \phi} \qquad (13.25)$$

and from Equation (13.12),

$$j\omega\varepsilon\vec{E}_r = \frac{1}{r}\frac{\partial \vec{H}_z}{\partial \phi} + j\beta\vec{H}_\phi \qquad (13.26)$$

$$-j\omega\varepsilon\vec{E}_\phi = \frac{\partial \vec{H}_z}{\partial r} + j\beta\vec{H}_r \qquad (13.27)$$

$$j\omega\varepsilon\vec{E}_z = \frac{1}{r}\frac{\partial}{\partial r}\left(r\vec{H}_\phi\right) - \frac{1}{r}\frac{\partial \vec{H}_r}{\partial \phi} \qquad (13.28)$$

By eliminating variables these equations can be rewritten such that when $\vec{E}_z$ and $\vec{H}_z$ are known, the remaining transverse components $\vec{E}_r$, $\vec{E}_\phi$, $\vec{H}_r$, and $\vec{H}_\phi$ can be determined.

$$\vec{E}_r = -\frac{j}{q^2}\left(\beta\frac{\partial \vec{E}_z}{\partial r} + \frac{\mu\omega}{r}\frac{\partial \vec{H}_z}{\partial \phi}\right) \qquad (13.29)$$

$$\vec{E}_\phi = -\frac{j}{q^2}\left(\frac{\beta}{r}\frac{\partial \vec{E}_z}{\partial \phi} - \mu\omega\frac{\partial \vec{H}_z}{\partial r}\right) \qquad (13.30)$$

$$\vec{H}_r = -\frac{j}{q^2}\left(\beta\frac{\partial \vec{H}_z}{\partial r} - \frac{\varepsilon\omega}{r}\frac{\partial \vec{E}_z}{\partial \phi}\right) \qquad (13.31)$$

$$\vec{H}_\phi = -\frac{j}{q^2}\left(\frac{\beta}{r}\frac{\partial \vec{H}_z}{\partial \phi} + \varepsilon\omega\frac{\partial \vec{E}_z}{\partial r}\right) \qquad (13.32)$$

where

$$q^2 = \omega^2\varepsilon\mu - \beta^2 = k^2 - \beta^2$$

The substitution of Equations (13.31) and (13.32) into Equation (13.28) results in the wave equation in cylindrical coordinates as

$$\frac{\partial^2 \vec{E}_z}{\partial r^2} + \frac{1}{r}\frac{\partial \vec{E}_z}{\partial r} + \frac{1}{r^2}\frac{\partial^2 \vec{E}_z}{\partial \phi^2} + q^2 \vec{E}_z = 0 \tag{13.33}$$

and substitution of Equations (13.28) and (13.29) into Equation (13.25) leads to

$$\frac{\partial^2 \vec{H}_z}{\partial r^2} + \frac{1}{r}\frac{\partial \vec{H}_z}{\partial r} + \frac{1}{r^2}\frac{\partial^2 \vec{H}_z}{\partial \phi^2} + q^2 \vec{H}_z = 0 \tag{13.34}$$

Equations (13.33) and (13.34) contact either $\vec{E}_z$ or $\vec{H}_z$ only. This appears to imply that the longitudinal components of E and H are uncoupled and can be chosen arbitrarily provided that they satisfy Equations (13.33) and (13.34).

If the boundary conditions do not lead to coupling between the field components, mode solutions can be obtained in which either $\vec{E}_z = 0$ or $\vec{H}_z = 0$. When $\vec{E}_z = 0$, the mode is called transverse electric or *TE* mode, and when $\vec{H}_z = 0$, a transverse magnetic or *TM* mode results. Hybrid modes exist if both $\vec{E}_z$ and $\vec{H}_z$ are nonzero. These are designated as *HE* or *EH* modes, depending on whether $\vec{H}_z$ or $\vec{E}_z$, respectively, makes a larger contribution to the transverse field.

Wave Equations for Step-Index Fibers

A standard mathematical procedure for solving equations such as Equation (13.33) is to use the separation of variables method, which assumes a solution of the form

$$\vec{E}_z = AF_1(r)F_2(\phi)F_3(z)F_4(t) \tag{13.35}$$

The time and z-dependent factors are given by

$$F_3(z)F_4(t) = e^{j(\omega t - \beta z)} \tag{13.36}$$

since the wave is sinusoidal in time and propagates in the z-direction. The circular symmetry of the waveguide and each field component must not change when the coordinate ϕ is increased by 2π. We thus assume a periodic function of the form.

$$F_2(\phi) = e^{jv\phi} \tag{13.37}$$

The constant v can be positive or negative, but it must be an integer, since the fields must be periodic if ϕ with a period of 2π.

Substituting Equation (13.36) into Equation (13.35), the wave equations for $\vec{E}_z$, Equation (13.33) becomes

$$\frac{\partial^2 F_1}{\partial r^2} + \frac{1}{r}\frac{\partial F_1}{\partial r} + \left(q^2 - \frac{v^2}{r^2}\right)F_1 = 0 \tag{13.38}$$

Equation (13.38) is a well-known *differential equation for Bessel functions*. An identical equation can be derived for $\vec{H}_z$. Equation (13.38) is solved for the regions inside and outside the core.

For the inside region the solutions for the guided modes must remain finite as $r \to 0$, whereas on the outside the solutions must decay to zero as $r \to \infty$.

Thus for $r < a$ the solutions are Bessel functions of the first kind of order v. For these functions we use the common designation $J_v = (ur)$. Here $u^2 = k_1^2 - \beta^2$ with $k_1 = \dfrac{2\pi n_1}{\lambda}$, n_1 is the refractive index of the core.

The expressions for $\vec{E}_z$ and $\vec{H}_z$ inside the core are thus

$$\vec{E}_z(r < a) = AJ_v(ur)e^{jv\phi}e^{j(\omega t - \beta z)} \tag{13.39}$$

$$\vec{H}_z(r < a) = BJ_v(ur)e^{jv\phi}e^{j(\omega t - \beta z)} \tag{13.40}$$

where A and B are arbitrary constants.

Outside of the core the solutions of Equation (13.38) are given by modified Bessel functions of the second kind $K_v(\omega r)$, where $\omega^2 = \beta^2 - k_2^2$ with $k_2 = \dfrac{2\pi n_2}{\lambda}$, n_2 = refractive index of cladding.

The expressions for $\vec{E}_z$ and $\vec{H}_z$ outside the core are, therefore,

$$\vec{E}_z(r > a) = CK_v(\omega r)e^{jv\phi}e^{j(\omega t - \beta z)} \tag{13.39}$$

$$\vec{H}_z(r > a) = DK_v(\omega r)e^{jv\phi}e^{j(\omega t - \beta z)} \tag{13.40}$$

where C and D are arbitrary constants.

From the definition of the modified Bessel functions, it is seen that $K_v(\omega r) \to e^{-\omega r}$ as $\omega r \to \infty$. Since $K_v(\omega r)$ must go to zero as $r \to \infty$, it follows that $\omega > 0$. This, in turn, implies that $\beta \geq k_2$, which represents a cutoff condition. The cutoff condition is the point at which a mode is no longer bound to the core region. A second condition on β can be deduced from the behavior of $J_v(ur)$. Inside the core the parameter u must be real for F_1 to be real, from which it follows that $k_1 \geq \beta$.

The permissible range of β for bound solutions is, therefore,

$$n_2 k = k_2 \leq \beta \leq k_1 = n_1 k \tag{13.41}$$

Where $k = \dfrac{2\pi}{\lambda}$ is the free-space-propagation constant.

Modal Equation

The solutions for β must be determined from the boundary conditions. The boundary conditions require that the tangential components $\vec{E}_\phi$ and $\vec{E}_z$ of $\vec{E}$ inside and outside of the dielectric interface at $r = a$ must be the same and, similarly, for the tangential components $\vec{H}_\phi$ and $\vec{H}_z$.

Consider first the tangential components of $\vec{E}$. For the z-components from Equation (13.39) at the inner core-cladding boundary $\left(\vec{E}_z = \vec{E}_{z_1} \right)$ and from Eqn. (13.39) at the outside of the boundary $\left(\vec{E}_z = \vec{E}_{z_2} \right)$, that is

$$\begin{aligned}
\vec{E}_{z_1} - \vec{E}_{z_2} &= A J_v(ur) e^{jv\phi} e^{j(\omega t - \beta z)} - C K_v(\omega r) e^{jv\phi} e^{j(\omega t - \beta z)} \\
&= A J_v(ur) C K_v(\omega r) = 0
\end{aligned} \tag{13.42}$$

Inside the core the factor q^2 is given by $q^2 = u^2 = k_1^2 - \beta^2$ where $k_1 = \dfrac{2\pi n_1}{\lambda} = \omega\sqrt{\varepsilon_1 \mu}$

Outside the core the factor w^2 is given by $w^2 = \beta^2 - k_2^2$ where $k_2 = \dfrac{2\pi n_2}{\lambda} = \omega\sqrt{\varepsilon_2 \mu}$.

Substituting Equations (13.39) and (13.40) into Equation (13.30) to find $\vec{E}_\phi$, and similarly using Equations (13.39) and (13.40) to determine $\vec{E}_{\phi_2}$ yield at $r = a$.

$$\begin{aligned}
\vec{E}_{\phi_1} - \vec{E}_{\phi_2} &= -\frac{j}{u^2}\left(A \frac{jv\beta}{a} J_v(ua) - B\omega\mu u J_v'(ua) \right) \\
&\quad - \frac{j}{w^2}\left(C \frac{jv\beta}{a} K_v(wa) - D\omega\mu w K_v'(wa) \right) = 0
\end{aligned} \tag{13.43}$$

where prime indicates differentiation with respect to the argument. Similarly, for the tangential components of $\vec{H}$, it is readily shown that at $r = a$.

$$\vec{H}_{z_1} - \vec{H}_{z_2} = B J_v(ua) - D K_v(wa) = 0 \tag{13.44}$$

and

$$
\vec{H}_{\phi_1} - \vec{H}_{\phi_2} = -\frac{j}{u^2}\left(B\frac{jv\beta}{a}J_v(ua) + A\omega\varepsilon_1 uJ'_v(ua) \right)
$$
$$
-\frac{j}{w^2}\left(D\frac{jv\beta}{a}K_v(wa) + C\omega\varepsilon_2 wK'_v(wa) \right) = 0
$$

(13.45)

Equations (13.44) to (13.45) are a set of four equations with four unknown coefficients A, B, C, and D. A solution to these equations exists only if the determinant of these coefficients is zero:

$$
\begin{bmatrix}
J_v(ua) & 0 & -K_v(wa) & 0 \\
\dfrac{\beta v}{au^2}J_v(ua) & \dfrac{jw\mu}{u}J'_v(ua) & \dfrac{\beta v}{aw^2}K_v(ua) & \dfrac{jw\mu}{w}K'_v(ua) \\
0 & J_v(ua) & 0 & -K_v(wa) \\
-\dfrac{jw\varepsilon_1}{u}J'_v(ua) & \dfrac{\beta v}{au^2}J_v(ua) & -\dfrac{jw\varepsilon_2}{w}K'_v(wa) & \dfrac{\beta v}{aw^2}K_v(wa)
\end{bmatrix} = 0
$$

(13.46)

Evaluation of this determinant yields the following eigen value equation for β:

$$
(J_v + K_v)(k_1^2 J_v + k_2^2 K_v) = \left(\frac{\beta v}{a}\right)^2\left(\frac{1}{u^2} + \frac{1}{w^2}\right)
$$

(13.47)

where

$$
J_v = \frac{J'_v(ua)}{uJ_v(ua)} \quad \text{and} \quad K_v = \frac{K'_v(wa)}{wK_v(wa)}
$$

(13.48)

Solving Equation (13.47) for β, it will be found that only discrete values restricted to the range given by Equation (13.41) will be allowed. Equation (13.47) is a complicated transcendental equation which is generally solved by numerical techniques; its solution for any particular mode will provide all the characteristics of that mode.

For $v = 0 \rightarrow \begin{cases} TE\text{(Transverse Electric) modes} \\ TM\text{(Transverse magnetic) modes} \end{cases}$

For $v \neq 0 \rightarrow$ hybrid modes (HE, EH)

TE, TM modes correspond to meridional rays, hybrid modes correspond to skew rays.

Normalized Frequency (V) or V-number or V-parameter

An important parameter connected with the cutoff condition is the normalized frequency V or V-number defined by

$$V^2 = \left(u^2 + w^2\right)a^2 = \left(\frac{2\pi a}{\lambda}\right)^2 \left(n_1^2 - n_2^2\right) \tag{13.49}$$

$$V = \frac{2\pi}{\lambda}a\sqrt{n_1^2 - n_2^2} = \frac{2\pi}{\lambda}an_1\sqrt{2\Delta} \tag{13.50}$$

λ is the free space wavelength of the light beam.

V is a dimensionless number that determines how many modes a fiber can support. The number of modes that can exist in a waveguide as a function of V may be conveniently represented in terms of a normalized propagation constant b defined by

$$b = \frac{a^2 w^2}{V^2} = \frac{\left(\frac{\beta}{k}\right)^2 - n_2^2}{n_1^2 - n_2^2} \tag{13.51}$$

Figure 13.6 shows that each mode can exist only for values of V that exceed a certain limiting value. The modes are cut off when $\dfrac{\beta}{k} = n_2$.

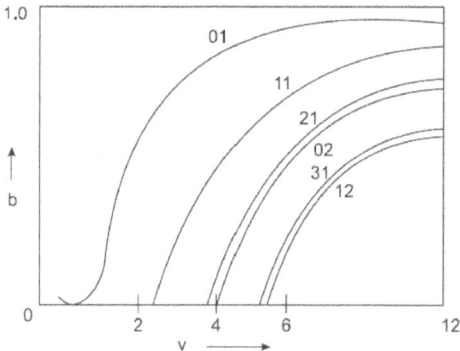

FIGURE 13.6. The curve number v_m designates the $HE_{v+1,m}$ and $EH_{v-1,m}$ for $v=1$, the curve number v_m gives the HE_{2m}, TE_{0m}, TM_{0m}.

The HE_{11} mode has no cutoff and ceases to exist only when the core diameter is zero. This is the principle on which the single-mode fiber is based.

By appropriately choosing a, n_1, and n_2 so that

$$V = \frac{2\pi a}{\lambda}\sqrt{\left(n_1^2 - n_2^2\right)} \geq 2.405$$

which is the value at which the lowest-order Bessel function J_o is zero, all modes except the HE_{11} mode are cut off.

The parameter V can also be related to the number of modes M in a multimode fiber when M is large. An approximate relationship for step-index fibers can be derived from ray theory.

The numerical aperture is given as

$$NA = \sin\theta = \sqrt{\left(n_1^2 - n_2^2\right)}$$

For practical numerical aperture $\sin\theta$ is small so that $\sin\theta \approx \theta$. The solid acceptance angle for the fiber is therefore

$$\Omega = \pi\theta^2 = \pi\left(n_1^2 - n_2^2\right) \tag{13.52}$$

For electromagnetic radiation of wavelength λ emanating from a laser or a waveguide, the number of modes per unit solid angle is given by $\dfrac{2A}{\lambda^2}$, where A is the area the mode is leaving or entering. The area A in this case is the core cross-section πa^2. The factor 2 comes from the fact that the plane wave can have two polarization orientations.

The total number of modes M entering the fiber is thus given by

$$M \simeq \frac{2A}{\lambda^2}\Omega = \frac{2\pi^2 a^2}{\lambda^2}\left(n_1^2 - n_2^2\right) = \frac{V^2}{2} \tag{13.53}$$

Example 13.4

Let $n_1 = 1.53, n_2 = 1.50, \lambda = 1\mu\text{m}, a = 50\mu\text{m}$. Find the total number of modes M entering the fiber.

Solution:

$$V = 94.72, \ M = \frac{V^2}{2} = \frac{(94.72)^2}{2} = 4486$$

13.4 PULSE DISPERSION IN STEP-INDEX FIBERS

The simplest type of optical fiber consists of a thin cylindrical structure of transparent glassy material of uniform refractive index n_1 surrounded by a cladding of another material of uniform but slightly lower refractive index n_2. These fibers are referred to as step-index fibers due to the step discontinuity of the index profile at the core-cladding interface.

In digital communication systems, information to be sent is first coded in the form of pulses, and then these pulses of light are transmitted from the transmitter to the receiver where the information is decoded. The larger the number of pulses that can be sent per unit time and still be resolvable at the receiver end, the larger is the transmission capacity of the system. A pulse of light sent into a fiber broadens in time as it propagates through the fiber; this phenomenon is known as pulse dispersion and happens because of the different times taken by different rays to propagate through the fiber, as shown in Figure 13.7.

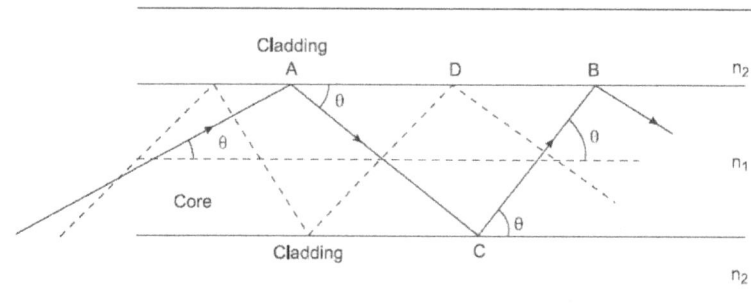

FIGURE 13.7. Pulse dispersion in step-index fiber.

Rays making larger angles with the axis take a longer time to traverse the length of the fiber. Consequently, the pulse broadens as it propagates through the fiber. Hence, even though two pulses may be well resolved at the input end, because of broadening of the pulses they may not be so at the output end. Where the output pulses are not resolvable, no information can be retrieved. Thus, the smaller the pulse dispersion, the greater the information-carrying capacity of the system.

Calculate the amount of dispersion in a step-index fiber as follows:

For a ray making an angle θ with the axis, the distance AB is traversed in time.

$$t = \frac{AC + CB}{\left(\dfrac{C}{n_1}\right)} = \frac{n_1(AB)}{C\cos\theta} \tag{13.54}$$

where $\left(\dfrac{C}{n_1}\right)$ represents the speed of light in a medium of refractive index n_1, C being the speed of light in free space. Since the ray path will repeat itself, the time taken by a ray to traverse a length L of the fiber will be

$$t = \frac{n_1 L}{C \cos\theta} \tag{13.55}$$

The previous expression shows that the time taken by a ray is a function of the angle made by the ray with the z-axis, which leads to pulse dispersion. If we assume that all rays between O and θ_c are present, then the time taken by rays corresponding respectively to $\theta = 0$ and $\theta = \theta_c = \cos^{-1}\left(\frac{n_2}{n_1}\right)$ will be given $t_{min} = \frac{n_1 L}{C}$, shortest ray congruence paths (the fundamental mode).

Since $\cos 0° = 1$, $t_{max} = \frac{n_1 L}{C\left(\frac{n_2}{n_1}\right)} = \frac{n_1^2 L}{C n_2}$, longest ray congruence paths (the highest-order mode).

Hence, if all the input rays were excited simultaneously, the rays would occupy a time interval at the output end of duration.

$$\delta T = t_{max} - t_{min} = \left(\frac{n_1^2 L}{C n_2}\right) - \left(\frac{n_1 L}{C}\right) = \frac{n_1^2 L}{C n_2}\left(1 - \frac{n_2}{n_1}\right) = \frac{n_1^2 L}{C n_2}\Delta, \tag{13.56}$$

$$\Delta = \frac{n_1 - n_2}{n_1} \tag{13.57}$$

where Δ is the relative refractive index difference.

This derivation considers only pulse broadening owing to meridional rays and does not take into account skew rays.

Where $\Delta \ll 1$, then the relative refractive index difference may also be given approximately by:

$$\Delta \simeq \frac{n_1 - n_2}{n_2} \tag{13.58}$$

Hence, Equation (13.56) becomes

$$\delta T_s = \frac{n_1^2 L}{C n_2}\left(\frac{n_1 - n_2}{n_1}\right) = \frac{n_1 L}{C}\left(\frac{n_1 - n_2}{n_2}\right) \simeq \frac{n_1 L \Delta}{C}, \tag{13.59}$$

Also, $NA = n_1\sqrt{2\Delta}$ the previous expression becomes

$$\delta T_s \simeq \frac{L}{C}\frac{(NA)^2}{2n_1} \tag{13.60}$$

where *NA* is the numerical aperture for the fiber.

The approximate expressions for the delay difference given in Equations (13.59) and (13.60) are usually employed to estimate the maximum pulse broadening in time due to intermodal dispersion in multimode step-index fibers.

The previous time delay is the characteristic of the fiber and independent of the wavelength of the light. It is of the order of nanoseconds per km. However, this type of delay is absent in single-mode fibers, as it has only one mode.

This phenomenon in graded-index fibers is at a minimum when profile shape parameter α is close to the value $2(1 - 1.2\Delta) = \alpha_{opt}$.

Using a ray theory approach,

$$\delta T_g \simeq \frac{Ln_1\Delta^2}{2C} \simeq \frac{(NA)^4}{8n_1^3C} \tag{13.61}$$

Using electromagnetic mode theory,

$$\delta T_g = \frac{Ln_1\Delta^2}{8C} \tag{13.62}$$

It is observed that the graded-index fibers have much lower intermodal dispersion than that of the step-index fiber.

Graded-index-α-profiles

$$n(r) = n_0\sqrt{\left(1 - 2\Delta\left(\frac{r}{a}\right)^\alpha\right)} \text{ for } r \le a \tag{13.63}$$

$$n(r) = n_0\sqrt{(1 - 2\Delta)} = n_c \text{ for } r \ge a \tag{13.64}$$

Refractive-index profiles ranging from the triangular profile, $\alpha = 1.0$, to the step-index profile, $\alpha = \infty$, are shown in Figure 13.8, with profiles for $\alpha = 1.7, 2.0, 2.3,$ and 6.0 lying in between. The profile parameter, α, may take on values between zero and infinity.

It may be expected that a light pulse with a given width and amplitude injected into one end of a fiber should arrive at its other end with the shape and width unchanged, and only its amplitude is reduced by losses. If the losses are extremely large, the pulse amplitude at the receiving end will be too small to be detected, and a repeater has to be included to boost up the signal level before it enters into the next section.

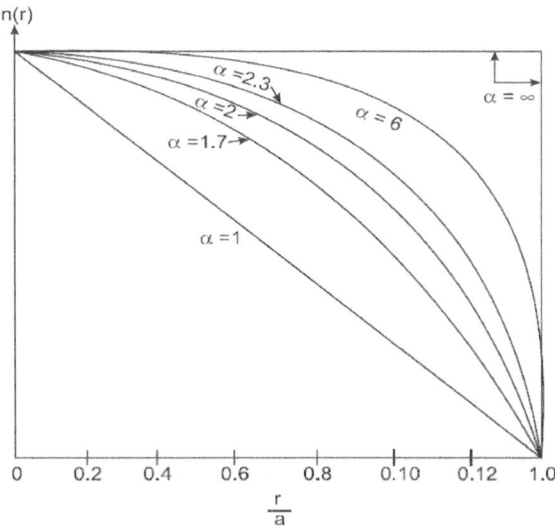

FIGURE 13.8. Refractive-index profiles ranging from the triangular profile, $\alpha = 1.0$, to the step-index profile, $\alpha = \infty$.

Several dispersion effects are encountered by the pulse of light propagating through a fiber, and these act to spread out the pulse in the time domain, changing the shape of the pulse. As a result, the pulses may become merged, with the previous or the succeeding pulses becoming indistinguishable at the receiver, as shown in Figure 13.9. This effect is known as inter symbol interference (ISI). The pulses may be separated by increasing the time interval between pulses, but thereby the maximum bit rate will be reduced.

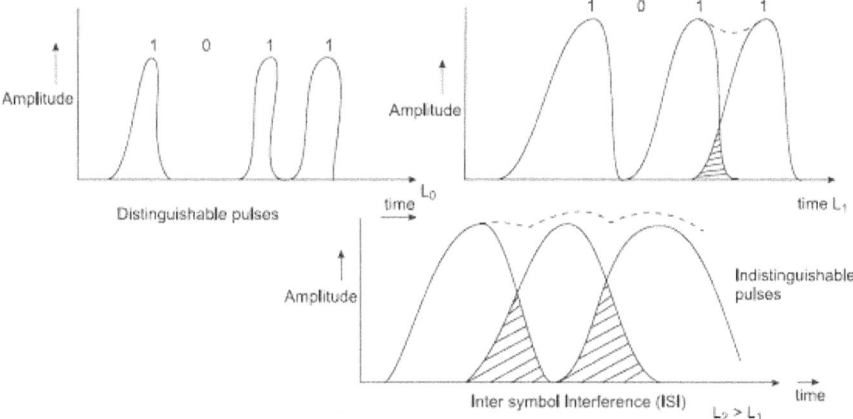

FIGURE 13.9. The pulse merged in the previous or in the succeeding pulses becoming indistinguishable at the receiver.

Dispersion Effect of Dispersion on Pulse Transmission

There are three kinds of dispersion due to three separate mechanisms existing in the fiber. These are:

1. Intermodal Dispersion or multi-path dispersion

2. Material or chromatic dispersion

3. Waveguide dispersion

13.5 MATERIAL DISPERSION

The refractive index of a core glass of fiber is not the same for lights of different wavelengths, but it depends on the wavelength of light in a complicated way. A pulse of light transmitted contains components of several wavelengths centered about a center wavelength λ_0, as shown in Figure (13.10).

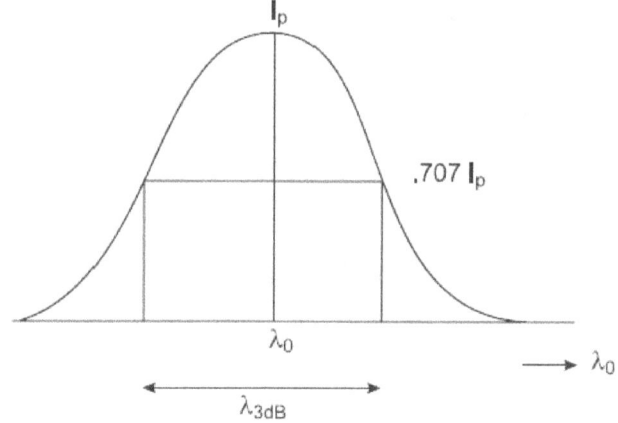

FIGURE 13.10. A pulse of light transmitted contains components of several wavelengths centered about a center wavelength λ_0.

The pulse component having shorter wavelengths will experience more delay that those of larger wavelengths. As a result, there is an effective time dispersion Δt of the pulse at the receiving end of the fiber.

Material dispersion depends only on the composition of the material, as shown in Figure 13.11. The material dispersion is given by

$$\Delta \tau_m - \frac{Z}{C} \lambda_0^2 \frac{d^2 n_1}{d\lambda_0^2} \left(\frac{\Delta \lambda_0}{\lambda_0} \right)$$

$$(13.65)$$

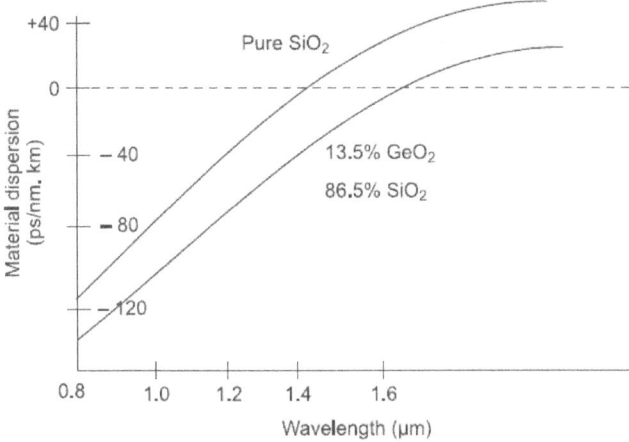

FIGURE 13.11. Material dispersion and wavelengths of different composite materials.

13.6 WAVEGUIDE DISPERSION

The effect of waveguide dispersion on pulse spreading can be approximated by assuming that the refractive index of the material is independent of wavelength, as shown in Figure 13.12.

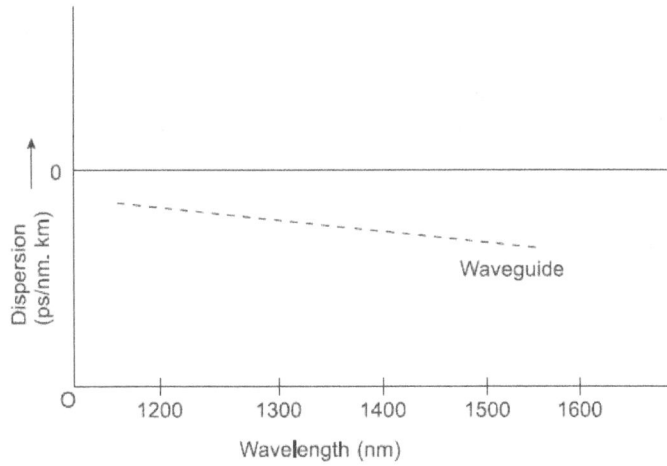

FIGURE 13.12. The effect of waveguide dispersion on pulse spreading.

When a light pulse is launched into a fiber, it is distributed among many guided modes. These various modes arrive at the fiber end at different times depending on their group delay, so that a pulse spreading results.

For multimode fibers the waveguide dispersion is generally very small compared with material dispersion and can therefore be neglected.

Waveguide dispersion depends on the core radius, the refractive-index difference, and the shape of the refractive-index profile. Thus, the waveguide dispersion can vary dramatically with the fiber design parameters.

The waveguide dispersion is given by

$$\Delta \tau_w = -L \frac{n_2 \Delta}{C} \left(\frac{\Delta \lambda_0}{\lambda_0} \right) V \frac{d^2 (bV)}{dV^2} \tag{13.66}$$

where

$$b \approx \frac{\left(\frac{\beta}{k - n_2} \right)}{n_1 - n_2}, \Delta = \frac{n_1 - n_2}{n_1}, V = ka\sqrt{(n_1^2 - n_2^2)} \approx kan_2 \sqrt{2\Delta}$$

13.7 DISPERSION SHIFTED AND DISPERSION FLATTENED FIBERS

The loss of signal through the fiber due to dispersion is reduced by two important methods.

i. Dispersion Shifted Fiber

The dispersion shifted fibers are made by increasing the negative waveguide dispersion as shown in Figure 13.13. This negative waveguide dispersion is exactly canceled by positive material dispersion, and the result is given by dispersion shifted fiber.

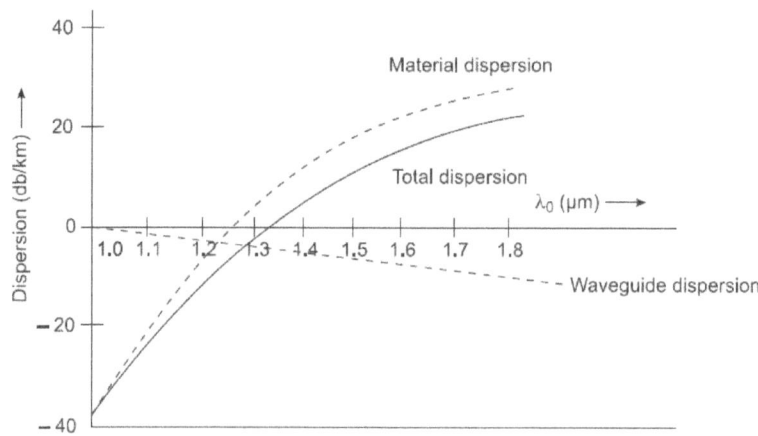

FIGURE 13.13. The dispersion shifted fiber.

Since for the single mode fiber the V-parameter should be

$0 < V < 2,408.0$

and the cutoff wavelength for the single mode fiber should be such that

$$\lambda_0 > \lambda_{c_0} = \frac{2\pi a \sqrt{\left(n_1^2 - n_2^2\right)}}{2.4048}$$

when λ_{c_0} is the cutoff wavelength in the step index fiber.
The material dispersion is given by

$$\Delta\tau_m = -\frac{z}{C}\lambda_0^2 \frac{d^2 n_1}{d\lambda_0^2}\left(\frac{\Delta\lambda_0}{\lambda_0}\right) \tag{13.67}$$

i.e., proportional to $\dfrac{d^2 n_1}{d\lambda_0^2}$ and spectral width $\Delta\lambda_0$.

Also, the waveguide dispersion is given by Equation (13.68).
Now we take the operated wavelength $\lambda_0 \simeq 1.3\ \mu m$ and we calculate the material dispersion. Hence

$$\lambda_0 \simeq 1.3\ \mu m,\ \frac{d^2 n}{d\lambda_0^2} = -5.5\times 10^{-4}\ \mu m^{-2}.$$

So that,

$$\frac{\Delta\tau_m}{L\Delta\lambda_0} \simeq 2.4\ \text{ps/km nm}$$

Again, we consider a step-index single-mode fiber for
$a = 5.6\mu m,\ \Delta = 0.00117,\ n_2 = 1.45$

So, V-parameter is so chosen that

$V \simeq 1.9$ at $\lambda_0 = 1.3\mu m$

Hence,

$$\frac{\Delta\tau_w}{L\Delta\lambda_0} \simeq 2.4\ \text{ps/km}\ \mu m$$

Thus, the waveguide dispersion and the material dispersion cancel each other to give zero dispersion. The explanation is shown by Figure 13.14. The same calculation done by the wavelength is $\lambda_0 = 1.55\ \mu m$.

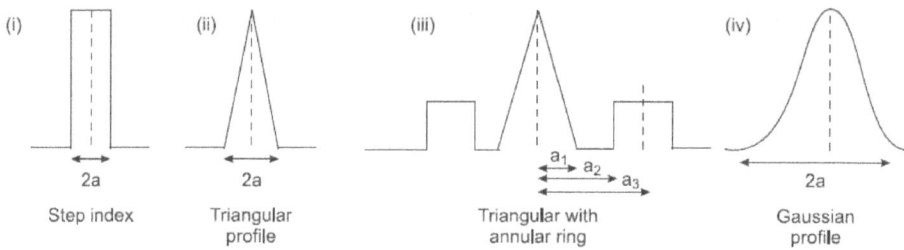

FIGURE 13.14. The refractive index profile for the dispersion shifted fiber.

ii. Dispersion Flattened Fiber

This is the second method by which we reduce the loss of the fiber. In this method, we show that by reducing the core radius, the zero-dispersion wavelength can be shifted to the minimum loss wavelength window. This can also be done by suitably grading the refractive index profile of the core.

Figure 13.15 shows the variation of total dispersion as a function of λ_0 in the step-index fiber.

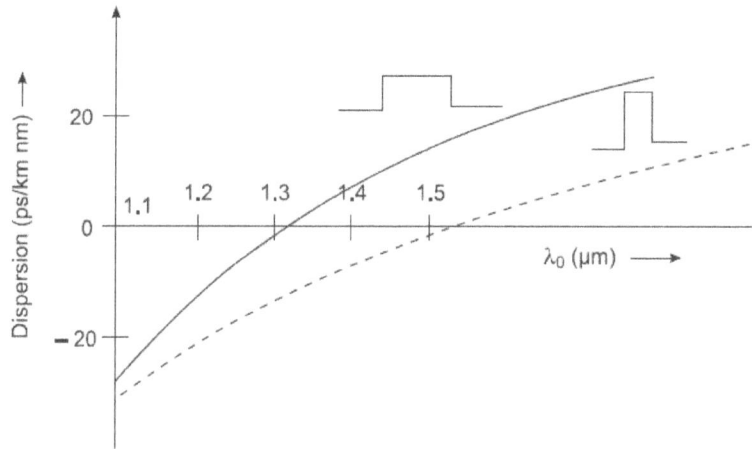

FIGURE 13.15. The variation of total dispersion as a function of λ_0 in the step-index fiber.

The solid line is for $\Delta = 0.0027$, $a = 4.1\mu m$, $\lambda_{c_0} = 1.13\mu m$, while the dashed curve is for $\Delta = 0.0075$, $a = 2.3\mu m$, $\lambda_{c_0} = 1.06\mu m$.

The variation of total dispersion for different classes is shown in Figure 13.16. Curve (i) corresponds to conventional fiber. Curve (ii) corresponds to the dispersion shifted fiber. Curves (iii) and (iv) correspond to dispersion flattened fiber.

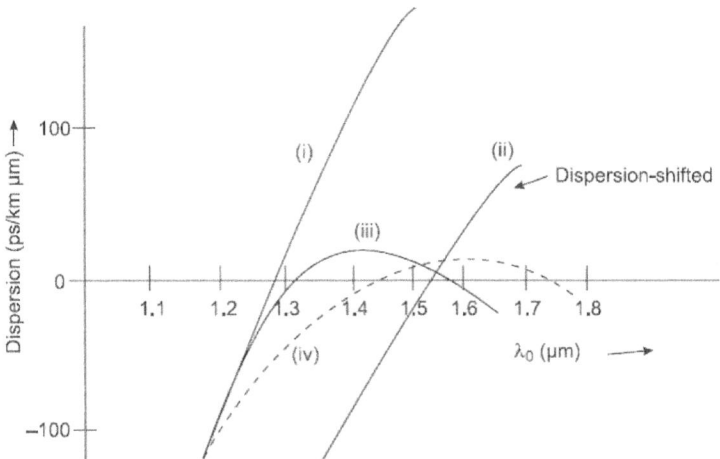

FIGURE 13.16. The variation of total dispersion for different classes.

The refractive index profile for the flattened fiber for corresponding (i), (ii), (iii), and (iv) are illustrated in Figure 13.17.

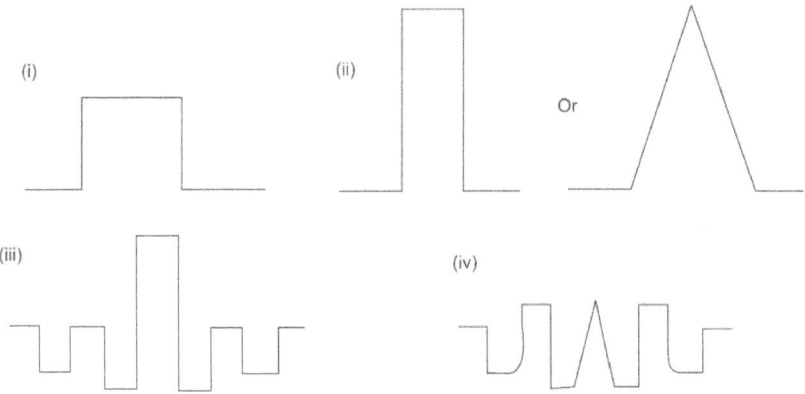

FIGURE 13.17. The refractive index profile for the flattened fiber for corresponding (i), (ii), (iii), and (iv).

13.8 ATTENUATION IN OPTICAL FIBERS

Attenuation in fibers means "loss of optical power" in the fiber itself. This loss may arise from different sources, for example,

1. Material absorption (or impurity losses)

2. Rayleigh scattering loss

3. Absorption loss

4. Leaky modes

5. Bending losses

6. Radiation induced losses

7. Defective construction losses

8. Inverse square law losses

9. Transmission losses

10. Temperature dependence of fiber losses

11. Core and cladding losses

Decibel (dB)

The decibel is referred to as ratios or relative power units:

$$\text{Attenuation} = 10 \log \left(\frac{P_2}{P_1} \right) dB$$

Attenuation in fibers means loss of optical power in the fiber itself or in a pair of connectors or splice. It is expressed in "decibels" or "dB," where $-10dB$ means a reduction in the power by 10 times, $-20dB$ means another 10 times, or 100 times overall, obviously $-30dB$ means 1000 times loss overall, and so on. Doubling the power means a 3dB gain.

The dBm

Decibel (dB) gives no indication of the absolute power level. The derived unit for doing this in optical communications is the dBm. dBm is decibel measure of power relative to 1mw. The power in dBm is an absolute value of power and it is given by

$$\text{Power level (in dBm)} = 10 \log \left(\frac{P}{1mw} \right)$$

In this relationship 0dBm = 1 mw.

Losses are expressed in decibels per kilometer (dB/km). The normal range of attenuation is from 0.154 dB/km at operating wavelength 1550 nm for a single-mode fiber to over 10 dB/m for plastic fiber.

The loss is expressed as in terms of a particular length L for a fiber.

$$P_0 = P_{in} 10^{-\frac{\alpha L}{10}} \qquad (13.68)$$

where
P_0 = power at a distance L from the input
P_{in} = amount of power coupled into the fiber
α = fiber attenuation in dB/km

Hence, attenuation in the fiber is defined as the ratio of the optical power output P_0 obtained from a fiber of length L to the optical power P_{in} fed to the input of the fiber.

$$\alpha = \frac{10}{L} \log\left(\frac{P_{in}}{P_{out}}\right) \text{ in dB/km} \qquad (13.69)$$

In the case of an ideal fiber, there is no loss of optical power and $P_{in} = P_0$. In other words, for an ideal fiber, the attenuation is 0dB. But practically this is not possible, because all fiber has a certain amount of loss.

$$\frac{P_0}{P_{in}} = 10^{-\frac{\alpha L}{10}} \rightarrow P_0 = P_{in} 10^{-\frac{\alpha L}{10}}$$

$$P_0(dBm) = 10\log\left(\frac{P_0}{1 \text{ m watt}}\right) + 10\log\left(10^{-\frac{\alpha L}{10}}\right) \qquad (13.70)$$

The Equation (13.68) can also be written in decibels,

$$P_0(dBm) = 10\log\left(\frac{P_{in}}{1 \text{ m watt}}\right) + 10\log\left(10^{-\frac{\alpha L}{10}}\right) = P_{in}(dBm) - \alpha L \qquad (13.71)$$

Material or Impurity Losses

Material absorption is a loss mechanism related to the material composition and the fabrication process for the fiber, which results in the dissipation of some of the transmitted optical power as heat in the waveguide. The absorption of the light may be intrinsic (caused by the interaction with one or more of the major components of the glass) or extrinsic (caused by impurities within the glass).

In the fabrication of various types of fibers, we use GeO_2, P_2O_5, B_2O_3, and so on as dopants in silica, in order to modify the refractive index. While B_2O_3 produces strong absorption peaks at 3.2 μm, P_2O_5 produces the same at 3.8 μm. However, in both these cases, absorption-tails extend below 1.3 μm. That is why Boron- and Phosphorous-based dopants are not used in the low loss, single-mode fiber. The introduction of P_2O_5 serves as a channel for the gradual buildup of OH– ions over time. In fiber using a high dose of P_2O_5, loss increases up to 0.9 dB/km at 1.3 μm over a period of 3 years. For zero-P single-mode fiber, the loss is less than 0.007 dB/km over a period of 25 years. Low-P multimode fiber has a projected loss of 0.07 dB/km over the same span of time.

Experiments have shown that the loss increases considerably when the operating wavelength is beyond 1.55 μm. This is an inherent property of all silica-based fiber. This is due to vibrational resonances of the silica tetrahedron molecules.

Rayleigh Scattering Loss

The glass which is used in the fabrication of fibers has many microscopic inhomogeneities and material density fluctuations of the silica material contents. As a result, a portion of light passing through the glass fiber gets scattered. This phenomenon is called *Rayleigh scattering*. The losses due to this scattering effect vary inversely with the fourth power of the wavelength. These inhomogeneities are frozen into the silica matrix when the fiber is formed and depend on the glass-forming temperature. The lower the glass-forming temperature, the lower the density fluctuation, and hence scattering. This scattering is most prominent between the wavelength range 500 nm to 1550 nm.

The expected minimum loss at 1550 nm is approximately 0.13 dB/km. It produces a $\dfrac{1}{\lambda^4}$ attenuation dependence; it is also a function of glass-forming temperature.

The scattering loss due to density fluctuation can be expressed as:

$$\alpha_{RS} = \frac{359.094}{\lambda^4} \left(\mu^2 - 1 \right)^3 K K_T K_F \qquad (13.72)$$

where λ = operating wavelength, K = Boltzmann constant, T_F = temperature at which the density fluctuations are frozen into the glass as it solidifies, K_T = Isothermal compressibility of the material. This is also can be expressed as

$$\alpha_{RS} = \frac{359.094}{\lambda^4} \mu^8 K_P K K_T K_F \qquad (13.73)$$

where K_p = Photoelastic coefficient.

The Rayleigh scattering coefficient is related to the transmission loss factor (transmissivity) of the fiber (P)

$$P = e^{-\alpha_{RS}L} \tag{13.74}$$

where L is the length of the fiber.

Absorption Loss

This form of loss is caused by the very nature of the core material and varies inversely to the transparency of the material. Absorption losses are not uniform across the light spectrum, and are regarded to be wavelength-sensitive, as shown in Figure 13.18.

Ion-resonance absorption, ultra-violet absorption, and infrared absorption are the three separate mechanisms which contribute to total absorption losses in glass fibers.

In pure fused silica, valence electrons can be ionized into conduction electrons by light with energy about 9 eV, and then *uv* absorption takes place. Thus, there is a loss of light energy due to this ionization.

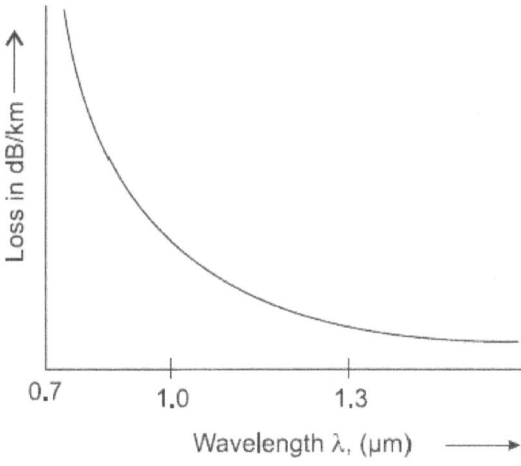

FIGURE 13.18. Absorption loss.

When photons of light energy are absorbed by the atoms within the glass molecules, infrared absorption takes place and is converted to the random mechanical vibrations typical of heating.

The nearest vibrational resonance of the $S_i - O$ bond is in the *IR* region at 9 μm. Some oxides which are doped in silica for changing the refractive index in the core (e.g., GeO_2, P_2O_5, B_2O_3) have similar resonances in *IR*. The tails of these absorption bands in *IR* may contribute in the wavelength region of interest 1.4–1.6 μm, depending on the doping materials.

Leaky Modes

The losses due to leaky modes arise due to irregularities in waveguide geometry, so these can be regarded as waveguide scattering. These leaky modes have radiative components that result in cladding power losses. So, these can be minimized by surrounding the thin cladding layer on the fiber-core by a third layer of pure silica, which has an index of refraction higher than that of the cladding but lower than that of the core. This third additional silica coating will not only give additional strength to the fiber, but side-by-side also removes the partial refraction rays from the leaky modes, as well as removes the passed rays from cutoff modes by total refraction.

Bending Losses

Whenever a fiber deviates from a straight-line path, radiative losses occur. These losses are prominent for improperly installed single-mode optical cable. Micro-bending and macro-bending are two types of bending losses.

Micro-bending Losses

Micro-bends have small random deviations about a nominal straight-line position. The micro-bending losses occur due to the fact that the small bends act as scattering, which causes mode-coupling to take place. The energy of the guided mode is cross-coupled into leaky modes. This cross-coupling leads to the loss through cladding.

Micro-bend loss for single mode fiber is expressed as

$$\alpha_{mic} = 0.05\alpha_m \mu^8 \frac{K^4 (Fd)^6 (NA)^4}{a^2} \tag{13.75}$$

where α_m = attenuation constant, K = wave vector $= \dfrac{2\pi}{\lambda}$, a = core radius, Fd = half of mode field diameter in the single mode fiber.

Figure (26.19) represents the spectral attenuation of a single-mode fiber with evidence of excessive micro-bending loss.

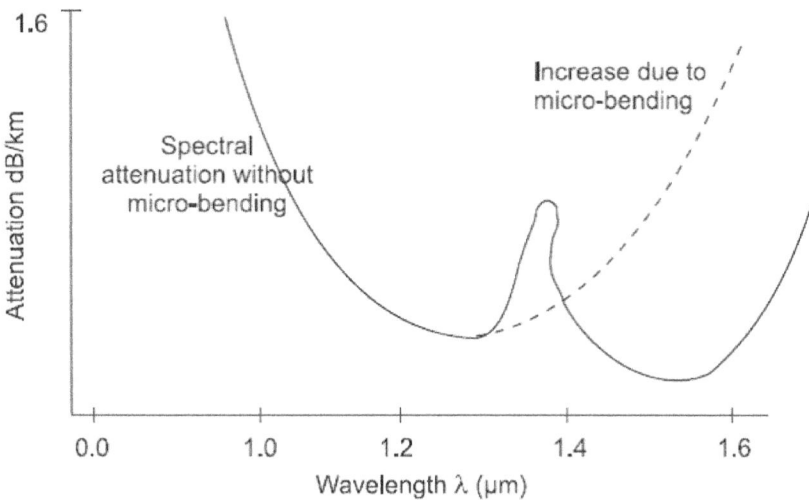

FIGURE 13.19. Spectral attenuation of a single-mode fiber with evidence of excessive micro-bending loss.

Macro-bending Losses

The larger the fiber-core radius and the smaller the bend radius, the greater the macro-bend loss. Macro-bend loss for multimode graded index fiber is given by the following expression:

$$\alpha_{mac} = -10\log\left(1 - \frac{p+2}{2p\Delta}\left[\frac{2a}{R} + \left(\frac{3}{2n_1 RK} \right)^{\frac{2}{3}} \right] \right) dB \qquad (13.76)$$

where p = index profile parameter for graded index fiber.

R = radius of curvature of bend, a = core radius,

$\Delta = \dfrac{n_1 - n_2}{n_1}$, n_1 = refractive index of the core, n_1 = refractive index of the cladding,

K = wave vector = $\dfrac{2\pi}{\lambda}$.

Figure 13.20 represents the variation of macro-bend loss for multimode step-index fiber with core radius in the range $200\,\mu$m to $800\,\mu$m. Figure 13.21 shows the fundamental mode field in a curved optical wave guide.

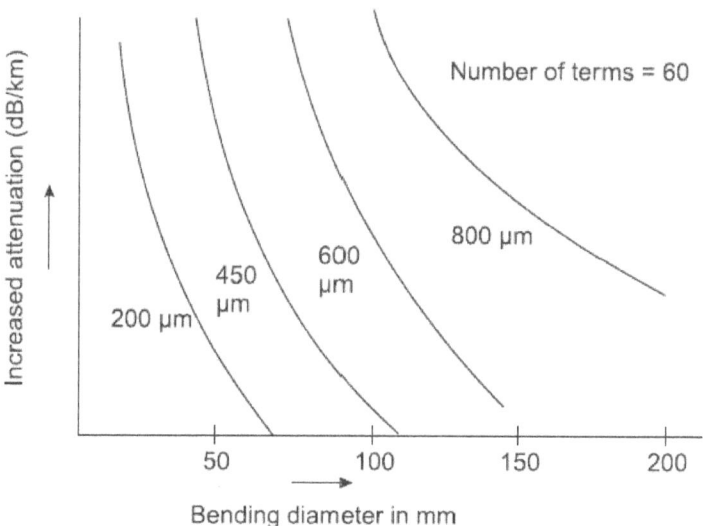

FIGURE 13.20. Variation of macro-bend loss for multimode step-index fiber with core radius in the range 200 μm to 800 μm.

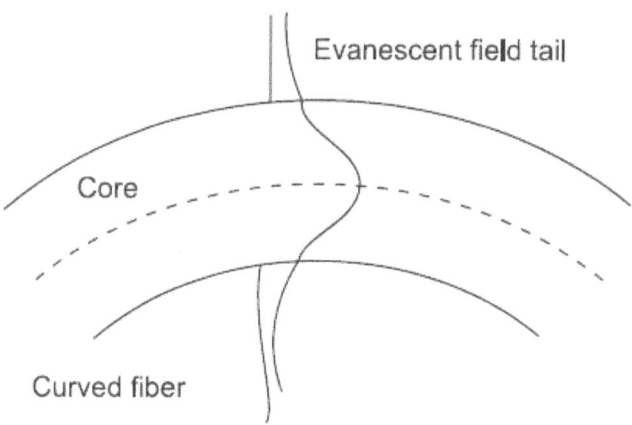

FIGURE 13.21. Fundamental mode field in a curved optical wave guide.

Radiation Induced Losses

When the glass molecular matrix interacts with electrons, neutrons, gamma rays, and X-rays, the structure of the glass molecules is altered and the fiber darkens. This introduces additional losses which increase with amount, type, dose, and exposure time of the radiation.

Inherent Defect Losses

Figure 13.22 shows several possible sources of loss due to defects in the fiber itself. First, in unclad fibers, surface defects (nicks or scratches) that breach the integrity of the surface allow light to escape. In other words, not all of the light is propagated along the fiber. Second (also in unclad fiber), grease, oil, or other contaminants on the surface of the fiber may form an area with an index of refraction different from what is expected and cause the light direction to change. There is always the possibility of inclusions, that is, objects, specks, or voids in the material making up the optical fiber. Inclusions can affect both clad and unclad fibers. When light hits the inclusion, it tends to scatter in all directions, causing a loss. Some of the light rays scattered from the inclusion may recombine either destructively or constructively with the main ray, but most do not.

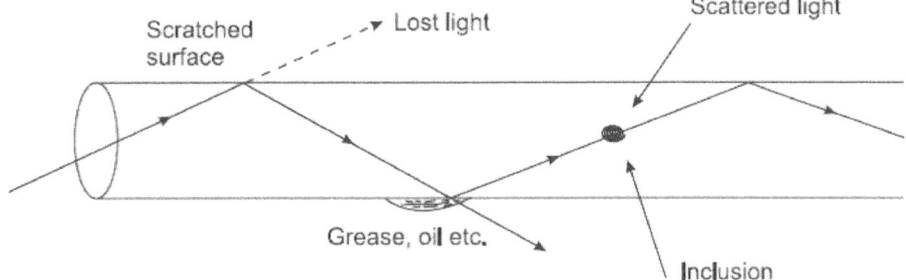

FIGURE 13.22. Several possible sources of loss due to defects in the fiber itself.

Inverse Square Law Losses

In all light systems, there is the possibility of losses caused by spreading (divergence) of the beam. If we take a flashlight and point it at a wall, measure the illuminance per unit area at the wall at a distance of, say, one meter, and then back off to twice the distance (two meters) and then measure again, we will find that the illuminance had dropped to one-fourth. In other words, the illuminance per unit area is inversely proportional to the square of the distance.

Transmission Losses

These losses are caused by light which is caught in the cladding material of clad optical fibers. This light is either lost to the outside or is trapped in the cladding layer and is thus not available for propagation in the core of the fiber.

Temperature Dependence of Fiber Losses

Temperature extremes have an adverse effect on fiber losses. Tight tube designs and clad made of plastic are normally usable down to –10°C. If the temperature is below –10°C, differential, thermal expansion between polymer coatings and glass causes stresses which cause microbending losses. Loose tube and glass-clad are used between –10 to –50°C. The central element of the fiber determines the amount of cable contraction with temperature and is called *stiffering*.

Core and Cladding Losses

In a fiber the core and cladding have different refractive indices, as they have different compositions. So the core and the cladding have different attenuation coefficients (α_c and α_d). If we neglect the modal coupling, then the loss associated with a mode (v, m) in *SI* fiber is expressed as:

$$\alpha_{vm} = \alpha_c \frac{P_C}{P_T} + \alpha_d \frac{P_d}{P_T} \tag{13.77}$$

where the fraction of powers $\dfrac{P_C}{P_T}$, $\dfrac{P_d}{P_T}$, P_T is the total power in the mode (n).

Misalignment Losses

In any optical fiber telecommunication link, one or more splices/joints in the fiber cable is required. The predominant method for connecting optical fibers involves a butt-joint connection. Any butt-joint requires three fundamental operations: fiber and preparation, fiber alignment to micron precision, and alignment retention. Demountable connections retain alignment mechanically while permanent connections retain alignment through melting and fusing of the fiber ends in a fusion-splicing machine. In any fiber joint, the fiber ends must be prepared smooth and perpendicular to the fiber axis. The next step of aligning the fiber ends is very crucial, because any kind of misalignment would lead to a transmission loss. Loss at a fiber splice could originate from either or a combination of the following possible misalignments, as shown in Figure 13.23:

1. Transverse offset between the fiber ends.

2. Angular tilt between the fiber ends.

3. Longitudinal end-separation between the fiber end faces.

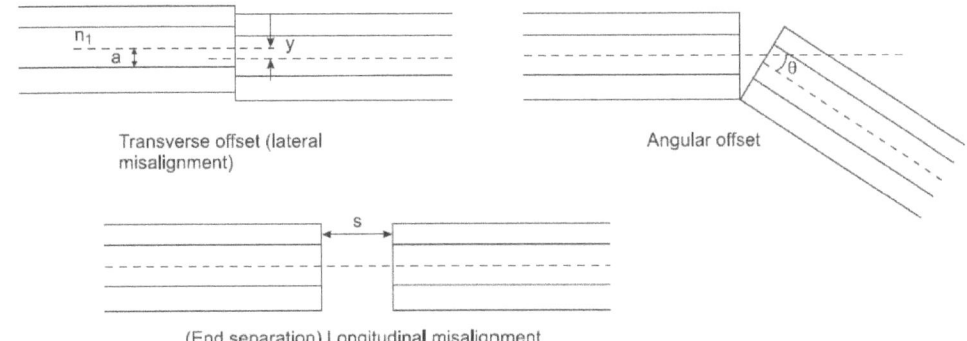

FIGURE 13.23. Possible misalignment losses of fiber.

$$\text{Loss}_{ang} = -10 \log_{10} \eta_{ang} dB \qquad (13.78)$$

$$\eta_{ang} \simeq \frac{16\left(\dfrac{n_1}{n}\right)^2}{\left(1+\dfrac{n_1}{n}\right)^4}\left(1 - \frac{n\theta}{\pi n_1 \sqrt{(2\Delta)}}\right) \qquad (13.79)$$

Δ is the refractive index difference for the fiber, n_1 is the core refractive index, and n is the refractive index of the medium between the fibers.

Angular Offset

$$\text{Loss}_{lat} = -10 \log_{10} \eta_{lat} dB \qquad (13.80)$$

$$\eta_{lat} \simeq \frac{16\left(\dfrac{n_1}{n}\right)^2}{\left(1+\dfrac{n_1}{n}\right)^4}\frac{1}{\pi}\left(2\cos^{-1}\sqrt{\left(\frac{y}{2a}\right) - \left(\frac{y}{a}\right)\left(1-\left(\frac{y}{2a}\right)^2\right)}\right) \qquad (13.81)$$

for the step-index fibers.

$$\text{Loss}_{long} = -10 \log_{10} \eta_{log} dB \qquad (13.82)$$

$$\eta_{long} \simeq \left(1 - \frac{s}{a}\frac{2}{\pi (NA)^2}\left(\sin^{-1}(NA) - NA\sqrt{1-(NA)^2}\right)\right) \text{ for } \frac{s}{a} \ll 1 \qquad (13.83)$$

Either or a combination of the following may also result in a joint loss, and these are sources of intrinsic loss:

i. Fresnel loss due to Fresnel reflection, as shown in Figure 13.24:

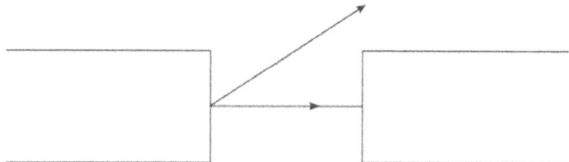

FIGURE 13.24. Fresnel loss due to Fresnel reflection.

A major consideration with all types of fiber–fiber connection is the optical loss encountered at the interface. Even when the two jointed fiber ends are smooth and perpendicular to the fiber axes, and the two fiber axes are perfectly aligned, a small proportion of the light may be reflected back into the transmitting fiber causing attenuation at the joint. This phenomenon is known as *Fresnel reflection*. The magnitude of this partial reflection of the light transmitted through the interface may be given by

$$r = \left(\frac{n_1 - n}{n_1 + n} \right)^2 \tag{13.84}$$

where r is the fraction of the light reflected at a single interface, n_1 is the refractive index of the fiber core, and n is the refractive index of the medium between the two jointed fibers.

The loss in decibels due to Fresnel reflection at a single interface is given by

$$\text{Loss}_{Fresnel} = -10 \log_{10} (1 - r) \tag{13.85}$$

This loss can be reduced to a very low level through the use of an index matching fluid in the gap between the jointed fibers, as shown in Figure 13.25.

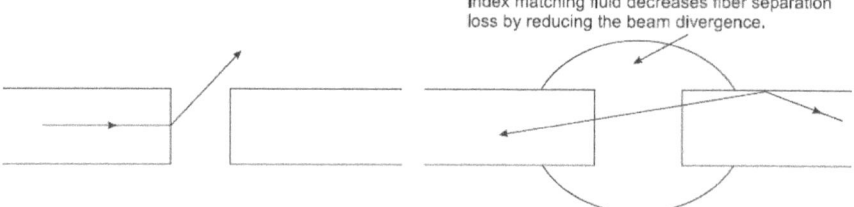

FIGURE 13.25. Loss through the use of an index matching fluid due to Fresnel reflection at a single interface.

In addition to mechanical misalignment,

ii. Any deviations in the geometrical and optical parameters of the two optical fibers which are jointed will affect the optical attenuation (insertion loss) through the connection. It is not possible within any particular connection technique to allow for all these variations. Hence, there are inherent connection problems when jointing fibers with, for instance:

a. Different core and/or cladding diameters

b. Different numerical apertures and/or relative refractive index differences

c. Different refractive index profiles

d. Fiber faults (core ellipticity, core cocentricity, etc.)

The losses caused by the previous factors together with those of Fresnel reflection are usually referred to *as intrinsic joint losses*.
Figure 13.26 illustrates the core diameter mismatch.

$$\text{Loss}_{CD} = -\log_{10}\left(\frac{a_2}{a_1}\right)^2 \quad (dB) \quad a_2 < a_1$$

$$\text{Loss}_{CD} = 0 \quad\quad\quad (dB) \quad a_2 \geq a_1$$

(13.86)

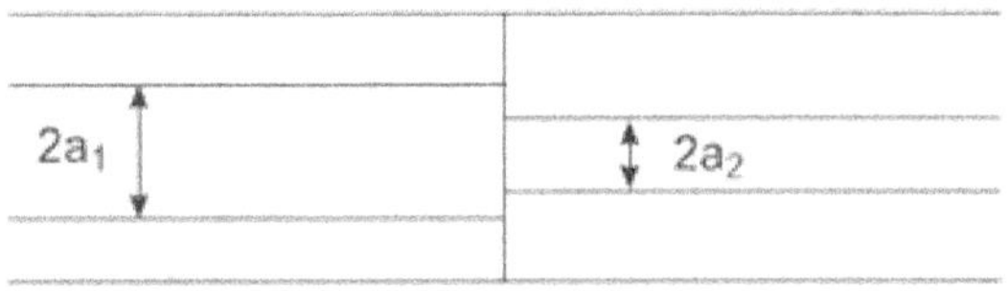

FIGURE 13.26. Core diameter mismatch.

Figure 13.27 shows the numerical aperture mismatch.

$$\text{Loss}_{NA} = -\log_{10}\left(\frac{NA_2}{NA_1}\right)^2 \quad (dB) \quad NA_2 < NA_1$$

$$\text{Loss}_{NA} = 0 \quad\quad\quad (dB) \quad NA_2 \geq NA_1$$

(13.87)

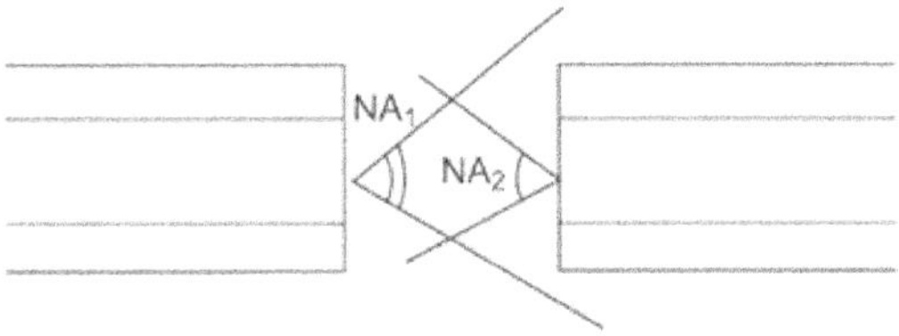

FIGURE 13.27. Numerical aperture mismatch.

Figure 13.28 illustrates refractive of index profile difference

$$\text{Loss}_{RI} = -\log_{10}\left(\frac{\alpha_2(\alpha_1+2)}{\alpha_1(\alpha_2+2)}\right)^2 \quad (dB) \quad \alpha_2 < \alpha_1$$

$$\text{Loss}_{RI} = 0 \qquad\qquad\qquad\qquad (dB) \quad \alpha_2 \geq \alpha_1$$

(13.88)

α_1 = Profile parameters for the transmitting fibers
α_2 = Profile parameters for the receiving fibers

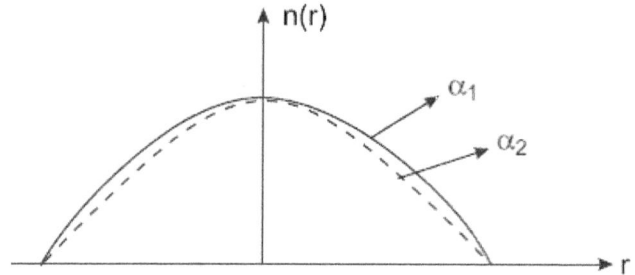

FIGURE 13.28. Refractive of index profile difference.

Typical results of splice losses due to various offsets between two fibers are shown in Figure 13.29.

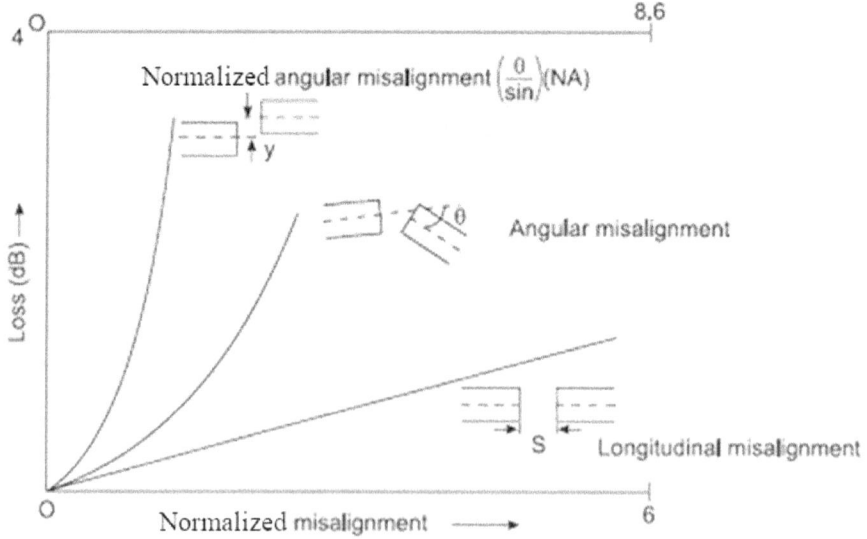

FIGURE 13.29. Splice losses due to various offsets between two fibers.

13.9 ERBIUM DOPED FIBER FOR OPTICAL AMPLIFIER

The expanding bandwidth demand in telecommunication networks which arises due to tremendous growth in the Internet and other communication services is being met by an erbium doped fiber optical amplifier (EDFA) in a WDM network. Er-ions when doped in low concentration in the core of the fiber are found to boost optical signal at the wavelength of 1550 nm when pumped by 980 nm light.

The advantage of EDFA is that it can simultaneously amplify all incoming wavelength (1530 to 1565 nm) channels to nearly the same output power. With additional care in dispersion compensation and nonlinear effects, the broadband gain flattering can be extended up to 1620 nm for dense wavelength division multiplexing (DWDM) applications. An energy band transition diagram shows the pump band as well as the metastable level of erbium ion in a silica host, as shown in Figure 13.30.

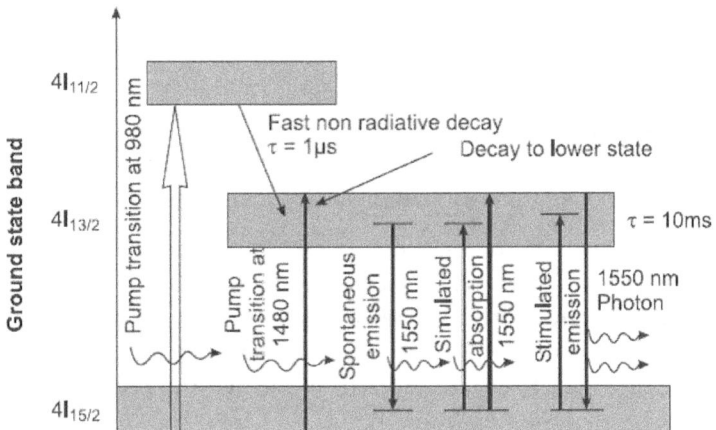

FIGURE 13.30. Energy level diagram of Er in silica host.

13.10 FIBER OPTIC SENSORS

In many industrial processes there is a need to monitor (such quantities as) displacement, pressure, temperature, flow rate, liquid level, chemical composition, and so on. Ideally, the measurement technique should be reliable, robust, corrosion-resistant, intrinsically safe, and free from external interference. Optical fibers have the potential for making a significant contribution in this area.

Generally, a transducer (or sensor) is used to convert one physical variable into another. Mostly these sensors are electrical in construction and provide convenient and controllable electrical signals for (a) amplifying weak signal, (b) measuring a process, (c) automatically recording these measurements and using them when needed, and (d) providing a signal that can be used to control another system or circuit. Thus, a transducer is an essential component of automation, wherever introduced.

In the case of optical fibers, these characteristics themselves provide an innovative approach to the design of an optical transducer/sensor.

The optical fibers have following characteristics:

1. Optical fibers are non-conducting, immune to electromagnetic and radio-frequency interference, and safe, even in explosive environments.

2. Their low attenuation coefficient allows the monitoring station to be located at a safe place, away from the hazardous environments (but where sensitizing has to be done)

3. Fiber optic sensors are not so sensitive to temperature changes.

4. Optical fibers are easily available to suit any design. (Thus, design-engineers naturally prefer to use optical fibers as the basic material for their sensor design.) Optical fiber sensors themselves can be divided into two main categories, namely: active (intrinsic) and passive (or extrinsic).

In the passive, the modulation takes place outside the optical fiber, which acts merely as a convenient transmission channel for the radiation.

In the active sensors, the quantity to be measured acts directly on the fiber itself to modify the radiation passing down the fiber. It is possible to modulate either the amplitude (or intensity), the phase, or the state of polarization of the radiation in the fiber.

In multimode fibers, however, mode coupling and the usually random relationships between the phases and polarization states of the propagating modes generally preclude the use of either phase or polarization modulation. Thus, active multimode fiber sensors almost invariably involve amplitude modulation. On the other hand, with single-mode fibers, both phase and polarization modulation become possible. Phase modulation opens up the possibility of fiber-based interferometric sensors that can offer exceptionally high sensitivities.

The extrinsic (passive) type sensors configured to allow the measurements to change the coupling characteristics between the feed and return fiber are shown in Figure 13.31:

FIGURE 13.31. The extrinsic type sensors allow the measurements to change the coupling characteristics between the feed and return fiber.

Intrinsic (Active) Fiber Sensors

A popular technique for the realization of an intrinsic multimode fiber sensor involves micro-bending of the fiber in the modulation region. The sensing mechanism for this technique is shown in the following figure. Deformation of the fiber on a small-scale causes light to be coupled from the guided optical modes propagating in the fiber core into the cladding region where they are lost through radiation into the surrounding region. When the spatial wavelength of the deformation L is correctly chosen, the power coupled from the fiber into radiation modes is high, providing a very high sensitivity to pressure applied to the deformer in a direction perpendicular to the fiber core axis as shown in Figure 13.32.

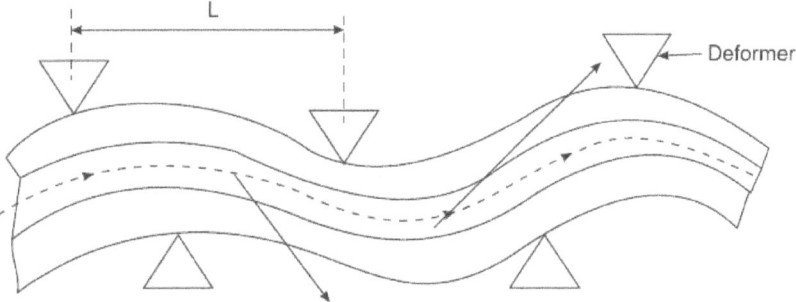

FIGURE 13.32. Intrinsic multimode fiber sensor involves micro-bending of the fiber in the modulation region.

Furthermore, if the deformation is caused by the measurand (e.g. pressure, vibration, sound, etc.), then the fluctuation in intensity of either the core or cladding light is directly proportional to the measurand for small deformations. Thus, monitoring the intensity of either the fiber core or cladding allows detection of the measurand.

Also, as is common with all fiber intensity modulation sensors, inaccuracies may occur due to source, detector, and fiber cable instabilities. These so-called common-mode variations usually necessitate the transmission of a separate optical reference signal which is not modulated by the measurand.

In this way, any optical intensity variations can be removed from the returned measurand signal by comparison with the returned reference signal at the receive terminal.

Flow Sensor

Figure 13.33 shows an intrinsic (Active) optical fiber flow sensor mechanism. In this device a multimode optical fiber is inserted across a pipe such that the liquid flows past the transversely stretched fiber. The turbulence resulting from the fiber's presence causes it to oscillate at a frequency roughly proportional to the flow rate. This results in a corresponding oscillation in the mode power distribution within the fiber, giving a similarly modulated intensity profile at the optical receiver. The technique has been used to measure flow rates from 0.3 to 3 m/sec.

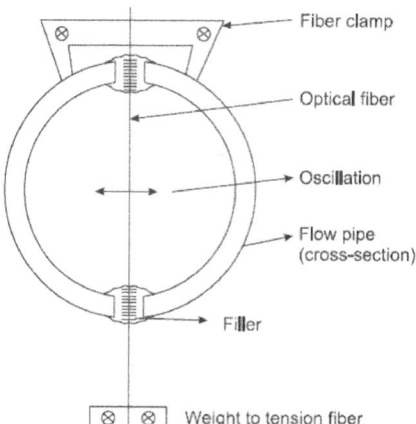

FIGURE 13.33. Intrinsic optical fiber flow sensor mechanism.

Extrinsic (Passive) Fiber Sensors

Numerous extrinsic optical fiber sensor mechanisms have been proposed and investigated, but to date relatively few practical commercial devices have emerged. A technique which has been realized as a commercial product is shown in Figure 13.34. This shows the operation of a simple optical fluid level switch—when the fluid, which has a refractive index greater than the glass forming the optical dipstick, reaches the chamfered end, total internal reflection ceases and the light is transmitted into the fluid. Hence, an indication of the fluid level is obtained at the optical detector. Although this system is somewhat crude and will not provide a continuous measurement of a fluid level, it is simple and safe for use with flammable liquids.

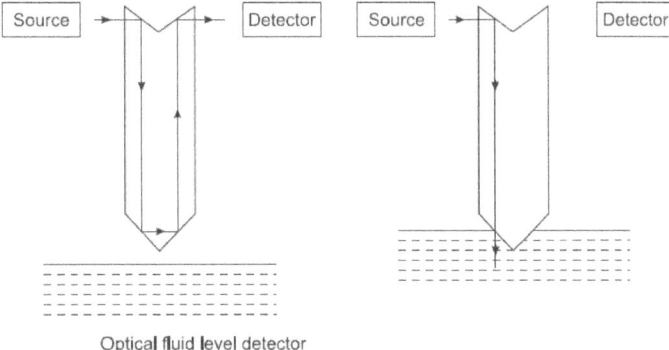

Optical fluid level detector

FIGURE 13.34. Extrinsic optical fiber sensor mechanisms.

Intensity modulation of the transmitted light beam is utilized in the extrinsic reflective or fotonic optical sensor, as shown in Figure 13.35, to give a measurement of displacement. Light reflected from the target is collected by a return fiber and is a function of the distance between the fiber ends and the target. Hence, the position or displacement of the target may be registered at the optical detector.

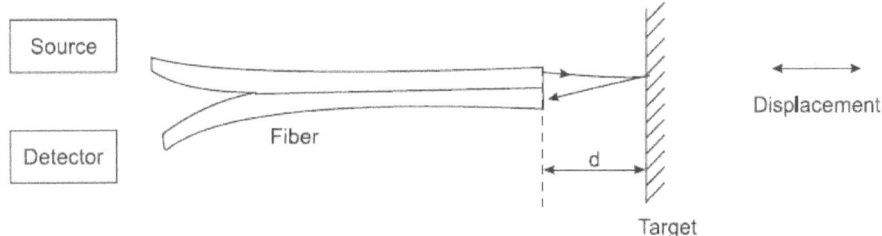

FIGURE 13.35. Intensity modulation of the transmitted light beam is utilized in the extrinsic reflective or fotonic optical sensor.

Furthermore, the sensitivity of this sensor may be improved by placing the axes of the feed and return fibers at an angle to one another and to the target. Unfortunately, this technique exhibits the drawback mentioned previously with regard to the stability of the optical components, which is a feature of intensity modulated fiber sensors.

A multimode fiber sensor which provides measurement of pressure is illustrated in Figure 13.36.

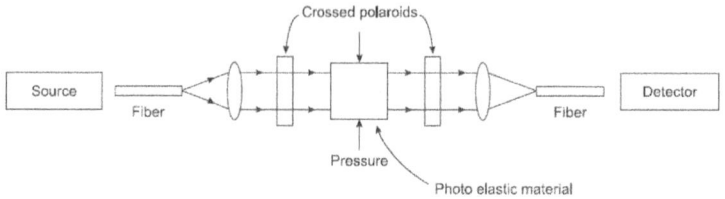

FIGURE 13.36. A multimode fiber sensor which provides measurement of pressure.

In this device the photoelastic effect induced by mechanical stress on a photoelastic material (e.g. piezo-optic glass, polyurethane, epoxy resin) is utilized to rotate the optical polarization between a pair of crossed polarizers. This phenomenon known as birefringence occurs with the application of mechanical stress to the transparent isotropic material, whereby it becomes optically anisotropic, giving a variation in transmitted light through the sensor. An advantage of this technique is that the stress may be induced directly without the need for an intermediate mechanism. A drawback, however, is that the fine fringence exhibited by photoelectric materials is often temperature-dependent, making measurement of a single parameter difficult.

Phase Modulated Sensors

In single-mode fiber sensors, we are mainly dealing with the effect of the external quantities to be measured on the phase (or mode velocity) of the light within the fiber.

One way in which we may detect phase changes is to construct an interferometric system where the phase of the beam through a sensing fiber is compared with that of a reference beam, as shown in Figure 13.37.

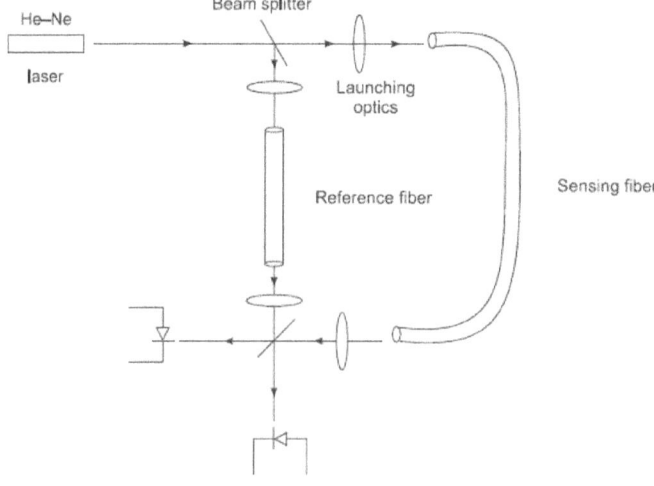

FIGURE 13.37. Interferometric system with He-Ne laser.

A beam splitter is used to divide a laser beam into two parts that are then launched into the sensing and reference fibers respectively. The outputs from these fibers meet at another beam splitter, and the two resulting beams are then allowed to fall onto suitable photo detectors. It is not strictly necessary to use both output beams, but there are advantages to be gained from doing so.

A more compact practical version may be constructed where the He-Ne laser is replaced by a semiconductor laser, and the beam splitters by "3 dB couplers," as shown in Figure 13.38.

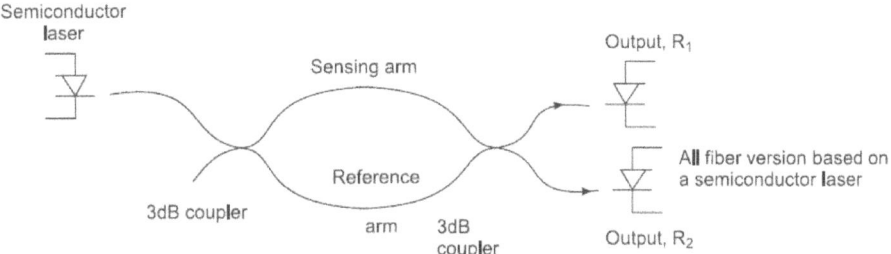

FIGURE 13.38. Interferometric system with a semiconductor laser.

Let us consider the form of the output expected from the interferometer. We write the electric fields of the two beams at one of the detectors as $A_S e^{i(wt+\Delta\phi)}$ and $A_r e^{iwt}$, where $\Delta\phi$ represents the phase difference between the two. In general, $\Delta\phi$ will be made up of a static differential phase term $\Delta\phi$ and a signal term, $\Delta\phi_s$, where $\Delta\phi = \Delta\phi_d + \Delta\phi_s$. The total signal amplitude is the sum of the two fields, and the detector output (R_1) is proportional to the product of this sum and its own complex conjugate, that is,

$$R_1 \alpha \left(A_S e^{i(wt+\Delta\phi)} + A_r e^{iwt} \right)\left(A_S e^{-i(wt+\Delta\phi)} + A_r e^{-iwt} \right)$$

i.e.

$$R_1 \alpha A_r^2 + A_S^2 + A_r A_S e^{i\Delta\phi} + e^{-i\Delta\phi} \alpha A_r^2 + A_S^2 + 2A_r A_S \cos\Delta\phi$$

Assuming for simplicity that $A_r = A_S = A$, then

$$R_1 \alpha 2A^2 \left(1 + \cos\Delta\phi \right) \tag{13.89}$$

Figure 13.39 illustrates the relative output of the interferometer as a function of the phase difference between the two arms $(\Delta\phi)$.

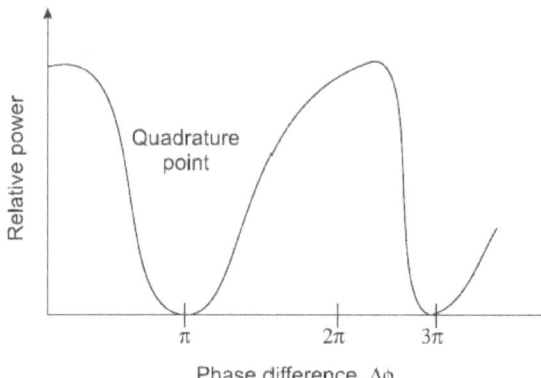

FIGURE 13.39. Relative output of interferometer as a function of the phase difference between the two arms ($\Delta\phi$).

This response as a function of $\Delta\phi$ is shown in the figure, illustrating the basic problem with any interferometric sensor, that its output is periodic in the phase difference. It is also evident that the greatest sensitivity to small variations in signal is obtained when the system is operating at points halfway between the maximum and minimum values of R. These are known as the quadrature points. Midway between these maximum sensitivity points, the sensitivity falls off to zero.

Thus, even if we could ensure that we started off operating at a quadrature point, it is quite likely that changes in ambient conditions around the reference arm would cause $\Delta\phi$ to change and hence cause a drift away from the quadrature point.

So far, we have considered the output from only one of the signal detectors (R_1). It may be shown that there is a phase difference between the beams falling on the two detectors of π. Thus, the response of the second detector (R_1) may be written as:

$$R_1 \alpha 2A^2 \left(1 + \cos\left(\Delta\phi + \pi\right)\right)$$

or

$$R_1 \alpha 2A^2 \left(1 - \cos\left(\Delta\phi\right)\right)$$

Thus,

$$\left(R_1 - R_2\right) \alpha 4A^2 \cos\left(\Delta\phi\right) \tag{13.90}$$

In effect, therefore, by taking the difference between the two output signals, we double the sensitivity of the interferometer. There are also advantages to be gained when the amplitudes of the beams from the signal and reference arms are not equal.

Fiber Optic Gyroscope

Another common single-mode fiber interferometric sensor which is finding wide-scale application is the fiber gyroscope. This device is based on the classical sagnac ring interferometer, a fiber version of which is given in Figure 13.40.

The sagnac effect is the phase shift induced between two light beams traveling in opposite directions around a fiber coil when the coil is rotating about an axis perpendicular to the plane of the coil.

In this device light entering the multi turn fiber coil is divided into two counter propagating waves which will return in phase after traveling along the same path in opposite directions. When the fiber coil is rotating about an axis perpendicular to the plane of the coil, however, then the path lengths between the counter propagating waves differ. This difference produces a phase shift which in turn can be measured by interferometric techniques in order to obtain the rotation.

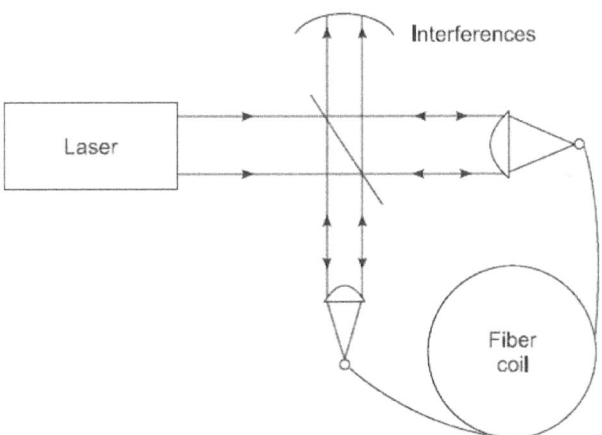

FIGURE 13.40. Fiber gyroscope.

To calculate the phase shift expected, we assume we have a single circular turn of fiber of radius R. It turns out that we must also assume that both light beams are effectively traveling in a vacuum. If there is no rotation, then both beams will return to their starting point in a time t, where $t = \dfrac{2\pi R}{C}$. If, however, the ring is rotating in a clockwise direction at a rate of Ω rad/sec, then the counterclockwise beam will arrive at its starting point sooner.

The effective velocity of the beam will be $C + R\Omega$ and the time taken, t', will be given by

$$t' = \frac{2\pi R}{C + R\Omega} \tag{13.91}$$

Similar arguments for the clockwise beam give a transit time of t'', where

$$t'' = \frac{2\pi R}{C - R\Omega} \tag{13.92}$$

The difference between the two transit times $\Delta t_S = t'' - t'$ is then

$$\Delta t_S = 2\pi R \left(\frac{1}{C - R\Omega} - \frac{1}{C + R\Omega} \right) = \frac{4\pi \Omega R^2}{C^2 - (R\Omega)^2} \tag{13.93}$$

Since $C^2 \gg R^2\Omega^2$
Thus,

$$\Delta t_S = \frac{4\Omega A}{C^2} \tag{13.94}$$

where $A = \pi R^2$ is the ring area.

In terms of a phase shift

$$\Delta\phi_S = 2\pi\Delta t_S f \tag{13.95}$$

where f is the frequency of the light, or putting the value of Δt_S from the previous equation we get

$$\Delta\phi_S = \frac{2\pi \Omega A}{\lambda_0 C} \tag{13.96}$$

This expression is independent of the medium in which the light is traveling. Most optical fiber sagnac interferometers are constructed with many turns (N ray), wrapped around a circular form. We then have

$$\phi_S = \frac{8\pi \Omega A N}{\lambda_0 C} \tag{13.97}$$

The basic arrangement of the sagnac fiber interferometer is shown in Figure 13.41. The form of the output signal from the detector as a function of the rotation rate (the phase change) will also be as shown in Figure 13.41.

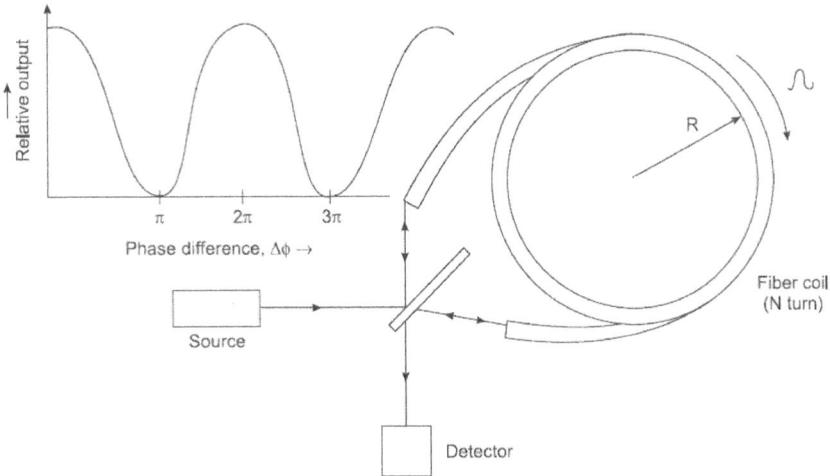

FIGURE 13.41. Basic arrangement of the sagnac fiber interferometer.

At low rotation rates, the sensitivity will be very low and will reach maximum values at the quadrature point $\left(\text{i.e., when } \phi_S = 2n + 1\left(\dfrac{\pi}{2}\right)\right)$

Another problem to be surmounted involves what is known as *reciprocity*. This concerns the necessity to ensure that the two counter-propagating beams travel absolutely identical paths, since any path difference is a potential source of phase noise.

In fact, the simple arrangement shown in Figure 13.41 is not reciprocal; the clockwise beam is reflected twice at the beam splitter, whereas the counter-clockwise beam is transmitted twice through it. The reciprocity requirements can be very stringent. For example, in a system capable of measuring rotation rates as low as $10^{-4} \deg^{-1} h$, the paths must be reciprocal to 1 part in 10^{16}.

Thus, although the two counter-propagating beams with identical polarizations may be launched in the fibers, energy may be coupled from one polarization mode into the orthogonal one as the beams traverse the fiber. The energy in the orthogonal modes will not be able to interfere with the original beams and a reduction in output will result, which may be interpreted as a phase signal.

Polarization Fiber Sensors

Modulation of the polarization state of light within a fiber may also be utilized to take a physical measurement. A successful implementation of this technique is displayed in the Faraday rotation current monitor as shown in Figure 13.42. This device consists of a single polarization maintaining fiber which passes up from the earth to loop around the current-carrying conductor before passing back to the earth.

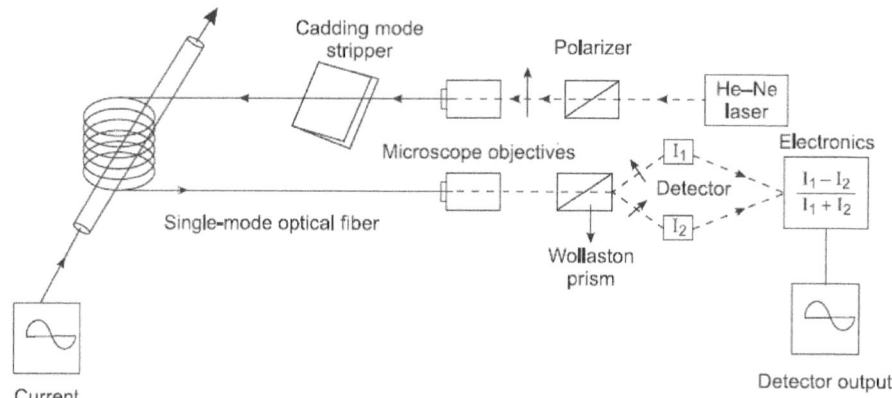

FIGURE 13.42. Single-mode optical fiber sensor for current measurement.

A He-Ne laser beam is linearly polarized and launched into the fiber, which is then stripped of any cladding modes. The direction of polarization of the light in the fiber core is rotated by the longitudinal magnetic field around the loop via the action of the Faraday magneto-optic effect. A Wollaston prism is used to sense the resulting rotation and resolves the emerging light into two orthogonal components. These components are separately detected with a photodiode prior to generation of a sum and difference signal of the two intensities (I_1 and I_2). The difference signal normalized to the sum gives a parameter which is proportional to the polarization rotation P.

$$PK = \frac{I_1 - I_2}{I_1 + I_2} \tag{13.98}$$

where K is a constant which is dependent on the properties of the fiber.

Hence, a current measurement (either d.c. or a.c.) may be obtained which is independent of the received light power. One area where such a sensor has proved valuable is in the monitoring of large currents in electricity generating stations. Currents up to 5000 A have been measured with an accuracy of about ±1%. Consider a single turn of fiber of radius r which passes around a wire carrying a current of I Amp.

Ampere's circuital theorem gives $H \times 2\pi r = I$ and hence $B = \dfrac{\mu_0 \mu_r I}{2\pi r}$. The amount of polarization rotation is given by

$$\theta_r = VB2\pi r \tag{13.99}$$

or

$$\theta_r = \mu_0 \mu_r IV \tag{13.100}$$

where μ_r is the relative permeability of the fiber and V is its verdict constant. For n turns of fiber, the amount of rotation will be n times greater, thus:

$$\theta_r = \mu_0 \mu_r nIV \tag{13.101}$$

This result is independent of r, and indeed in general θ_r is independent of the size or shape of the loop and the position of the conductor within the loop. This is a useful result, since it indicates that the device will be insensitive to any vibrations.

13.11 EXERCISES

1. Define fiber optic?

2. How do optical fibers carry light from end to end?

3. What are the three common types of fiber optic cables?

4. What are the advantages of optical fibers?

5. What are the three kinds of dispersion in fiber?

6. What are the three separate mechanisms that contribute to total absorption losses in glass fibers?

7. Consider a fiber with $n = 1.48$ in the core and $n = 1.46$ in the cladding. Determine the critical angle for reflections on the inner surface of the core.

8. A fiber has a core index of 1.455, and the cladding has an index of 1.483. Determine the NA.

9. Find the numerical aperture NA of a step-index fiber having $n_1 = 1.46$ and $n_2 = 1.43$.

10. An optical fiber 600 m long has an input power 9.7 μW and an output power of 8.6 μW. Find the fiber loss.

11. An optical fiber with an input power of 8.7 μW has a loss of -1.45 dB/km. If the fiber is 2600 m long, find the output power.

12. What determines the wavelengths that can be amplified in an optical fiber amplifier?

13. A light source for a fiber optic system has an output power -12 dBm. Determine the power in watts.

14. A power of 0.36 mW is launched into a fiber with an overall loss of 13 dB. What is the output power?

15. What causes the absorption loss in optical fiber?

16. What are the three kinds of dispersion due to three separate mechanisms existing in fiber?

17. What is the meaning of attenuation in fiber?

18. Provide five reasons for loss of power in fiber optics.

19. Find the minimum core diameter of a multimode fiber designed to carry IR signals at a wavelength $\lambda \simeq 1600$ nm.

20. A fiber with $n_1 = 1.45$ and $\Delta = 0.02$ has $NA \approx 0.29$. Let $\lambda_0 = 0.87$ μm and the core radius $a = 23$ μm. Determine the V-parameter.

21. A fiber has an attenuation of 3.2 dB/km at 820 nm. Let 0.4 mW of optical power be launched into the fiber. Determine the power level after 5km.

22. A fiber has lost 80% of its power after traversing 400 m of fiber. Find the loss in dB/km of fiber.

The Greek Alphabet

Names	Upper Case Letters	Lower Case Letters	Names	Upper Case Letters	Lower Case Letters
Alpha	A	α	Nu	N	ν
Beta	B	β	Xi	Ξ	ξ
Gamma	Γ	γ	Omicron	O	o
Delta	Δ	δ	Pi	Π	π
Epsilon	E	ε	Rho	P	ρ
Zeta	Z	ζ	Sigma	Σ	σ
Eta	H	η	Tau	T	τ
Theta	Θ	θ	Upsilon	Y	υ
Iota	I	ι	Phi	Φ	ϕ
Kappa	K	κ	Chi	X	χ
Lambda	Λ	λ	Psi	Ψ	ψ
Mu	M	μ	Omega	Ω	ω

The International System of Units (SI) Prefixes

Power	Prefix	Symbol	Power	Prefix	Symbol
10^{-1}	deci	d	10^{24}	yotta	Y
10^{-2}	centi	c	10^{21}	zetta	Z
10^{-3}	milli	m	10^{18}	exa	E
10^{-6}	micro	μ	10^{15}	peta	P
10^{-9}	nano	n	10^{12}	tera	T

(continued)

(continued)

Power	Prefix	Symbol	Power	Prefix	Symbol
10^{-12}	pico	p	10^9	giga	G
10^{-15}	femto	f	10^6	mega	M
10^{-18}	atto	a	10^3	kilo	k
10^{-21}	zepto	z	10^2	hecto	h
10^{-24}	yocto	y	10^1	deka	da
10^{-35}	stringo	-	10^0	-	-

Logarithmic Identities

$\log_e a = \ln a$ (natural logarithm)

$\log_{10} a = \log a$ (common logarithm)

$\log ab = \log a + \log b$

$\log \dfrac{a}{b} = \log a - \log b$

$\log a^n = n \log a$

Vector Derivatives and Coordinates:

1. Cartesian Coordinates

Coordinates	(x, y, z)
Vector	$\vec{A} = A_x \vec{a}_x + A_y \vec{a}_y + A_z \vec{a}_z$
Gradient	$\nabla \vec{A} = \dfrac{\partial A}{\partial x} \vec{a}_x + \dfrac{\partial A}{\partial y} \vec{a}_y + \dfrac{\partial A}{\partial z} \vec{a}_z$
Divergence	$\nabla \cdot \vec{A} = \dfrac{\partial A_x}{\partial x} + \dfrac{\partial A_y}{\partial y} + \dfrac{\partial A_z}{\partial z}$
Curl	$\nabla \times \vec{A} = \begin{vmatrix} \vec{a}_x & \vec{a}_y & \vec{a}_z \\ \dfrac{\partial}{\partial x} & \dfrac{\partial}{\partial y} & \dfrac{\partial}{\partial z} \\ A_x & A_y & A_z \end{vmatrix}$ $= \left(\dfrac{\partial A_z}{\partial y} - \dfrac{\partial A_y}{\partial z} \right) \vec{a}_x + \left(\dfrac{\partial A_x}{\partial z} - \dfrac{\partial A_z}{\partial x} \right) \vec{a}_y + \left(\dfrac{\partial A_y}{\partial x} - \dfrac{\partial A_x}{\partial y} \right) \vec{a}_z$
Laplacian	$\nabla^2 \vec{A} = \dfrac{\partial^2 A}{\partial x^2} + \dfrac{\partial^2 A}{\partial y^2} + \dfrac{\partial^2 A}{\partial z^2}$

2. Cylindrical Coordinates

Coordinates	(ρ, ϕ, z)
Vector	$\vec{A} = A_\rho \vec{a}_\rho + A_\phi \vec{a}_\phi + A_z \vec{a}_z$
Gradient	$\nabla \vec{A} = \dfrac{\partial A}{\partial \rho} \vec{a}_\rho + \dfrac{1}{\rho} \dfrac{\partial A}{\partial \phi} \vec{a}_\phi + \dfrac{\partial A}{\partial z} \vec{a}_z$
Divergence	$\nabla \cdot \vec{A} = \dfrac{1}{\rho} \dfrac{\partial}{\partial \rho}(\rho A_\rho) + \dfrac{1}{\rho} \dfrac{\partial A_\phi}{\partial \phi} + \dfrac{\partial A_z}{\partial z}$
Curl	$\nabla \times \vec{A} = \dfrac{1}{\rho} \begin{vmatrix} \vec{a}_\rho & \rho \vec{a}_\phi & \vec{a}_z \\ \dfrac{\partial}{\partial \rho} & \dfrac{\partial}{\partial \phi} & \dfrac{\partial}{\partial z} \\ A_\rho & \rho A_\phi & A_z \end{vmatrix}$ $= \left(\dfrac{1}{\rho} \dfrac{\partial A_z}{\partial \phi} - \dfrac{\partial A_\phi}{\partial z} \right) \vec{a}_\rho + \left(\dfrac{\partial A_\rho}{\partial z} - \dfrac{\partial A_z}{\partial \rho} \right) \vec{a}_\phi + \dfrac{1}{\rho} \left(\dfrac{\partial}{\partial x}(\rho A_\phi) - \dfrac{\partial A_\rho}{\partial \phi} \right) \vec{a}_z$
Laplacian	$\nabla^2 \vec{A} = \dfrac{1}{\rho} \dfrac{\partial}{\partial \rho} \left(\rho \dfrac{\partial A}{\partial \rho} \right) + \dfrac{1}{\rho^2} \dfrac{\partial^2 A}{\partial \phi^2} + \dfrac{\partial^2 A}{\partial z^2}$

3. Spherical Coordinates

Coordinates	(r, θ, ϕ)
Vector	$\vec{A} = A_r \vec{a}_r + A_\theta \vec{a}_\theta + A_\phi \vec{a}_\phi$
Gradient	$\nabla \vec{A} = \dfrac{\partial A}{\partial r} \vec{a}_r + \dfrac{1}{r} \dfrac{\partial A}{\partial \theta} \vec{a}_\theta + \dfrac{1}{r \sin \theta} \dfrac{\partial A}{\partial \phi} \vec{a}_\phi$
Divergence	$\nabla \cdot \vec{A} = \dfrac{1}{r^2} \dfrac{\partial}{\partial r}(r^2 A_r) + \dfrac{1}{r \sin \theta} \dfrac{\partial}{\partial \theta}(A_\theta \sin \theta) + \dfrac{1}{r \sin \theta} \dfrac{\partial A_\phi}{\partial \phi}$
Curl	$\nabla \times \vec{A} = \dfrac{1}{r^2 \sin \theta} \begin{vmatrix} \vec{a}_r & r\vec{a}_\theta & (r \sin \theta)\vec{a}_\phi \\ \dfrac{\partial}{\partial r} & \dfrac{\partial}{\partial \theta} & \dfrac{\partial}{\partial \phi} \\ A_r & rA_\theta & (r \sin \theta)A_\phi \end{vmatrix}$ $= \dfrac{1}{r \sin \theta} \left(\dfrac{\partial}{\partial \theta}(A_\phi \sin \theta) - \dfrac{\partial A_\theta}{\partial \phi} \right) \vec{a}_r + \dfrac{1}{r} \left(\dfrac{1}{\sin \theta} \dfrac{\partial A_r}{\partial \phi} - \dfrac{\partial}{\partial r}(rA_\phi) \right) \vec{a}_\theta$ $+ \dfrac{1}{r} \left(\dfrac{\partial}{\partial r}(rA_\theta) - \dfrac{\partial A_r}{\partial \theta} \right) \vec{a}_\phi$
Laplacian	$\nabla^2 \vec{A} = \dfrac{1}{r^2} \dfrac{\partial}{\partial r} \left(r^2 \dfrac{\partial A}{\partial r} \right) + \dfrac{1}{r^2 \sin \theta} \dfrac{\partial}{\partial \theta} \left(\sin \theta \dfrac{\partial A}{\partial \theta} \right) + \dfrac{1}{r^2 \sin^2 \theta} \dfrac{\partial^2 A}{\partial \phi^2}$

Vector Identity

$$\vec{A}\cdot\left(\vec{B}\times\vec{C}\right) = \vec{B}\cdot\left(\vec{C}\times\vec{A}\right) = \vec{C}\cdot\left(\vec{A}\times\vec{B}\right)$$

$$\vec{A}\times\left(\vec{B}\times\vec{C}\right) = \vec{B}\left(\vec{A}\cdot\vec{C}\right) - \vec{C}\left(\vec{A}\cdot\vec{B}\right)$$

$$\nabla(fg) = f(\nabla g) + g(\nabla f)$$

$$\nabla(\vec{A}\cdot\vec{B}) = \vec{A}\times(\nabla\times\vec{B}) + \vec{B}\times(\nabla\times\vec{A}) + (\vec{A}\cdot\nabla)\vec{B} + (\vec{B}\cdot\nabla)\vec{A}$$

$$\nabla\cdot(f\vec{A}) = f(\nabla\cdot\vec{A}) + \vec{A}\cdot(\nabla f)$$

$$\nabla\cdot(\vec{A}\times\vec{B}) = \vec{B}\cdot(\nabla\times\vec{A}) - \vec{A}\cdot(\nabla\times\vec{B})$$

$$\nabla\times(f\vec{A}) = f(\nabla\times\vec{A}) - \vec{A}\times(\nabla f) = \nabla\times(f\vec{A}) = f(\nabla\times\vec{A}) + (\nabla f)\times\vec{A}$$

$$\nabla\times(\vec{A}\times\vec{B}) = (\vec{B}\cdot\nabla)\vec{A} - (\vec{A}\cdot\nabla)\vec{B} + \vec{A}(\nabla\cdot\vec{B}) - \vec{B}(\nabla\cdot\vec{A})$$

$$\nabla\cdot(\nabla\times\vec{A}) = 0$$

$$\nabla\times(\nabla f) = 0$$

$$\nabla\cdot(\nabla f) = \nabla^2 f$$

$$\nabla\times(\nabla\times\vec{A}) = \nabla(\nabla\cdot\vec{A}) - \nabla^2\vec{A}$$

$$\nabla(f + g) = \nabla f + \nabla g$$

$$\nabla\cdot\left(\vec{A}+\vec{B}\right) = \nabla\cdot\vec{A} + \nabla\cdot\vec{B}$$

$$\nabla\times\left(\vec{A}\times\vec{B}\right) = \nabla\times\vec{A} + \nabla\times\vec{B}$$

$$\nabla\left(\frac{f}{g}\right) = \frac{g(\nabla f) - f(\nabla g)}{g^2}$$

$$\nabla f^n = nf^{n-1}\nabla f \quad (n = \text{integer})$$

Divergence Theorem: $\displaystyle\int_{volume} (\nabla\cdot\vec{A})dv = \oint_{surface} \vec{A}\cdot d\vec{s}$

3. Curl (Stokes) Theorem: $\displaystyle\int_{surface} (\nabla\times\vec{A})\cdot d\vec{s} = \oint_{line} \vec{A}\cdot d\vec{l}$

Approximate Common Optical Wavelengths Ranges of Light

Color	Wavelength, λ
Ultraviolet Region	10–380 nm
Visible Region	380–750 nm
violate	380–450 nm
blue	450–495 nm
green	495–570 nm
yellow	570–590 nm
orange	590–620 nm
red	620–750 nm
Infrared	750 nm–1mm

Common Optical Wavelengths Conversions

Wave-length, λ	Angstrom, Å	Nano-meter, nm	Micro-meter, μm	Centi-meter, cm	Meter, m
1 Å	1	10^{-1}	10^{-4}	10^{-8}	10^{-10}
1 nm	10	1	10^{-3}	10^{-7}	10^{-9}
1 μm	10^4	10^3	1	10^{-4}	10^{-6}
1 cm	10^8	10^7	10^4	1	10^{-2}
1 m	10^{10}	10^9	10^6	10^2	1

Index of Refraction for Common Substances

Substance	Index of Refraction, n
Air	1.000293
Diamond	2.24

(continued)

(continued)

Substance	Index of Refraction, n
Ethyl alcohol	1.36
Fluorite	1.43
Fused quartz	1.46
Crown Glass	1.52
Flint Glass	1.66
Glycerin	1.47
Ice	1.31
Polystyrene	1.49
Rock salt	1.54
Water	1.33

Physical Constants

Quantity	Best Experimental Value	Approximate Value for Problem Work
Speed of light (m/s)	2.9997×10^8	3×10^8
Electron charge (C)	-1.6022×10^{-19}	-1.6×10^{-19}
Electron mass (kg)	9.1066×10^{-31}	9.1×10^{-31}
Proton mass (kg)	1.67248×10^{-27}	1.67×10^{-27}
Neutron mass (kg)	1.6749×10^{-27}	1.67×10^{-27}
Planck's constant (J. s)	6.6261×10^{-34}	6.62×10^{-34}
Boltzmann constant (J/k)	1.38047×10^{-23}	1.38×10^{-23}
Permeability of free space (H/m)	$4\pi \times 10^{-7}$	12.6×10^{-7}
Intrinsic impedance of free space (Ω)	376.6	120π
Speed of light in vacuum (m/s)	2.9979×10^8	3×10^8

(continued)

(continued)

Quantity	Best Experimental Value	Approximate Value for Problem Work
Acceleration due to gravity (m/s^2)	9.8066	9.8
Avogadro's number (/kg-mole)	6.0228×10^{26}	6×10^{26}
Universal constant of gravitation (m^2/kg . s^2)	6.658×10^{-11}	6.66×10^{-11}
Electron-volt (J)	1.6030×10^{-19}	1.6×10^{-19}
Gas constant (J/mol K)	8.3145	8.3

INDEX